Applications of Numerical Methods in Molecular Spectroscopy

Chemometrics Series

Series Editor

Steven D. Brown
University of Delaware

Advisory Editors

Lloyd A. Currie
National Institute of Standards and Technology

Paul Geladi
University of Umeå

Barry K. Lavine
Clarkson University

Paul J. Gemperline
East Carolina University

Phillip K. Hopke
Clarkson University

Titles in the Series

Sequential Simplex Optimization
Parker, Deming, Morgan, and Walters

Factor Analysis of Chromatographic Data
Gemperline and Ramos

Applications of Numerical Methods in Molecular Spectroscopy
Pelikán, Čeppan, and Liška

Pattern Recognition: Data Analysis Via Multivariate Statistical Methods
B. Lavine

Introduction to Chemometrics
Rohrback, Ramos, and Tayler

Applications of Numerical Methods in Molecular Spectroscopy

Peter Pelikán, D.Sc.
Slovak Technical University
Bratislava, Slovakia

Michal Čeppan, Ph.D.
Slovak Technical University
Bratislava, Slovakia

Marek Liška, Ph.D.
Slovak Academy of Sciences
Bratislava/Trenčín, Slovakia

Translated from the Slovak by Dana Marianová

CRC Press
Boca Raton Ann Arbor London Tokyo

Library of Congress Cataloging-in-Publication Data

Pelikán, Peter
Applications of numerical methods in molecular spectroscopy.
p. cm.
Includes bibliographical references and index.
ISBN 0-8493-7322-0
1. Molecular spectroscopy—Statistical methods. 2. Molecular spectroscopy—Data processing. I. Čeppan, Michal, 1952– . II. Liška, Marek, 1951– . III. Title. IV. Series: Chemometrics series (Boca Raton, Fla.).
QD96.M65P45 1993
539′.6—dc20 92-46758
CIP

International Standard Book Number 0-8493-7322-0
Library of Congress Card Number 92-46758
Printed in the United States of America 1 2 3 4 5 6 7 8 9 0
Printed on acid-free paper

Series Editor's Preface

Modern spectrometers — of all sorts — commonly include a computer as an integral part of the spectroscopic hardware. Many also come with sophisticated chemometrics software to run on the computer. Getting the most from a set of spectroscopic data requires much more than just easy access to a computer and one or two chemometrics packages, however. Having a good understanding of spectroscopic theory and the way in which spectroscopic theory should be coupled to such chemometric methods as those for peak resolution, deconvolution, noise reduction, and multivariate calibration, can determine whether useful information or meaningless sets of numbers result from an experiment. Most texts deal exclusively with either the theory of spectroscopy or the chemometrics; few try to integrate the two. This text is unusual in that the authors focus the discussion on bridging that literature gap. The most useful mathematical techniques for reducing and analyzing spectroscopic data of all types are collected here, a feature that should make the book an important reference for students of chemometrics and spectroscopy, as well as practicing spectroscopists.

The CRC Chemometrics Series has as its goal the publishing of texts that will make the conversion of data to results a little easier for students and practicing chemists. Those with an interest in the collection and analysis of data from chemical systems will find covered in this series both the theory and practical applications needed for the efficient use of the computing power available on their desks. The texts in this series will be useful not only to students of chemistry and practicing chemists, but also to those in related fields, such as food science, pharmacy, materials science, and environmental science, where the analysis of chemical measurements plays a central role.

Steven D. Brown
Newark, Delaware

Preface

Spectroscopic methods have undergone unprecedented developments during the past decades. These developments can be seen as a result of microelectronic applications in spectral measurement control and automation as well as an extensive utilization of various spectral data treatment methods using the powerful computing facilities of recent spectrophotometer systems. Application of computers and numerical methods to spectral data acquisition and processing creates certain unifying phenomena and requires background, basic ideas, and special approaches to numerical methods in spectroscopic experiments.

Spectral data processing methods generally can be classified into two groups. Methods in the first group are relatively simple from the mathematical point of view, with the goal of providing convenient and comfortable handling of spectral data. Methods in the second group are based on more complex mathematics and are aimed at extracting the maximal amount of useful chemical information contained in spectral data. Some of these methods are included in standard software delivered with recent spectrophotometers; some of them are parts of various commercial software packages. *Applications of Numerical Methods in Molecular Spectroscopy* provides a mathematical background, a theoretical overview, and a review of the second group of methods with an effort to uniformly apply numerical processing of all spectral data types.

The book is divided into seven chapters and four appendices. The first chapter is devoted to spectral data acquisition and basic processing (smoothing, deconvolution, numerical derivation, etc.). Various methods of complex spectral profile resolution into individual bands are discussed in Chapter 2. Chapter 3 introduces molecular spectroscopy aspects of factor analysis methods, while Chapter 4 describes methods of quantitative analysis. The area of reflectance spectroscopy, where progress in experimental equipment and techniques has enhanced data processing methods, is treated in Chapter 5. The last two chapters are devoted to mathematical methods in special areas of molecular spectroscopy — electron spin resonance (Chapter 6) and nuclear magnetic resonance (Chapter 7) spectroscopies. The appendices contain the mathematical background for topics discussed in the previous chapters.

The Authors

Peter Pelikán, D.Sc., was born in 1939 and took his M.Sc. in 1969, at the Faculty of Chemical Technology of the Slovak Technical University, Bratislava. In 1973 he finished his doctoral study at the same university and obtained his Ph.D. in Physical Chemistry. His first position was that of Lecturer at the Department of Physical Chemistry, Faculty of Chemical Technology, Slovak Technical University. In 1978, he was appointed Research Fellow; in 1989, Research Professor; in 1990, Associate Professor; and in 1991, Full Professor. In the fall of 1990, he was elected Dean of the Faculty of Chemical Technology. He is Editor in Chief of the journal *Chemical Papers,* published in Bratislava.

His main research interests are quantum and molecular spectroscopy, especially theoretical problems of coordination chemistry, stereochemistry, catalysis, radical chemistry, photochemistry, and ESR spectroscopy. He has written approximately 200 original papers, and was a member of the team of authors responsible for 6 monographs and handbooks — *Handbook of Chemistry and Chemical Physics,* Alfa, Bratislava, 1976; *Encyclopedia of Chemistry and Chemical Technology I,* Alfa, Bratislava, 1983 (English translation: *Comprehensive Dictionary of Physical Chemistry,* Ellis Horwood, Chichester, 1992); *Quantum Chemistry of Coordination Compounds,* Veda, Bratislava, 1987 (English translation: *Coordination Chemistry Review — Special Issue,* Volume 118, Elsevier, Amsterdam, 1992), *Radicals in Chemistry and Biochemistry,* CRC Press, New York, 1989; *Molecular Nitrogen Fixation,* Veda, Bratislava, 1989; and *Photochemical Processes,* Alfa, Bratislava, 1989.

Michal Čeppan, Ph.D., born in 1952, graduated in 1976 with an M.Sc. from the Faculty of Chemical Technology, Slovak Technical University, Bratislava. In 1983, he earned his Ph.D. in Technology of Macromolecular Compounds at the same institution. In 1981, he was appointed Lecturer at the Department of Printing Technology and Applied Photochemistry, Faculty of Chemical Technology, Slovak Technical University.

Michal Čeppan's research interests include photochemistry and its applications, as well as UV-VIS and IR spectroscopy, with emphasis on utilization of these methods in photochemical reaction studies and in applications of numerical algorithms to spectral data processing. He has written approximately 20 papers, and was a member of the team of authors who wrote *Photochemical Processes,* Alfa, Bratislava, 1989; and *Encyclopedia of Chemistry and Chemical Technology I,* Alfa, Bratislava, 1983 (English translation: *Comprehensive Dictionary of Physical Chemistry,* Ellis Horwood, Chichester, 1992).

Marek Liška, born in 1951, took his M.Sc. in 1974 and his Ph.D. in Inorganic Chemistry in 1978 at the Faculty of Chemical Technology, Slovak Technical University, Bratislava. His first position was as Lecturer at the Department of Inorganic Chemistry at this university. In 1983, he was appointed Research Fellow at the Laboratory for Glass Research of the Slovak Academy of Sciences. At present he is Head of this Laboratory.

Marek Liška has devoted his research to quantum chemistry. The main area of his present studies is the relation between structure and properties of glasses and glass melts. He has written more than 60 papers, and was a member of the team of authors for *Encyclopedia of Chemistry and Chemical Technology I,* Alfa, Bratislava, 1983 (English translation: *Comprehensive Dictionary of Physical Chemistry,* Ellis Horwood, Chichester, 1992).

Contents

Applications of Numerical Methods in Molecular Spectroscopy

Chapter 1

Acquisition and Processing of Spectral Data

SCANNING AND ACQUISITION OF SPECTRAL DATA

A digitized spectrum is the starting point of spectral data processing. Generally, an ordered set of N pairs of values (y_i, x_i), $i = 1, 2, \ldots N$, of independent variable x (frequency, wavenumber, wavelength, incident angle, magnetic field induction, and so on) and dependent variable y (absorbance, transmittance, radiant flux, etc.) can be considered a digitized spectrum. The particular experimental arrangement used to make the spectral measurement, as determined by the type of spectroscopy, is not important for the study of methods for numerical processing.

The common basic characteristics of each spectral measurement,[1] schematically illustrated in Figure 1.1, are important. The quantifying abstract quantity (e.g., the ability to absorb, emit, or reflect the electromagnetic radiation of a certain frequency) is designated by the symbol A in the scheme. This quantity results from the interaction of the physical object studied with the general measuring probe (i.e., irradiation, thermal excitation, etc.) and is converted (transduced) by the measuring system to the electrical signal M, which gives the resulting experimental spectrum y. Practically, the measurements are performed so that the independent variable x changes with the time t, i.e.,

$$x = x(t) \tag{1.1}$$

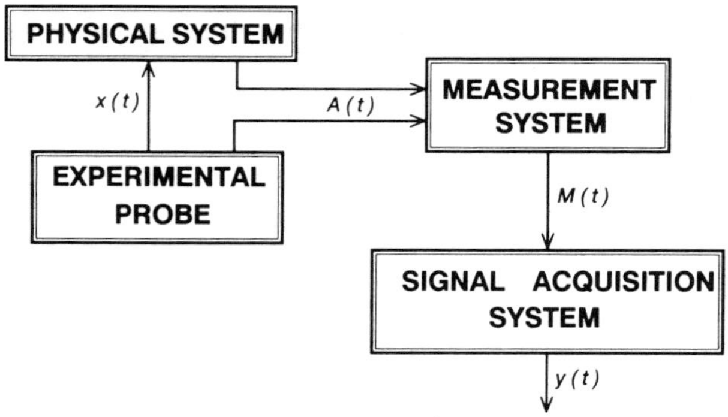

Figure 1.1 The simplified scheme of spectral scanning.

The digitized spectrum (y_i, x_i) can be generally understood as the time series (y_i, t_i) allowing for the possibility of the application of numerical methods developed within the universal theory of time series.[2-4]

Besides the instrument function, the set of random influences causing the stochastic noise involved in the spectrum $y(t)$ appears in the spectral measurement. Occasionally, interaction of the measured system with the measuring probe can change the physical characteristics of the object investigated.

Numerical methods in spectroscopy are used on two levels. First, the best approximation to the so-called true spectrum $A(t)$ (filtering, deconvolution, and elimination of background) is obtained by the application of these methods to a measured spectrum $y(t)$. The following step is the analysis of the true spectrum with the goal of extracting detailed information about the structure, composition, and properties of the studied sample.

Using older spectral devices, the measured spectrum was obtained graphically. Its digitization was performed manually by reading the coordinates of individual points, or semiautomatically by applying different digitizers connected with the computer. Modern spectrometers are mostly controlled by microprocessors, and on-line connections with computers enable the acquisition of spectral data while recording on external storage media (magnetic disk or tape).

The fact that the whole continuous function is not used for numerical processing, but only the part represented by the digitized points (y_i, t_i), $i = 1, 2, \ldots N$, influences the amount and quality of information obtained from the digitized spectrum. Let us assume that a spectrum $y(t)$ (Figure 1.2a) was digitized with a constant time step T. The digitized spectrum $y_d(t)$ (Figure 1.2c) can be expressed as the product of the original continuous spectrum with the series of the unit pulses $\delta_T(t)$ spaced equidistantly from T (Figure 1.2b).

$$y_d(t) = y(t)\,\delta_T(t) = \sum_{k=-\infty}^{\infty} y(kT)\,\delta(t-kT) \tag{1.2}$$

where $\delta(t)$ is the Dirac function (see Appendix D).

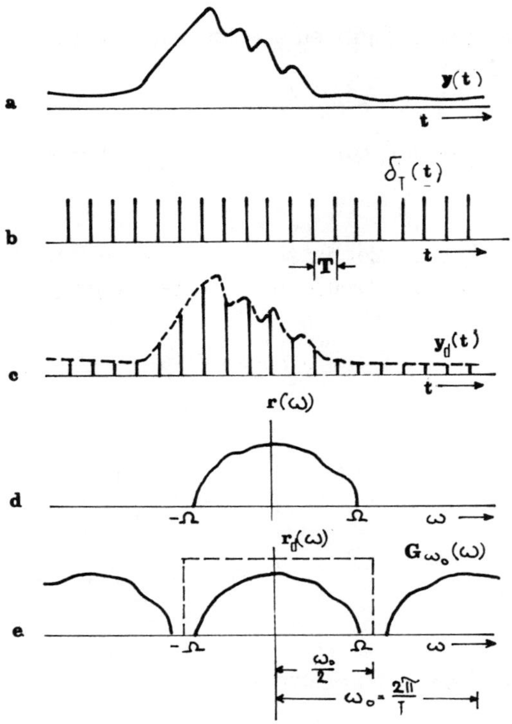

Figure 1.2 The relationship between the character of the measured and digitized spectrum in time t and frequency ω domain: **(a)** measured spectrum $y(t)$; **(b)** the series of the unit pulses $\delta_T(t)$; **(c)** the digitized spectrum $y_d(t)$; and **(d)** the Fourier projection $r(\omega)$ of the digitized spectrum and the rectangular low-frequency filter $G_{\omega_0}(\omega)$. (From Blass, W. E., *Appl. Spectrosc. Rev.*, 11, 57, 1976. With permission.)

Let the Fourier projection of the spectrum $y(t)$ be labeled $r(\omega)/2\pi$. From Figure 1.2d, the limitation of this projection by the maximal angle frequency Ω, or by the maximal frequency $f = \Omega/2\pi$ is supposed. The frequency limit is determined from the frequency bandpass of the amplifier used, and those other electronic elements detecting and processing the signal. If an accurate representation of the spectrum $y(t)$ by the function $y_d(t)$ is required, the sampling interval must be sufficiently small. For the Fourier projection (see Appendix D) of the impulse series $\delta_T(t)$, the following relationship can be derived:

$$\mathcal{F}\left\{\delta_T(t)\right\} = \omega_0\,\delta_{\omega_0}(\omega) = \begin{Bmatrix} 0 & \text{for } \omega \neq 0 \\ \omega_0 & \text{for } \omega = 0 \end{Bmatrix} \qquad \textbf{(1.3)}$$

where $\omega_0 = 2\pi/T$. If the convolution in the Fourier projection space corresponding to the product in Equation 1.2 is considered, the relationship of the

Fourier projection to the digitized spectrum is obtained from Equations 1.2 and 1.3 (Figure 1.2e):

$$r_d(\omega) = \frac{1}{T} r(\omega) * \delta_{\omega_0}(\omega) = \frac{1}{T} \sum_{k=-\infty}^{\infty} r(\omega - k\omega_0) \tag{1.4}$$

It is obvious from Figure 1.2e that the Fourier projection $r(\omega)$ can be obtained from the projection $r_d(\omega)$ by application of the low-frequency rectangle filter $G_{\omega_0}(\omega)$:

$$r(\omega) = T\, r_d(\omega)\, G_{\omega_0}(\omega) \tag{1.5}$$

where

$$G_{\omega_0}(\omega) = \left\{ \begin{vmatrix} 1 \text{ for } |\omega| \le \dfrac{\omega_0}{2} \\ \\ 0 \text{ for } |\omega| > \dfrac{\omega_0}{2} \end{vmatrix} \right\} \tag{1.6}$$

After the inverse Fourier transformation

$$y(t) = T\, y_d(t) * \text{sinc}\left(\omega_0 \frac{t}{2}\right) \tag{1.7}$$

and rewriting the convolution product

$$y(t) = \sum_{k=-\infty}^{\infty} y_d(kT)\, \text{sinc}\left[\omega_0 \frac{(t - kT)}{2}\right] \tag{1.8}$$

is obtained, where (see Appendix D)

$$\text{sinc}(x) = \frac{\sin(x)}{x} \tag{1.9}$$

From the treatment above, it is apparent that the spectrum $y(t)$ can be obtained from the digitized spectrum only if the following inequality is fulfilled:

$$\frac{T}{2\pi} \le \frac{1}{2\Omega} \tag{1.10}$$

where

$$T \le \frac{1}{2f} \tag{1.11}$$

The value T satisfying the equality in Equation 1.10 is called the Nyquist interval.[5]

If the value T is higher than $1/2f$, multiple representations of the spectrum $r(\omega)$ projection mutually overlap in the projection $r_d(\omega)$, and by applying Equation 1.8 a false result is obtained. This effect is known as aliasing.[6]

If $T \ll 1/2f$, the digitization is uselessly dense (the spectrum is strongly oversampled), and the numerical processing becomes wasteful of both time and memory.

In the work of Kelly and Horlick[7] the effects of sampling interval, sampling duration, quantization, digitization time, aperture time, and random variations in sampling interval were examined. The maximum sampling interval and minimum number of samples needed to digitize triangular, exponential, Lorentzian, and Gaussian peaks for given values of maximum absolute errors were tabulated.

SPECTROPHOTOMETER CALIBRATION

The individual values of the dependent variable y and independent variable x are determined from the scales of the spectrometer. In the recording devices these scales are usually printed on registered paper, whereas for spectrometers connected on-line with a computer, the introduced values are emitted from the spectrometer in digital form through the interface between spectrometer and computer. The calibration of both scales is necessary for precise measurements. The means of calibration depends on the type of spectroscopic method; generally, in calibration, precise values obtained from spectra of suitable chosen standards are compared with data given by the measuring device.

The task of numerical methods at spectrometer calibration is the analytical expression of the functional relationships

$$x^{\text{prec}} = f\left(x^{\text{mea}}\right) \tag{1.12}$$

and

$$y^{\text{prec}} = g\left(y^{\text{mea}}\right) \tag{1.13}$$

where upper indices (prec and mea) represent the precise and measured values. After choosing suitable types of functions f and g, the problem is reduced to the standard problem of regression analysis (see Appendix C), starting from the sets of value pairs (x_i^{prec}, x_i^{mea}), and (y_i^{prec}, y_i^{mea}) obtained by measurements of standards.

The functions f and g are often expressed in polynomial form. However, the disadvantages of polynomial functions are their relative inflexibility for lower degree polynomial applications and, on the other hand, the oscillation properties of higher degree polynomials. Therefore, application of spline functions[8] (see Spectral Smoothing, below) has been recommended recently. A greater number of calibration standards, uniformly distributed in the measured area, is necessary to enable the suitable distribution of the knot points of the spline function.

For example, at precise measurements in infrared spectroscopy[8-14] wavenumber calibration is based on the use of known rotational-vibrational spectra of gaseous compounds. A more dense calibration grid can be obtained by recording interference spectra of the empty measuring cell with a precisely determined thickness (Figure 1.3).

Devices with rotating segments that produce a precise reduction of the radiant flux or the measuring (reference) flux can be used for calibration of the photometric scale.

SPECTRAL SMOOTHING

The procedure by which noise is suppressed from the digitally represented experimental data is designed as smoothing, or digital filtering. Recently, a relatively large number of numerical methods have been developed for spectral smoothing.[5,15] Certain undesirable deformation of spectral information always appears with the application of any filtering method. Noise suppression depends not only on the method of smoothing used but also on the character of the spectral data. This fact must be

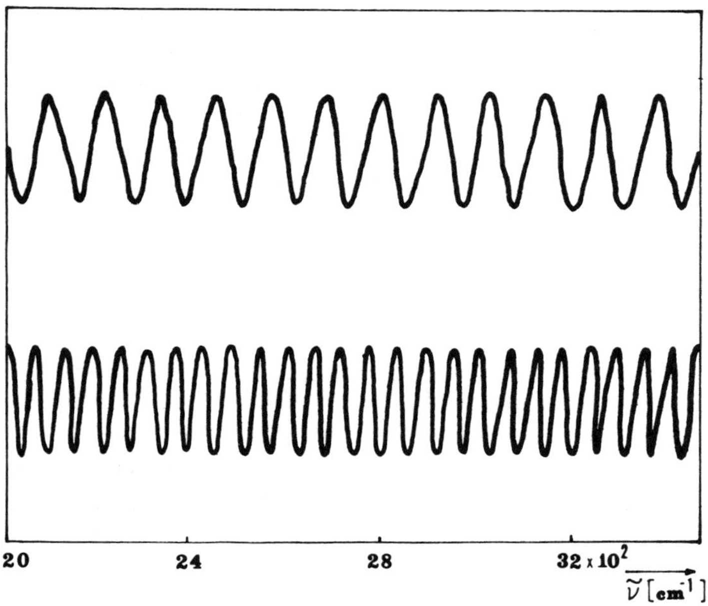

Figure 1.3 The IR spectrometer record of NaCl cells' interference bands: **(A)** for the thickness of the measuring cell, 0.004 cm; **(B)** for the cell thickness, 0.01 cm.

taken into consideration when evaluating the suitability of any numerical smoothing method.

One of the most frequently applied procedures for spectral smoothing was developed by Savitzky and Golay.[16] The method is applicable not only for all types of spectra but for any experimental data digitized with a constant step $\Delta\nu$ of an independent variable ν (wavenumber, wavelength, and so on) as well. The smoothed functional value $y_i^* = y^*(\nu_i)$ for the ith point of a spectrum is obtained from approximation of $2m + 1$ values of $y_j (j = i - m, i - m + 1, \ldots i + m)$ of the original spectrum by the kth degree polynomial, with $k < 2m$. The constant spectral digitization is used to introduce the transformed independent variable:

$$x_j = \frac{\nu_j - \nu_i}{\Delta\nu} = j - i \quad \textbf{(1.14)}$$

Then, for values $j = i - m, i - m + 1, \ldots i + m$, a new independent variable acquires the values $-m, -m + 1, \ldots m$, independent of the chosen value i, when $x_i = 0$. Generally, the polynomial of approximation can be expressed by the relationship

$$f(x) = \sum_{n=1}^{k+1} b_n \, x^{n-1} \quad \textbf{(1.15)}$$

where b_n are unknown coefficients, whose values are determined by minimizing the sum of the squared deviations (see Appendix C):

$$S(\vec{b}) = \sum_{j=i-m}^{i+m} \left[y_j - \sum_{n=1}^{k+1} b_n \, x_j^{n-1} \right]^2 \quad \textbf{(1.16)}$$

The solution of the minimization can be presented in matrix form:

$$\vec{b} = \bar{\bar{A}}^{-1} \vec{c} \quad \textbf{(1.17)}$$

where

$$A_{rs} = \sum_{j=i-m}^{i+m} x_j^{r+s-2} = \sum_{j=-m}^{m} j^{r+s-2} \quad \textbf{(1.18)}$$

$$c_r = \sum_{j=i-m}^{i+m} y_j \, x_j^{r-1} = \sum_{j=-m}^{m} y_{i-j} \, j^{r-1} \quad \textbf{(1.19)}$$

for $r, s = 1, 2, \ldots k + 1$.

The fact that the elements of the symmetrical matrix $\bar{\bar{A}}$ (Equation 1.18) depend only on a value m and do not depend on concrete values of ν_i is the advantage of the transformation introduced (Equation 1.14) for the independent variable. Moreover, elements of a matrix $\bar{\bar{A}}$ defined by the sum of unpaired powers j have values of zero, and so simplify the calculation of an inverse matrix $\bar{\bar{A}}^{-1}$ (see Appendix B). Rewriting Equation 1.17, the following expression for the individual coefficients of the approximation polynomial will be obtained:

$$b_r = \sum_{s=1}^{k+1} \left(\bar{\bar{A}}^{-1} \right)_{rs} c_s$$

$$= \sum_{j=i-m}^{i+m} \left(\sum_{s=1}^{k+1} \left(\bar{\bar{A}}^{-1} \right)_{rs} \cdot (j-1)^{s-1} \right) y_j \quad \textbf{(1.20)}$$

for $r = 1, 2, \ldots k + 1$. Substituting in Equation 1.15 yields the next equation for the value y_i^*:

$$y_i^* = f(x_i) = f(0) = b_1 \quad \textbf{(1.21)}$$

If the labeling

$$c_{rj} = \sum_{s=1}^{k+1} \left(\bar{\bar{A}}^{-1} \right)_{rs} \cdot (j-1)^{s-1} \quad \textbf{(1.22)}$$

is introduced, then

$$y_i^* = y^*(\nu_i) = \sum_{j=i-m}^{i+m} c_{1j}\, y_j \quad \textbf{(1.23)}$$

results, i.e., the smoothed spectrum can be expressed as the convolution of an original spectrum with the function

$$g(\nu) = \sum_{j=-m}^{m} \delta(\nu - j\Delta\nu) c_{1j} \quad \textbf{(1.24)}$$

i.e.,

$$y^*(\nu_i) = \sum_j g(\nu_i - \nu_j)\, y(\nu_j) \quad \textbf{(1.25)}$$

The coefficients c_{rj} are tabulated for various values of k and m in Reference 16. The approximate polynomial (Equation 1.15) also can be used for calculation of the spectral derivatives. For the nth derivation of the smoothed spectrum y_i^*:

$$y_i^{*(n)} = f^{(n)}(x_i) = f^{(n)}(0) = n!\, b_{n+1} \tag{1.26}$$

is obtained for $n \le k$, by application of Equations 1.20 and 1.22:

$$y_i^{*(n)} = n! \sum_{j=i-m}^{i+m} c_{n+1,j}\, y_j \qquad \text{for } n \le k \tag{1.27}$$

i.e., the spectral derivatives also are expressed as a certain type of convolution of original spectral values. An example using this method for calculation of the first four derivatives of the spectrum is described in Chapter 2.

The whole method of spectral smoothing can be repeated several times with the same or different values of m and k. If τ repetitions with the same values of m and k are supposed, the method of smoothing is uniquely determined by the triplet of values k, m, and τ. This fact accounts for the wide range of applications of the Savitzky-Golay (S-G) filtering, and also the need for universal optimization of the selection of values k, m, and τ. Many papers[16-28] deal with the investigation of the general properties, various modifications, and ways of applying S-G filtering.

The best results of calculations of the spectral derivatives are obtained using preliminary smoothing by an S-G filter of the fourth degree and followed by differentiation using a sixth degree S-G filter,[18] as shown in the model examples. The bandwidth $(2n + 1)\Delta\nu$ of both filters must be close to the widths of the spectral bands at half of their maximal height.

The polynomial approximation of the S-G method has the character of the convolution. The division of the spectrum into a sequence of overlapping segments, in which the spectrum is approximated by polynomials, forcing continuity of the functional values and some of the derivatives, is one benefit of polynomial approximation. The condition of continuity connects some of the coefficients of the adjacent polynomials and so reduces the total number of unknowns. The approximation by such functions is generally similar to approximation by spline functions.[29]

Gans and Gill[30] studied spectral smoothing by B-splines. The polynomial spline of the degree h and of the order k ($k = h + 1$) consists of the series of $m + 1$ segments forming hth degree polynomials in the independent variable ν connected at the m points. The values of ν at the connecting points are called the internal knots. If all internal knots have values different from one another, then the spline function y^* and its first $(h - 1)$ derivatives are continuous over the whole range of ν. The derivative of the hth degree, $y^{*(h)}$, is discontinuous at the internal knots; between them it has a constant value. The spline function of order h with m internal knots is determined by the $k + m$ parameters. All spline functions of order k with m internal knots given can be expressed as linear combinations of the set $k + m$ basic splines (B-splines). B-splines are uniquely defined by the set of $2k + m$ points with values of $\nu = t_1, t_2,\ldots t_{2k+m}$, where in addition to m internal knots, $2k$ of the external arbitrarily defined knots are involved. The values of t_i must be strictly arranged in increasing order, i.e.,

$$t_i \le t_{i+1} \qquad \text{for } i = 1, 2,\ldots 2k + m - 1 \tag{1.28}$$

For simplicity in spectroscopic applications, the first k external knots are selected at the first point of the spectrum and the last k external knots at the last point of the spectrum, i.e.,

$$t_i = \nu_{\min} \qquad \text{for } i = 1, 2,\ldots k \tag{1.29}$$

$$t_i = \nu_{\max} \qquad \text{for } i = m + k + 1, m + k + 2,\ldots m + 2k \tag{1.30}$$

where $\nu_{\min}$ and $\nu_{\max}$ are the minimal and maximal values of the independent variable.

Let us label the value of the ith B-spline of the kth order at the point ν as $B_{ik}(\nu)$. The values of the individual B-splines can be evaluated with respect to the relationships[31-34]

$$B_{i1}(\nu) = \begin{Bmatrix} 0 \text{ for } x < t_i \\ 1 \text{ for } t_1 \le x < t_{i+1} \\ 0 \text{ for } x > t_{i+1} \end{Bmatrix} \tag{1.31}$$

$$B_{ik}(\nu) = \frac{\nu - t_i}{t_{i+k-1} - t_i} B_{i,k-1}(\nu) + \frac{t_{i+k} - \nu}{t_{i+k} - t_{i+1}} B_{i+1,k-1}(\nu) \qquad \text{for } k > 1 \tag{1.32}$$

Then, each B-spline has a nonzero value only inside the interval limited by the series $k + 1$ of the neighboring knots (Figure 1.4). On the contrary, at most k splines can have nonzero values at any point ν, which significantly simplifies the calculations. The resulting spline function is defined as the linear combination of the $k + m$ B-splines

$$y^*(\nu) = \sum_{i=1}^{k+m} c_i \, B_{ik}(\nu) \tag{1.33}$$

where the unknown coefficients c_i are determined by minimizing the sum of the squared deviations:

$$S(\vec{c}) = \sum_j \left[y(\nu_j) - \sum_i c_i \, B_{ik}(\nu_j) \right]^2 \qquad \textbf{(1.34)}$$

where the first summation is performed over all the points of the digitized spectrum.

By successive differentiation of Equations 1.33, 1.32, and 1.31, the following recurrent relationships can be obtained for the calculation of the nth derivative:

$$y^{*(n)}(\nu) = \sum_i c_i^{(n+1)} \, B_{i,k-n}(\nu) \qquad \textbf{(1.35)}$$

where

$$c_i^{(n+1)} = c_i \qquad \text{for } n = 0 \qquad \textbf{(1.36)}$$

and

$$c_i^{(n+1)} = \frac{\left[c_i^{(n)} - c_{i-1}^{(n)}\right][k-n]}{t_{i+k-n} - t_i} \qquad \text{for } n > 0 \qquad \textbf{(1.37)}$$

Gans and Gill[30] brought attention to the optimal choice of the knot locations as well. Based on theoretical considerations, the criterion of the maximal smoothness of the resulting spline can be proposed for the optimal knot distribution[29]. This criterion leads to the minimax task that requires obtaining a minimum of the maximal value of a scaled norm of a kth spectral derivative defined between the pairs of neighboring knots. The result is the requirement that the knots be placed so that

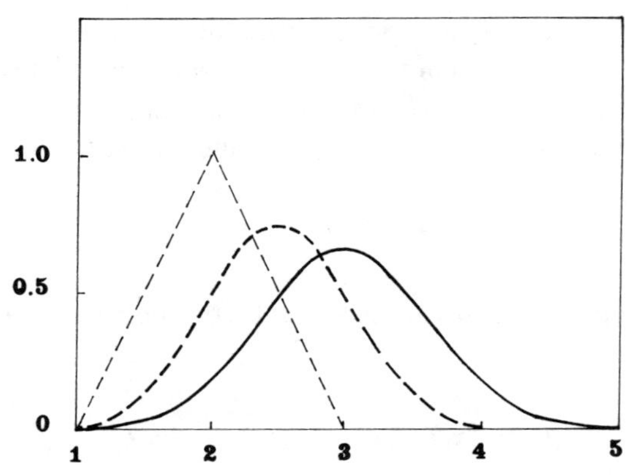

Figure 1.4 B-spline B_{lk} on the knots 1, 2, 3, 4, 5: $k = 2$ (–•–•–), $k = 3$ (– – – – –), $k = 4$ (———). (From Gans, P. and Gill, J. B., *Appl. Spectrosc.,* 38, 370, 1984. With permission.)

$$\sum_{\nu \geq a}^{\nu < b} \left| y^{(k)}(\nu) \right|^{\frac{1}{k\Delta\nu}} = \frac{1}{m+1} \sum_{\nu = t_k}^{t_{k+m}} \left| y^{(k)}(\nu) \right|^{\frac{1}{k\Delta\nu}} \tag{1.38}$$

where

$$a = t_i \tag{1.39}$$

$$b = t_{i+1} \qquad \text{for } k \leq i \leq k+m \tag{1.40}$$

and $\Delta\nu$ is the interval of the spectral data digitization. The summation on the left side is performed via all the spectral points lying between the knot points t_i and t_{i+1}. The summation on the right side is performed over all the spectral points. The values of the kth derivative must be estimated by some independent method, because the kth derivative of the kth order spline is identically zero. First of all, the right side of Equation 1.38 is calculated. If $i = k$, the maximal value t_{i+1}, for which the sum on the left side does not exceed the value of Equation 1.38's right side is found for the step $\Delta\nu$. The method is repeated for all $i = k + 1, k + 2, \ldots k + m - 1$. The optimal positions of the m internal knots are determined in this way. Figure 1.5 illustrates an example of this procedure on the model Gaussian and Lorentz functions.

Horne and Parker[35] proposed the spectral smoothing method based on the approximation of the spectrum by linear combination of Tchebyschev polynomials:

$$y^*(x) = \sum_{i=0}^{k} c_i \, T_i(x) \tag{1.41}$$

where $T_i(x)$ is a Tchebyschev polynomial of the ith degree and c_i are the constants of the linear combination, as determined by least squares. The independent variable x is obtained by a suitable linear transformation, $x = x(\nu)$, so that over the whole spectral range the following will be valid:

$$-1 \leq x(\nu) \leq 1 \tag{1.42}$$

The Tchebyschev polynomial values are determined recursively according to the relationships

$$\begin{aligned} T_i(x) &= 2x\, T_{i-1}(x) - T_{i-2}(x) \qquad \text{for } i \geq 2 \\ T_1(x) &= x \\ T_0(x) &= 1 \end{aligned} \tag{1.43}$$

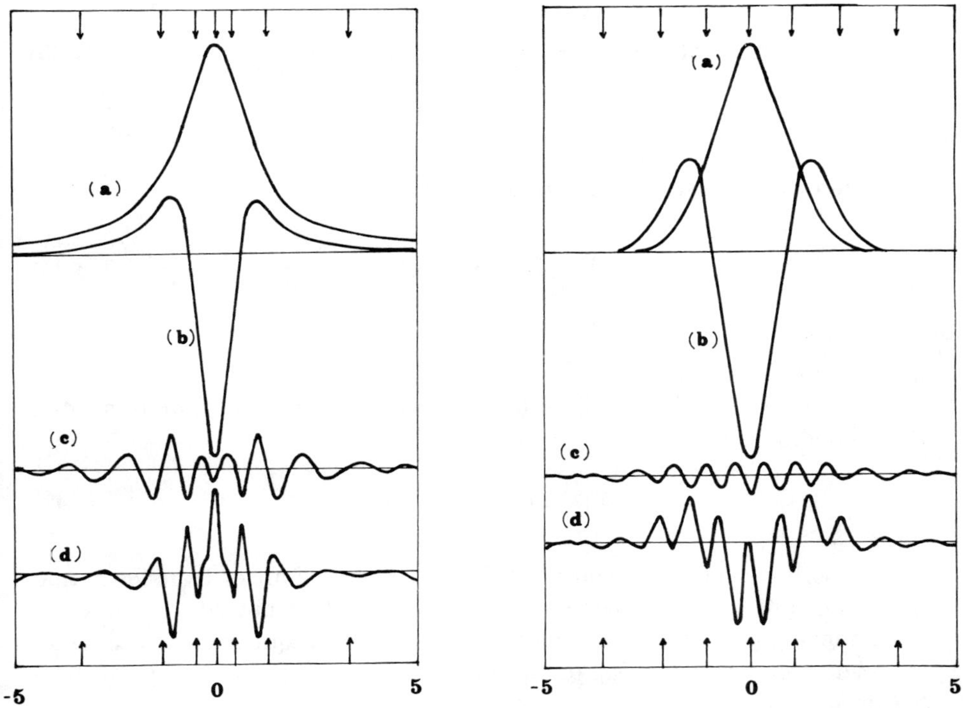

Figure 1.5 The optimized spline function of the fifth degree with the seven internal knots designed by arrows: **(a1)** the Lorentz curve; **(a2)** the Gaussian curve; **(b)** the second derivative calculated from the spline function; **(c)** the difference between the original and the spline functions multiplied by the factor 1000; and **(d)** the difference between the analytical second derivative and the second derivative of the spline function multiplied by the factor 50. (From Gans, P. and Gill, J. B., *Appl. Spectrosc.,* 38, 370, 1984. With permission.)

The nth derivative spectrum can be calculated according to the relationship

$$y^{*(n)}(\nu) = \sum_{i=0}^{k} c_i \frac{d^n T_i(x)}{dx^n} \left(\frac{dx}{d\nu} \right)^n \quad \textbf{(1.44)}$$

The advantage of this method is that about 25 to 30 coefficients are enough for the description of the whole spectrum. One disadvantage is the necessity of the transformation of the independent variable.

It is also necessary to mention here spectral smoothing by a Fourier transformation.[36-38] Besides the spectral information the experimental spectrum also contains superimposed high frequency noise. If these high frequency components in the Fourier projection of the experimental spectrum are eliminated by a suitable filter, the smoothed spectrum is obtained by the inverse Fourier transformation (see Chapter 2 and Appendix D).

The discrete Fourier transformation can be defined for a spectrum digitized using N equidistant points:

$$\mathcal{F}\{y(j)\} = Y(k) = \frac{1}{N}\sum_{j=0}^{N-1} y(j)\exp\left(-i\,\frac{2\pi jk}{N}\right) \quad \textbf{(1.45)}$$

for $k = 0, 1,\ldots,N - 1$.

Then the inverse Fourier transformation

$$\mathcal{F}^{-1}\{Y(k)\} = y(j) = \sum_{k=0}^{N-1} Y(k)\exp\left(i\,\frac{2\Pi jk}{N}\right) \quad \textbf{(1.46)}$$

is valid for $j = 0, 1,\ldots,N - 1$.

For smoothing with the discrete Fourier transformation (Equation 1.45), the following relationship is given:

$$Y(k) = 0 \qquad \text{for } k > k_{\max} \quad \textbf{(1.47)}$$

The smoothed spectrum is obtained by the inverse Fourier transformation (Equation 1.46). The fundamental assumption of this procedure is that all the components of the transformed spectrum with frequencies greater than the frequency corresponding to $k_{\max}$ are created by high frequency, stochastic noise.

The Fourier transformation (Equation 1.45) represents harmonic analysis of a periodic function with a period of N points, i.e., it is assumed that the spectrum with the period N is repeated so the last spectral point "confluents" with the first point and the whole spectrum is repeated like this an infinite number of times (Figure 1.6a). If values $y(0)$ and $y(N - 1)$ are different, the periodic function has discontinuities in the boundary points $\pm(N - 1)\mu$, where μ is any integer. The nonzero high frequency components $Y(k)$ (for $k > K_{\max}$) correspond to these discontinuities. If these components are suppressed in smoothing, a distorted spectrum is obtained after the inverse transformation (Equation 1.46). Therefore the spectrum $y(j)$ is transformed to the spectrum $\eta(j)$ by the subtraction of the line passing through the first and last spectral points, described by the equation

$$\eta(j) = y(j) - y(0) - j\,\frac{y(N-1) - y(0)}{(N-1)} \quad \textbf{(1.48)}$$

for $j = 1, 2,\ldots,N - 1$. The function $\eta(j)$ fulfills the condition

$$\eta(0) = \eta(N-1) = 0 \quad \textbf{(1.49)}$$

and with its periodic repetition eliminates the functional value discontinuity. Its first derivative is discontinuous at the boundary points. Therefore, Arnold et al.[36] proposed extending the function η to $2N - 1$ points by adding to $N - 1$ points obtained by the inversion around the last point of a spectrum $\eta(N-1)$. The new function $\xi(j)$ will be an odd function with the period $2N-1$:

$$\xi(j) = \begin{cases} \eta(j) & \text{for } j = 0, 1, \ldots N-1 \\ -\eta(2N-2-) & \text{for } j = N, N+1, \ldots 2(N-1) \end{cases} \tag{1.50}$$

The function ξ is smooth in the boundary points, and the Fourier coefficients are strictly imaginary, as a consequence of its odd properties:

$$\begin{aligned} \mathcal{F}\{\xi(j)\} &= X(k) \\ &= \frac{1}{2N-1} \sum_{j=0}^{2(N-1)} \xi(j) \exp\left[-i \frac{2\pi jk}{(2N-1)}\right] \\ &= \frac{-i}{2N-1} \sum_{j=0}^{2(N-1)} \xi(j) \sin\left[\frac{2\pi jk}{(2N-1)}\right] \end{aligned} \tag{1.51}$$

for $k = 0, 1, \ldots 2(N-1)$. The high frequency part of the Fourier projection (Equation 1.51) of the function ξ does not contain the components resulting

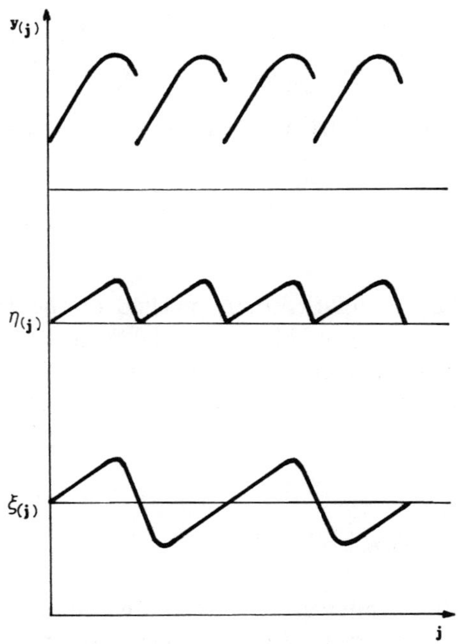

Figure 1.6 The mutual relationship of the functions $y(j)$, $\eta(j)$ and $\xi(j)$.

from the discontinuities in the boundary points, and its filtration will result in the smoothing of this function, i.e., elimination of the noise with minimal distortion of spectral information. The mutual relationship of the functions $y(j)$, $\eta(j)$, and $\xi(j)$ is illustrated in Figure 1.6.

As was shown by Arnold et al.,[35] this method also gives good results for the calculation of the derivative spectra. Coefficients $X^n(k)$ of the Fourier series of the nth spectral derivative are simply related to the Fourier series of the undifferentiated spectrum:

$$X^{(n)}(k) = (-i\,2\pi n k)^n\; X(k) \tag{1.52}$$

for $k = 0, 1, \ldots 2(N-1)$. For the calculation of the smoothed value of the nth derivative of the transformed spectrum $h^{*(n)}$, the following relationship can be derived:

$$\eta^{*(n)}(j) = \sum_{k=0}^{k_{max}} (-i\,2\pi k)^n\; X(k)\exp\left[i\,\frac{2\pi jk}{(2N-1)}\right] \tag{1.53}$$

for $j = 0, 1, \ldots N-1$ and for the nth derivative of the untreated, smoothed spectrum $y^{*(n)}$ from Equations 1.48 and 1.50:

$$y^{*(n)}(j) = \eta^{*(n)}(j) + \frac{d^n}{dj^n}\left\{y(0) + j\,\frac{[y(N-1) - y(0)]}{(N-1)}\right\} \tag{1.54}$$

results for $n = 0, 1, 2, \ldots$. Starting with the second derivative

$$y^{*(n)}(j) = \eta^{*(n)}(j) \tag{1.55}$$

is valid.

Finally, the Gasyn and Stillman method of the Fourier analysis of luminescence extinction should be mentioned.[39] In this case, after the Fourier transformation of the convolution product, the functions of the coefficients digitized with N points are extended to N other points with the functional value of zero.

SPECTRAL DECONVOLUTION

The dependence of the output signal S on the wavenumber ν' can be expressed for any dispersive spectrometer by the integral relationship:[5,40]

$$S(\nu') = \int\!\!\int I(\nu)\, G(\nu' - \nu)\, A(\nu' - \nu'')\,d\nu\, d\nu'' \tag{1.56}$$

where $I(\nu)$ describes the radiation source, $G(\nu' - \nu)$ is the instrument function, and $A(\nu' - \nu)$ is the measured absorption spectrum. Equation 1.56 describes a double convolution product (see Appendix D):

$$S(\nu) = I(\nu) * G(\nu) * A(\nu) \tag{1.57}$$

If the response characteristic of the detector, $E(\nu)$, and of the electronic circuits processing the signal, $S(\nu)$, are added to Equation 1.57, the following is obtained for the output spectrum:

$$y(\nu) = I(\nu) * G(\nu) * A(\nu) * E(\nu) \tag{1.58}$$

With a scanning spectrometer, there is a direct relation between the wavenumber ν and the time t ($\nu = \nu[t]$), so that Equation 1.58 can be rewritten to the time scale

$$y(t) = I(t) * G(t) * A(t) * E(t) \tag{1.59}$$

In fact, the signal $y_d\,(t)$ obtained by sampling the output with time step T

$$y_d(T) = y(T)\,\delta_T(t) = \sum_{k=-\infty}^{\infty} y(kT)\,\delta(t - kT) \tag{1.60}$$

is measured, where δ is the Dirac function (see Appendix D).

The convolution of the functions $I(t)$, $G(t)$, and $E(t)$ can be formally replaced by the general instrument function $F(t)$, where:[41]

$$F(t) = I(t) * G(t) * E(t) \tag{1.61}$$

and, for the sampled signal, the following is obtained:

$$y_d(t) = A(t) * F(t) * \delta_T(t) \tag{1.62}$$

The instrument function closely associated with the spectral slot width can be approximated, e.g., by Gaussian, trapezoidal, or triangular functions.[8] The function $F(t)$ can be determined experimentally[42] or by calculation.[43]

The aim of the convolution procedure is the elimination of the instrument function $F(t)$ from the spectral record $y_d(t)$ and the optimization of physical information $A(t)$, sometimes also called the "true" spectrum.

If the function $F(t)$ is understood to be a filter, then the deconvolution task is to find the inverse filter $F^{-1}(t)$ defined by the relationship

$$A_d(t) = y_d(t) * F^{-1}(t) = A(t)\,\delta_T(t) \tag{1.63}$$

Many deconvolution methods have been proposed for the calculation of this task. Their review and mutual comparison is introduced in the work of Frieden.[44] Further, special attention has been given to the Willson modification,[45] and to Jansson's iterative algorithm.[46,47] The van Citter iterative scheme,[1,48] according to which the nth approximation $a_n(t)$ of the true spectrum $A(t)$, is recursively defined by the following relationship

$$a_n(t) = a_{n-1}(t) + \left[y(t) - a_{n-1}(t) * F(t)\right] \qquad \textbf{(1.64)}$$

as the basis of the procedure; the starting estimation

$$a_0(t) = y(t) \qquad \textbf{(1.65)}$$

is applied for the first approximation.

Jansson[46,47] and Willson[45] extended the correction used in the van Citter scheme by adding the relaxation parameter $\alpha_n(q)$:

$$a_n(t) = a_{n-1}(t) + \alpha_n(q)\left[y(t) - a_{n-1}(t) * F(t)\right] \qquad \textbf{(1.66)}$$

where $q = a_{n-1}(t)$. The dependence of the Jansson relaxation parameter on the value q is shown in Figure 1.7. The limitation of the values $a_n(t)$ to the given interval is ensured by the choice of the $q_{\min}$ and $q_{\max}$ values. For example, by normalization of the absorption spectrum to the interval (0, 1) nonphysical, negative values in transmittance, or transparency values greater than 100%, can be eliminated. This algorithm has a complicated, nonlinear character, due to the dependence of the relaxation parameter a_n on the estimated spectral values. For special cases, with constant $\alpha_n = \alpha$ (e.g., $\alpha = 1$ for the van Citter method), the first n iterations are equivalent to the application of the linear inverse filter:[5]

$$F_n^{-1}(t) = \sum_{i=0}^{n} (-1)^i \left[\binom{n}{i} + \alpha\binom{n}{i+1}\right]\alpha^i f^i \qquad \textbf{(1.67)}$$

where

$$f^i = \frac{F(t) * F(t) * \ldots * F(t)}{i} \qquad \textbf{(1.68)}$$

Then for the nth estimation of $A(t)$ the following is valid:

$$a_n(t) = F_n^{-1}(t) * y(t) \qquad \textbf{(1.69)}$$

During the iterative refinement, according to Equation 1.66, the average quadratic correction

$$E_n = \left\{ \frac{1}{t_{max} - t_{min}} \int_{t_{min}}^{t_{max}} \alpha_n^2(q) \left[y(t) - a_{n-1}(t) * F(t) \right]^2 dt \right\}^{\frac{1}{2}} \quad \textbf{(1.70)}$$

is evaluated at every iteration.

Monitoring the decrease of E_n makes judging the convergence of the iterative cycle possible. The convergence speed is significantly influenced by the choice of the functional dependency given by $\alpha_n(q)$. The Jansson function $\alpha_n(q)$ is schematically illustrated in Figure 1.7; for $q_{min} = 0$ and $q_{max} = 1$ it has the following form:

$$\alpha_n(q) = \alpha_n\left[a_{n-1}(t)\right] = \alpha_{max}\left(1 - 2\left|a_{n-1}(t) - 0.5\right|\right) \quad \textbf{(1.71)}$$

where α_{max} is the constant.

Willson proposed the quadratic function:

$$\alpha_n(q) = \alpha_n\left[a_{n-1}(t)\right] = 4\alpha_{max}\left[a_{n-1}(t) - a_{n-1}^2(t)\right] \quad \textbf{(1.72)}$$

instead of the Jansson triangular function (Equation 1.71). The Willson function eliminates the oscillatory behavior sometimes observed with the application of the Jansson triangular function.

The general function of the type

$$\alpha_n\left[a_{n-1}(t)\right] = \alpha_{max}\left\{a_{n-1}(t)\left[1 - a_{n-1}(t)\right]\right\}^m \quad \textbf{(1.73)}$$

was tested for various values of m.

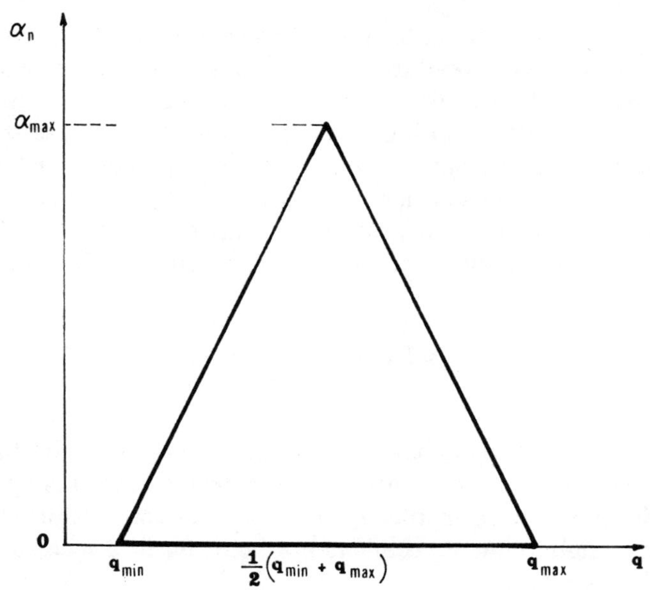

Figure 1.7 The course of the triangle Jansson function $\alpha_n(q)$.

The functions (Equations 1.71 to 1.73) assume the constraint of spectral values to the inequality

$$0 \le a_n(t) \le 1 \tag{1.74}$$

For the universal bound of the type

$$A \le a_n(t) \le B \tag{1.75}$$

Frieden[44] proposed the relationship

$$\alpha_n\left[a_{n-1}(t)\right] = C\left[1 - 2(B-A)^{-1}\left|\frac{a_{n-1}(t)-1}{2(A+B)}\right|\right] \tag{1.76}$$

where C is the constant.

The dependence of the relaxation parameter α on the value $a(t)$, obtained by the previous iteration, must be chosen carefully. Also, the investigation into how the value of the area P under the spectral curve is influenced by a choice of $\alpha_n(q)$, i.e., the value of the integral

$$P = \int a_n(t)\,dt \tag{1.77}$$

is necessary. In Figure 1.8, Jansson deconvolution is applied to the CHD_3 segment of the IR spectrum around 2140 cm^{-1} (the part of the Q-segment of the vibrational transition ν_2).[49]

The significant sensitivity to spectral noise is a common property of the iterative deconvolution procedures. Therefore, smoothing the experimental spectra is necessary before deconvolution can be attempted (see Section 1.5).

Recently, more attention has been given to application of the Fourier transformation for deconvolution due to the development of fast computers and algorithms.[50-53] The Fourier projection (see Appendix D) of the true spectrum can be obtained, in principle, by dividing the projection of the measured spectrum by the projection of the instrument function:

$$\mathcal{F}\{A(t)\} = \frac{\mathcal{F}\{y(t)\}}{\mathcal{F}\{F(t)\}} \tag{1.78}$$

However, many high frequency components, which must be filtered out before the backward transformation, are formed during this operation as artifacts. Suitable choice of the apodization function, ensuring accurate attenuation of high frequency components, is the basis of successful Fourier deconvolution (see Chapter 2).

Deutsch and Beniaminy[54] proposed deconvolution based on expression of the functions $y(t)$ and $A(t)$ by the cubic polynomial series with common

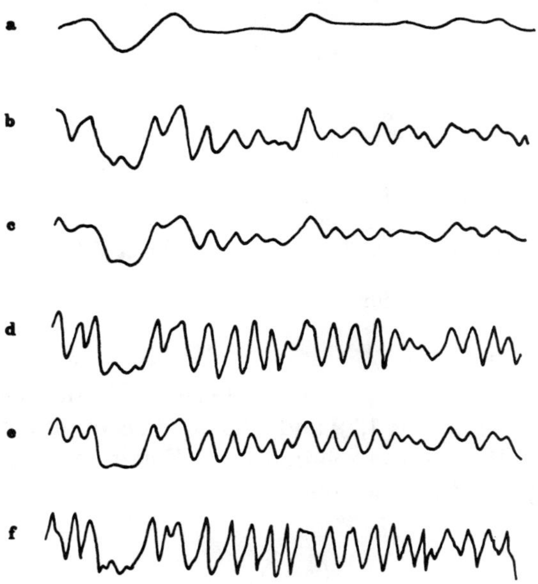

Figure 1.8 The deconvolution results of the IR spectrum of CHD_3 in the vicinity of 2140 cm^{-1} (the Q-branch of the vibration ν_2): the experimental spectra recorded with resolution of 0.055 and 0.022 cm^{-1} **(a, c, e)** and the result of the Jansson deconvolution after 50 iterations with the value α_n from the interval (1.0, 2.5).[49] (From Blass, W. E., *Appl. Spectrosc. Rev.*, 11, 57, 1976. With permission.)

knot points. The method is based on the expression of the true spectrum in the neighborhood of point t by the Taylor series:

$$A(t-s) = A(t) - A'(t)s + \frac{1}{2}A''(t)s^2 - \frac{1}{6}A'''(t)s^3 + \ldots \quad \textbf{(1.79)}$$

Replacing the convolution product (Equation 1.62) yields the following:

$$y(t) = A(t)\,M_0 - A'(t)\,M_1 + \frac{1}{2}A''(t)\,M_2 - \frac{1}{6}A'''(t)\,M_3 + \ldots \quad \textbf{(1.80)}$$

where M_m labels the mth moment of the instrument function:

$$M_m = \int_{-\infty}^{+\infty} t^m\, F(t)\, dt \quad \textbf{(1.81)}$$

In the case of the standardized instrument function, $M_0 = 1$, and the odd moments are zero for the symmetric instrument functions. The true spectrum can be expressed by the superposition of $n - 1$ cubic polynomials of the form

$$A(t) = \sum_{k=1}^{n-1} s_k(t) \tag{1.82}$$

where

$$s_k(t) = \begin{cases} 0 & \text{for } t < \xi_k \\ a_k t^3 + b_k t^2 + c_k t + d_k & \text{for } \xi_k \le t \le \xi_{k+1} \\ 0 & \text{for } t > \xi_{k+1} \end{cases} \tag{1.83}$$

and where values ξ_1, ξ_2,...ξ_n are suitable selected knot points. With such expressions $A(t)$, Equations 1.79 and 1.80 do not contain terms higher than the cubic terms. Placing the relationships (Equations 1.79 and 1.83) into Equation 1.80 yields the following:

$$y(t) = \sum_{k=1}^{n-1} s_k(t) \tag{1.84}$$

where

$$S_k(t) = \begin{cases} 0 & \text{for } t < \xi_k \\ A_k t^3 + B_k t^2 + C_k t + D_k & \text{for } \xi_k \le t \le \xi_{k+1} \\ 0 & \text{for } t > \xi_{k+1} \end{cases} \tag{1.85}$$

The coefficients of Equation 1.85 are related to the coefficients of Equation 1.83 by the relationships

$$\begin{aligned} a_k &= \frac{A_k}{M_0} \\ b_k &= \frac{B_k}{M_0} + \frac{3a_k M_1}{M_0} \\ c_k &= \frac{C_k}{M_0} + \frac{2b_k M_1}{M_0} - \frac{3a_k M_2}{M_0} \\ d_k &= \frac{D_k}{M_0} + \frac{c_k M_1}{M_0} - \frac{b_k M_2}{M_0} + \frac{a_k M_3}{M_0} \end{aligned} \tag{1.86}$$

The experimental spectrum $y(t)$ can be expressed by a cubic spline function,[55] i.e., the function of the type (Equation 1.84) keeping the continuity of the functional value and of the first two derivatives in the knot points. The third derivative in the point ξ_k is designed as δ_k. If the function $A(t)$ is calculated by the relationships illustrated in Equations 1.86 and 1.82, the discontinuous functional values and derivatives

$$\Delta\left[A(t)\right]_{t=\xi_k} = \delta_k\left(M_1^3 - M_1 M_2 + \frac{M_3}{6}\right)$$

$$\Delta\left[A'(t)\right]_{t=\xi_k} = \delta_k\left(M_1^2 - \frac{M_2}{2}\right)$$

$$\Delta\left[A''(t)\right]_{t=\xi_k} = \delta_k M_1$$

$$\Delta\left[A'''(t)\right]_{t=\xi_k} = \delta_k \qquad \textbf{(1.87)}$$

are obtained at the individual knots. From the given relationships, the functional values and the values of the second derivative will be discontinuous at all knot points of the symmetrical instrument function.

Deutsch and Beniaminy[54] analyzed the procedure for choosing the optimal knot points conditioned upon the instrument function and the peaks of the function $y(t)$. As the examples of various model functions showed, the proposed method of deconvolution gives better results for the correct choice of knots than procedures previously used.

NUMERICAL DIFFERENTIATION OF SPECTRA

Interpolation of spectra is convenient if derivatives with respect to the wavenumber and wavelength are used. One of the main advantages of derivative spectroscopy is the possibility of better resolution of small bands overlapping with larger bands, background, or noise (see Chapter 2).[56,57] For this reason, derivative spectroscopy is convenient for quantitative as well as qualitative analysis.

Spectral derivatives can be obtained by three different methods.[57] Frequency modulation, in which the frequency of the incident radiation is sinusoidally modulated around the main frequency and the amplitudes of individual harmonic components are determined on output after detection and amplifying, is used for optical differentiation. Low amplitudes of the higher harmonics restrict the application of this method to lower order derivatives. The second method of optical differentiation is the use of two wavelengths; the sample is simultaneously irradiated by two fluxes with a small difference in their wavelengths (1 to 2 nm). Electronic differentiation is used when spectral derivation is obtained by analog processing of the electrical signal from the third, independent group.

In this chapter attention is given to methods of numerical differentiation, which can be divided into two main groups. The procedures used for direct differentiation, i.e., obtaining the values of derivatives directly from experimental data using the differences, form the first group. The methods of curve fitting, i.e., approximating the spectral data by segments of analytical functions and derivatives obtained analytically by differentiation of the corresponding functional relationships, belong to the second group.

In principle, all procedures yielding derivatives tabulated from the functional values detailed in various handbooks of numerical mathematics[58-61] are direct differentiation methods.

The simplest way to calculate the derivative numerically is based on the differences of the functional values of the neighboring points of the spectrum $A(\lambda)$:

$$\left.\frac{dA}{d\lambda}\right|\bar{\lambda} = \frac{A_{i+1} - A_i}{\lambda_{i+1} - \lambda_i} \tag{1.88}$$

where

$$\bar{\lambda} = \frac{\lambda_{i+1} + \lambda_i}{2} \tag{1.89}$$

The calculation is extremely quick and simple, but the results are extraordinarily sensitive to errors in experimental data caused by high frequency noise. Therefore, variously defined direct or weighted averages of differences around the central point are used. In these procedures the information at the spectral boundaries is lost in each transition, so it is convenient to record the spectrum over a more extended range.

The method suitable for the spectra digitized with constant step wavelength (or wavenumber) $\Delta\lambda$ was proposed by Butler and Hopkins.[62] The differences of the values spaced at α intervals,

$$\left.\frac{dA}{d\lambda}\right|_{\frac{n+a}{2}} = \frac{A(\alpha + n) - A(n)}{\alpha} \tag{1.90}$$

are used to calculate the first derivative, and for simplicity, $\Delta\lambda = 1$ is supposed. The optimal value α is set high enough to decrease the noise influence, but, on the other hand, small enough so the spectral details will not be erased.

The value β, determining the number of intervals in the first derivative value, is chosen independently for the second derivative calculation:

$$\left.\frac{d^2A}{d\lambda^2}\right|_{\frac{n+(\alpha+\beta)}{2}} = \frac{\left[A(\alpha + \beta + n) - A(\beta + n)\right] - \left[A(\alpha + n) - A(n)\right]}{\alpha\beta} \tag{1.91}$$

Similarly, independent intervals γ and δ can be introduced to calculate the third and fourth derivatives. Generally, this procedure can be understood as a convolution defining a certain type of weighted average. Butler and Hopkins[62] studied the dependence of the signal/noise ratio in the fourth derivative on the choice of the values α, β, γ, and δ. The optimal results are obtained when these parameters have very similar, but unequal, values, and $\alpha + \beta \neq \gamma + \delta$. Also, it is necessary that the digitization interval of the spectrum be sufficiently small, so that the values α, β, γ, and δ could be higher than 10. Butler and Hopkins[62] found the following optimal values: of $\Delta\lambda = 0.125$ nm, $\alpha = 21$, $\beta = 19$, $\gamma = 18$, and $\delta = 17$.

The method of curve fitting used here approximates the spectrum, by the linear combination of the analytical functions $f_i(\lambda)$:

$$A(\lambda) \approx \sum_i c_i \, f_i(\lambda) \tag{1.92}$$

The unknown coefficients c_i are determined by the classical least squares method (see Appendix C), and derivatives are calculated analytically according to the relationship

$$\frac{d^n A}{d\lambda^n} = \sum_i c_i \frac{d^n f_i}{d\lambda^n} \tag{1.93}$$

If the number of coefficients c_i is lower than the number of digitized spectrum points, the simultaneous removal of noise (i.e., the spectral smoothing) also results from the curve fitting. Therefore, the most important applied procedures are characterized in the section dealing with spectral smoothing (see Spectral Smoothing, above).

ELIMINATION OF SPECTRAL BACKGROUND

The tails of the bands from the spectral range studied contribute to the spectral background. Besides this, it is necessary to consider measurement effects due to the uncompleted intensity compensations of the measured and reference rays by solvent absorption respective to the measuring cell material, radiation dispersion, reflection, and so on of the background. Also, very weak bands of higher harmonic and combination frequencies contribute to the spectral background in IR spectroscopy. According to the different physical properties of these contributions, the mathematical function describing the background is very complicated. Moreover, only some of these contributions can be determined experimentally.

For these reasons the background is determined empirically using different types of functions.[8] The simplest and most commonly applied function is the polynomial:

$$P(x) = \sum_{i=0}^{n} c_i \, x^i \tag{1.94}$$

If the spectral area studied is small, first-order polynomials adequately describe the background. Orthogonal Legendre polynomials are also used.[63]

Bulmer et al.[64] referred to the contradiction between the necessity of maintaining flexibility in the background function and the need to enhance spectral separation. Their comparative study showed that the exponentially quadratic function of the form

$$P(x) = c_1 \exp(c_2 x) + c_3 + c_4(c_6 - x) + c_5(c_6 - x)^2 \quad \textbf{(1.95)}$$

was the most suitable function. The cubic spline functions and polynomial series seemed less suitable.

Numerical elimination of the spectral background is performed simultaneously with the separation of the spectrum into bands (see Chapter 2) by methods based on nonlinear regression analysis (see Appendix C).

REFERENCES

1. Blass, W. E. and Halsey, G. W., *Deconvolution of Absorption Spectra,* Academic Press, New York, 1981.
2. Blackburn, J. A., Ed., Spectral analysis, in *Methods and Techniques,* Marcel Dekker, New York, 1990.
3. Cadzow, J. A. *Discrete-Time Systems,* Prentice Hall, Engelwood Cliffs, NJ, 1972.
4. Lathi, B. P., *An Introduction to Random Signals and Communication Theory,* Int. Textbook Comp., Scranton, PA, 1968.
5. Blass, W. E., *Appl. Spectrosc. Rev.,* 11, 57. 1976.
6. Vanassi, G. A. and Sakai, H., *Prog. Optics,* 6, 260, 1967.
7. Kelly, P. C. and Horlick, G., *Anal. Chem.,* 45, 518, 1973.
8. Horák, M. and Vítek, A., *Interpretation and Processing of Vibrational Spectra,* John Wiley & Sons, New York, 1978.
9. IUPAC Commission on Molecular Structure and Spectroscopy, *Tables of Wavenumbers for Calibration of Infrared Spectrometers,* Butterworths, London, 1961.
10. Rao, K. N., Humphreys, C. J., and Rank, D. H., *Wavelength Standards in the Infrared,* Academic Press, New York, 1966.
11. Blaine, L. R., Plyler, E. K., and Benedict, W. S., *J. Res. Nat. Bur. Stand., Sect. A,* 66, 223, 1962.
12. Silvera, I. F., Hardy, W. N., and McTague, J. P., *Rev. Sci. Instrum.,* 43, 58 1972.
13. Jones, R. N. and Madeau, A., *Pure Appl. Chem.,* 37, 649, 1974.
14. Jones, R. N. and Madeau, A., *Can. J. Spectrosc.,* 20, 33, 1975.
15. Ormsby, J. F. A., Numerical filtering in spectral analysis, in *Methods and Techniques,* Blackburn, J. A., Ed., Marcel Dekker, New York, 1970.
16. Savitzky, A. and Golay, M. J. E., *Anal. Chem.,* 36, 1627, 1964.
17. Bromba, M. U. A and Ziegler, H., *Anal. Chem.,* 53, 1583, 1981.

18. Bershtein, Ya., Mogilnitsky, M. M., and Komarov, E. V., *Anal. Chim. Acta,* 222, 335, 1989.

19. Korobkov, M. E., *Zh. Prikl. Spektrosk,* 30, 694, 1979.

20. Maden, H. H., *Anal. Chem.,* 50, 1383, 1978.

21. Wilson, P. D. and Edwards, T. H., *Appl. Spectrosc. Rev.,* 12, 1, 1976.

22. Khan, A., *Anal. Chem.,* 59, 654, 1987.

23. Dubrovkin, I. M., *Ž. Prikl. Spektrosk.,* 41, 73, 1984.

24. Gans, P. and Gill, J. B., *Appl. Spectrosc.,* 37, 515, 1983.

25. Bražnik, L. G., *Ž. Prikl. Spektrosk.,* 43, 605, 1985.

26. Arthur, J. W., *J. Raman Spectrosc.,* 5, 9, 1976.

27. Bromba, M. U. A. and Ziegler, H., *Anal. Chem.,* 55, 1299, 1983.

28. Nevius, T. A. and Pardue, H. L., *Anal. Chem.,* 56, 2249, 1984.

29. de Boor, C., *A Practical Guide to Splines,* Springer, New York, 1978.

30. Gans, P. and Gill, J. B., *Appl. Spectrosc.,* 38, 370, 1984.

31. Curry, H. B. and Schoenberg, I. J., *J. Anal. Math,* 17, 71, 1966.

32. Schoenberg, I. J., *Inequalities,* Shishda, O., Ed., Academic Press, New York, 1967.

33. de Boor, C. J., *Approx. Theory,* 6, 50, 1972.

34. Cox, M. G., *J. Inst. Math. Applic.,* 10, 134, 1972.

35. Horne, D. S. and Parker, T. G., *Biochem. Biophys. Acta,* 625, 18, 1980.

36. Arnold, R., Franz, K.-F., Heidecke, G., Krotz, J., and Stark, G., *Z. Anal. Chem.,* 322, 574, 1985.

37. Dubrovkin, I. M., *Zavod. Lab.,* 4, 434, 1978.

38. Aspens, D. E., *Anal. Chem.,* 47, 1181, 1975.

39. Gasyna, Z. and Stillman, M. J., *J. Photochem.,* 38, 1987.

40. Blass, W. E. and Nielsen, A. H., *Methods of Experimental Physics,* Vol. 3A, Williams, D., Ed., Academic Press, New York, 1974.

41. Miller, R. J. and Stace, B. C., Eds., Laboratory Methods, in *Infrared Spectroscopy,* Heyden, London, 1972.

42. Stokes, A. R., *Proc. Phys. Soc. London,* 1, 382, 1948.

43. Hendricks, R. W. and Schmidt, P. W., *Acta Phys. Austriaca,* 26, 97, 1967; 37, 20, 1973.

44. Frieden, B. R., *Picture Processing and Digital Filtering,* Huang, T. S., Ed., Springer, Berlin, 1975.

45. Willson, P. D., Ph.D. dissertation, Michigan State University, East Lansing, 1973.

46. Jansson, P. A., Hunt, R. H., and Plyler, E. K., *J. Opt. Soc. Am.,* 60, 596, 1970.

47. Jansson, P. A., *J. Opt. Soc. Am.,* 60, 184, 1970.

48. van Citter, P. H., *Z. Phys.,* 69, 298, 1931.

49. Halsey, G. W., Jennings, D. E., Boyd, J. W., and Blass, W. E., *29th Symposium on Molecular Structure and Spectroscopy,* Columbus, Ohio, 1972.

50. Mann, C. K. and Vickers, T. J., *Appl. Spectrosc.,* 43, 33, 1989.

51. Kauppinen, J. K., Moffatt, D. J., Mantsch, H. H., and Cameron, D. G., *Appl. Spectrosc.,* 35, 271, 1981.

52. Jones, R. N. and Shimokoshi, K., *Appl. Spectrosc.,* 37, 59, 1983.

53. Friesed, W. I. and Michaelian, E. H., *Appl. Spectrosc.,* 39, 484, 1985.

54. Deutsch, M. and Beniaminy, I., *Rev. Sci. Instrum.,* 53, 90, 1931.

55. Ahlberg, J. H., Nielson, E. N., and Walsh, J. L., *The Theory of Splines and their Application,* Academic Press, New York, 1967.

56. Dixit, L. and Ram, S., *Appl. Spectrosc. Rev.,* 21, 311, 1985.

57. Bridge, T. P., Fell, A. F., and Wardman, R. H., *J. Soc. Dyers Colour.,* 103, 17, 1987.

58. Ralston, A., *A First Course in Numerical Analysis,* McGraw-Hill, New York, 1965.

59. Rektorys, K., *Survey of Applied Mathematics,* Iliffe Books, London, 1969.

60. Kopchenova, N. V. and Maron, I. A., *Computational Mathematics,* MIR, Moskva, 1987.

61. Demidovich, B. P. and Maron, I. A., *Computational Mathematics,* MIR, Moskva, 1987.

62. Butler, W. and Hopkins, D., *Photochem. Photobiol.,* 12, 439, 1970

63. Lipkus, A. H., *Appl. Spectrosc.,* 42, 395, 1988.

64. Bulmer, J. T., Irish, D. E., Grossmann, F. W., Herriot, G., Tseng, M., and Weerheim, A. J., *Appl. Spectrosc.,* 29, 506, 1975

Chapter 2

Resolution of Spectra

Solving many tasks involved with various spectroscopic methods requires the decomposition of a measured spectrum to an additive sum of elementary spectral line shapes. Such elementary forms can be, e.g., various spectral bands of a compound, spectra of individual compounds contained in a mixture, the spectra of various isomers, or spectra from close relatives of a given compound. The decomposition is simplest if an analytical expression for the shape of every band in a spectrum is known, i.e., if the physical effect determining its shape is known. For example, in the vibrational-rotational spectra of gaseous compounds under normal conditions the line shapes of individual rotational components are specified by collisions of elements; therefore, they have a dispersed character well described by the Lorentz function.

However, spectra with unknown physical effects, line broadening, are often studied. Various approximation functions using parameters without any physical meaning, characterizing only the shape of a spectral curve, should be used in such cases. Because the precision of approximation usually increases monotonically with the assumed number of bands, it is necessary to know the number of elementary bands of a given type in a decomposed spectrum.

Recently, with the appearance of fast digital computers, many methods capable of describing spectra consisting of tens of overlapping bands have been reported; often these methods produce results that are more precise than those obtained from spectral devices. The results of these methods, however, are strongly dependent on the suitability of the chosen model and the suitability of the chosen approximation function or functions. In less favorable situations, e.g., a vibrational band with a nonsymmetric shape, often observed in systems with strong intermolecular interactions, the application of such a procedure is rather questionable.

TYPES OF SPECTRAL BANDS

Measuring the spectrum of a single molecule yields very narrow spectral bands. Spectral sets of compounds in a gaseous, liquid, or solid state, as a rule, give considerably broad individual bands. Many physical factors responsible for the broadening of lines are known.

The Heisenberg principle of uncertainty, according to which the interval of a measured energy is inversely proportional to the lifetime of the excited state of a particle, is the basic reason for the broadening of spectral lines. The broadening of spectral lines can also be caused by weak intermolecular interactions that change the electron structure of molecules only slightly, the existence of different energetically close conformations, the possibility of free rotation around individual bands, and so on. The broadening of the spectral bands in vibrational spectra in the gaseous and liquid phases is caused by degenerate rotational transitions within the given vibrational transitions. This is also the case for electronic spectra, often represented by an envelope of various vibrational states. The dynamics of fast chemical reactions can also lead to the broadening of spectral lines. The Doppler effect also usually causes broadening of the spectral lines. A spectral transition derives its characteristic shape according to the type and range of the broadening.

A large number of approximation functions,[1-13] representing, for better or worse, the experimental type of spectral lines, are used for the description of spectral bands. Individual types of approximation functions are characterized by the number of parameters (e.g., bandwidth, position of a band maximum, maximal height of a band, parameter of asymmetry, etc.) defining uniquely the given type of band.

The shapes of spectral line types used in spectroscopy are summarized in Tables 2.1 and 2.2. In spite of the relatively large number of various forms, Lorentz, Gaussian, and lognormal distribution functions (and the combination of Gaussian and Lorentz distribution functions), are the most used for practical applications. Therefore, these types of functions will be emphasized.

The Gaussian shape for a spectral line (Figure 2.1a):

$$Y = Y_0 \exp\left[-4(\ln 2)\frac{(x - x_0)^2}{\Delta x^2}\right] \tag{2.1}$$

is characterized by three parameters: Y_0 is the maximal intensity, x_0 is the position of a band maximum, and Δx is the halfwidth, i.e., the width at the point of half-maximum intensity.

It is convenient to use the Gaussian shape for a spectral line under conditions of strong Doppler broadening of curves or when the spectra consist of quasimonochromatic bands, with the instrument function of the Gaussian type (using the slit width close to the natural width).

The following equation describes the Lorentz shape line for a spectral band (Figure 2.1b):

$$Y = Y_0\left[1 + 4\frac{(x - x_0)^2}{\Delta x^2}\right]^{-1} \tag{2.2}$$

Table 2.1 Types of Symmetric Distribution Functions Used in Spectroscopy

Name of function	Mathematical expression	Number of parameters
Gaussian[1-4]	$Y = Y_0 \exp\left[-4(\ln 2)\dfrac{(x-x_0)^2}{\Delta x^2}\right]$	3
Lorentz (Cauchy)[1-4]	$Y = Y_0\left[1+4\dfrac{(x-x_0)^2}{\Delta x^2}\right]^{-1}$	3
Student[1]	$Y = Y_0\left[1+4\left(\sqrt{2}-1\right)\dfrac{(x-x_0)^2}{\Delta x^2}\right]^{-2}$	3
Reciprocal polynomials[4]	$Y = Y_0\left[\sum_{i=0}^{n} A_i\dfrac{(x-x_0)^{2i}}{\Delta x^{2i}}\right]^{-1}$	$n+2$
Voigt[2]	$Y = \int_{-\infty}^{+\infty}\dfrac{Y_0}{1+\left(\dfrac{Y-x_0}{\Delta x_1}\right)^2}\cdot\exp\left[-\dfrac{(Y-x_0)^2}{\Delta x^2}\right]dY$	4
Cauchy moduled by cosine[8]	$Y = Y_0\left[\cos\left(\pi\dfrac{x-x_0}{\Delta x_1}\right)\right]^k\cdot\left[1+4\dfrac{(x-x_0)^2}{\Delta x_2^2}\right]^{-1}$	5
Product of Lorentz and Gaussian[2]	$Y = Y_0\left[1+4\dfrac{(x-x_0)^2}{\Delta x_L^2}\right]^{-1}\cdot\exp\left[-4(\ln 2)\dfrac{(x-x_0)^2}{\Delta x_G^2}\right]$	4
Combination of Lorentz and Gaussian[2,4]	$Y = Y_0\left\{A\left[1+4\dfrac{(x-x_0)^2}{\Delta x_L^2}\right]^{-1}+(1-A)\exp\left[-4\ln 2\dfrac{(x-x_0)^2}{\Delta x_G^2}\right]\right\}$	4

Table 2.2 Types of Asymmetric Distribution Functions Used in Spectroscopy

Name of function	Mathematical expression	Number of parameters
Cauchy with two halfwidths[9]	$Y = Y_0\left[1 + 4\dfrac{(x-x_0)^2}{\chi^2}\right]^{-1}$ $\chi = \begin{cases}\Delta x_1 & \text{for } x \le x_0 \\ \Delta x_2 & \text{for } x > x_0\end{cases}$	4
Reciprocal polynomials[5]	$Y = Y_0\left[\sum_{i=0}^{n} A_i \dfrac{(x-x_0)^i}{\Delta x^i}\right]^{-1}$	$n+2$
lognormal[10-12]	$Y = Y_0 \exp\left\{-\dfrac{\ln 2}{(\ln\rho)^2}\left[\ln\left(\dfrac{x-x_0}{\Delta x}\cdot\dfrac{\rho^2-1}{\rho}+1\right)\right]^2\right\}$ for $x \ge x_0 - \Delta x\dfrac{\rho}{\rho^2-1}$ $Y = 0$ for $x < x_0 - \Delta x\dfrac{\rho}{\rho^2-1}$	4
Gaussian asymmetrical[13]	$Y = Y_0 \exp\left\{-\left[\dfrac{(x-x_0)\,x_0}{x\,\Delta x}\right]^2\right\}$	4
Lorentz with exponent	$Y = Y_0\left[1 + \left(2^{a^2}-1\right)\cdot\left(2\dfrac{x-x_0}{\Delta x}\right)^2\right]^{-1}$	4

The parameters have the same meaning as do the parameters of the Gaussian line. Compared to the Gaussian function, the Lorentz function is more narrow around the midpoint of a line, but the intensity decreases much more slowly from the half-height, so that the function is more broadened in the wings.

The Lorentz shape of a spectral line is convenient for conditions of radiative decay and for the collision-based line broadening. Subsequently, Δx correlates with the mean lifetime of an excited state, based on the mean free trajectory of a particle.

The combination of the Gaussian and Lorentz spectral bands is often used in the following form:

$$Y = Y_0 \left\{ (1-A)\exp\left[-8\ln 2 \frac{(x-x_0)^2}{\Delta x^2}\right] + A \cdot \left[1 + 4\frac{(x-x_0)^2}{\Delta x^2}\right]^{-1} \right\} A \in (0,1) \tag{2.3}$$

The dependence of a band shape on the mixing parameter A is shown in Figure 2.2.

Spectral bands often have a considerable shape asymmetry. The lognormal distribution function is frequently used in such situations:

$$Y = Y_0 \exp\left\{ -\frac{\ln 2}{(\ln \rho)^2} \left[\ln\left(\frac{x-x_0}{\Delta x} \frac{\rho^2 - 1}{\rho} + 1 \right) \right]^2 \right\} \tag{2.4}$$

for

$$x \geq x_0 - \frac{\Delta x \rho}{\rho^2 - 1} \tag{2.5}$$

and

$$Y = 0 \tag{2.6}$$

for

$$x < x_0 - \frac{\Delta x \rho}{\rho^2 - 1} \tag{2.7}$$

The parameters Y_0, x_0, and Δx are the intensity of the maximum, the position of the maximum, and the halfwidth. The asymmetry parameter ρ is defined by the following relationship (see Figure 2.1c):

$$\rho = \frac{x_R - x_0}{x_0 - x_L} \tag{2.8}$$

The dependence of the shape of the lognormal distribution function on the asymmetry parameter is shown in Figure 2.3.

PROBLEM OF OVERLAP OF SPECTRAL BANDS

If individual spectral bands strongly overlap each other, it is difficult to visually specify their parameters. The example of two almost identical bands of the Gaussian type is shown in Figure 2.4. This situation is more complicated if the individual spectral bands have considerably different intensities (Figure 2.5) or have different half-widths (Figure 2.6). Various numerical algorithms were developed for solving the problem of spectral band overlap; computerized spectral profile analysis uses the following procedures or combinations:

1. Initial non-optimization analysis working with original spectral data without previous indication of band parameters
2. Iterative optimization of band parameters obtained from the previous method

The first method gives estimates of the band parameters, which are inaccurate in the case of overlapping bands. The reconstruction of a profile shows that additional optimization is needed.

The first level of spectral profile analysis can be split into three categories: visual inspection, analog, and digital and analog methods. With sufficient experience, intuition, and logical ability a reasonable estimate of the parameters of individual bands often can be obtained from the visual analysis; these parameters are used in subsequent analysis as the initial

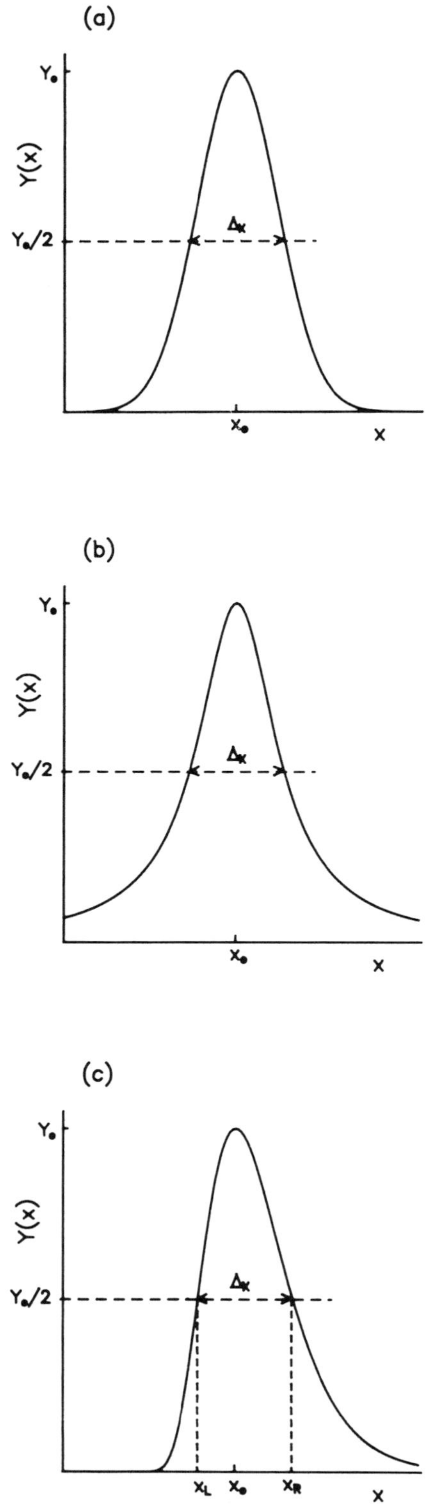

Figure 2.1 The Gaussian **(a)**, Lorentz **(b)**, and lognormal **(c)** shape of a spectral curve.

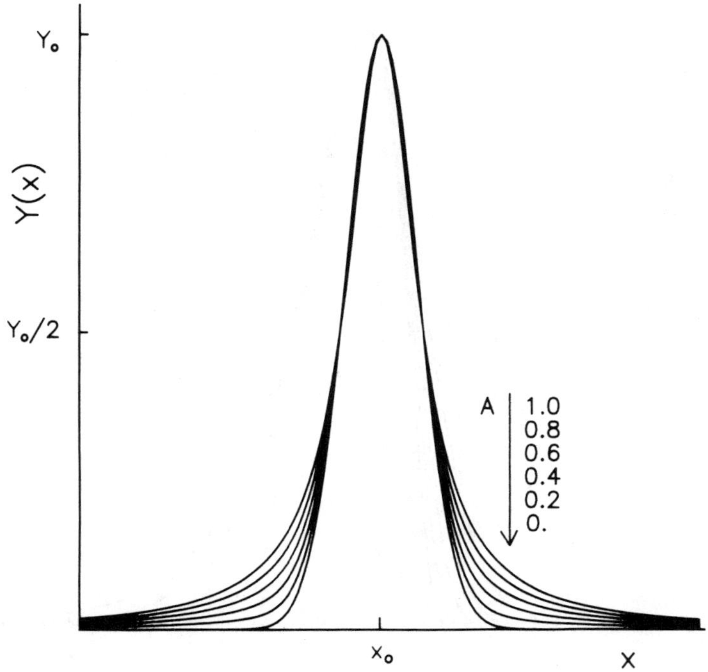

Figure 2.2 The linear combination of the Gaussian and Lorentz shape of a spectral curve (A is the Lorentz function portion, see Table 2.1).

parameters for iterative optimization techniques (see Resolution of Overlapped Spectral Bands Using Nonlinear Regression Analysis, below). The numerical methods used to estimate the band parameters utilize spectra defined as the coordinate set (see Logarithm-Differential Analysis, Determination of Spectral Band Parameters from Properties of Spectral Derivatives, Fourier Deconvolution of the Spectral Profile to Individual Spectral Bands, Combined Fourier Transformation and Least Squares, and Resolution of Overlapped Spectral Bands Using Nonlinear Regression Analysis, below).

ANALOG RESOLUTION OF SPECTRA

An analog resolution of spectra uses a line analyzer, the Du Pont line analyzer,[14,15] for example. The analyzed spectrum is placed on the operating desk of the apparatus. The tube with the high intensity cathode rays projects the synthetic spectrum through the mirror system directly onto the original spectrum. The synthetic spectrum is the sum of the individual bands yielded by the band generator. Simultaneously, up to 12 spectral bands of the Gaussian, Lorentz, and Voight types, with the given parameters, can be generated. The total synthetic absorption profile is compared with the original spectrum, and the band parameters are changed until visual correspondence is obtained. To decrease subjectivity in comparing

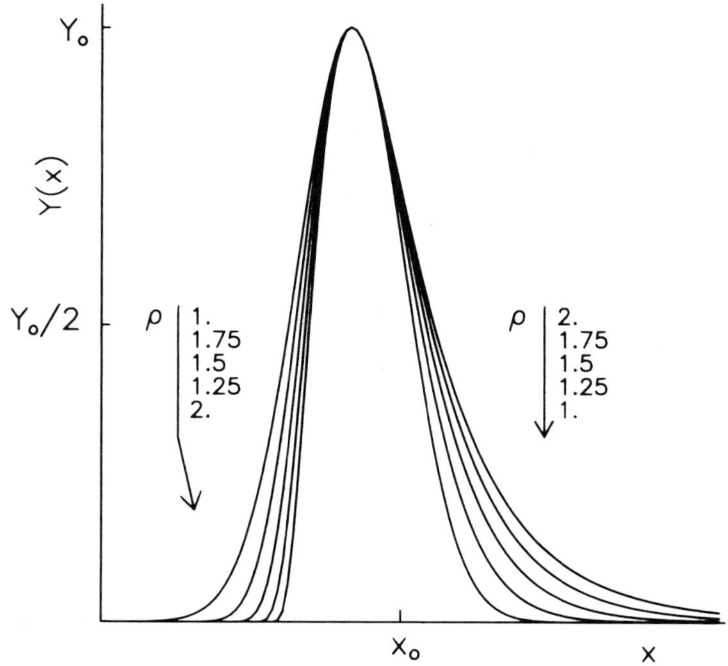

Figure 2.3 The lognormal spectral curve for various values of the asymmetry parameter ρ.

the synthetic spectrum with the original spectrum, the automatic integral measure is used, giving the percentage content of the nonoverlapped regions as well as the percentage representation of the individual bands in the total spectral profile.

The Du Pont line analyzer was used for analyzing various systems, for example, the Gaussian shape of the bands of the Raman spectra,[16] the absorption spectra of halogenides in the far UV region,[17] and the spectral analysis of the circulation dichroism of nucleotides in solution.[18]

Generation of spectral profiles on the graphics terminal of a computer is accomplished by a similar method. A joystick or light pen[19,20] is used to direct the decomposition of the total absorption profile into individual absorption bands.

The methods above are "controlled" by the operator and are relatively time consuming. The results are influenced by subjective errors of the operator when comparing the experimental and generated spectra. On the other hand, these methods are relatively simple and convenient, especially for the first estimation of band parameters for the minimization procedures.

LOGARITHM-DIFFERENTIAL ANALYSIS

The method of logarithm-differential analysis of spectral profiles is based on linearized Gaussian bands.[19]

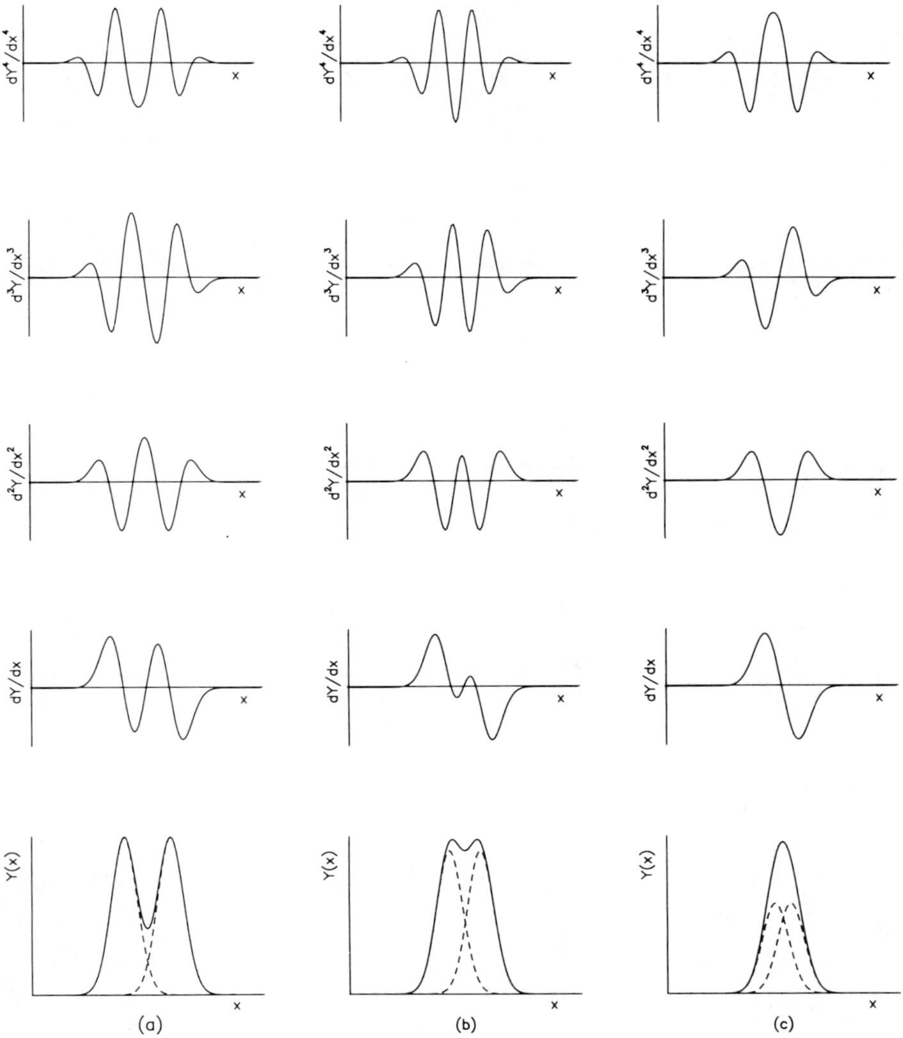

Figure 2.4 The overlap of two identical spectral curves of a Gaussian shape with the dependence of their maximal Δ distances: **(a)** $\Delta = 1.5\ \Delta x$, **(b)** $\Delta = 1.0\ \Delta x$, and **(c)** $\Delta = 0.5\ \Delta x$.

Let us suppose that a spectrum consists of the Gaussian bands centered at points x_0, with halfwidth Δx and height Y_0 (see Types of Spectral Bands, above). Let the quantity S be given by

$$S = \frac{\Delta x}{2\sqrt{2\ln 2}} \tag{2.9}$$

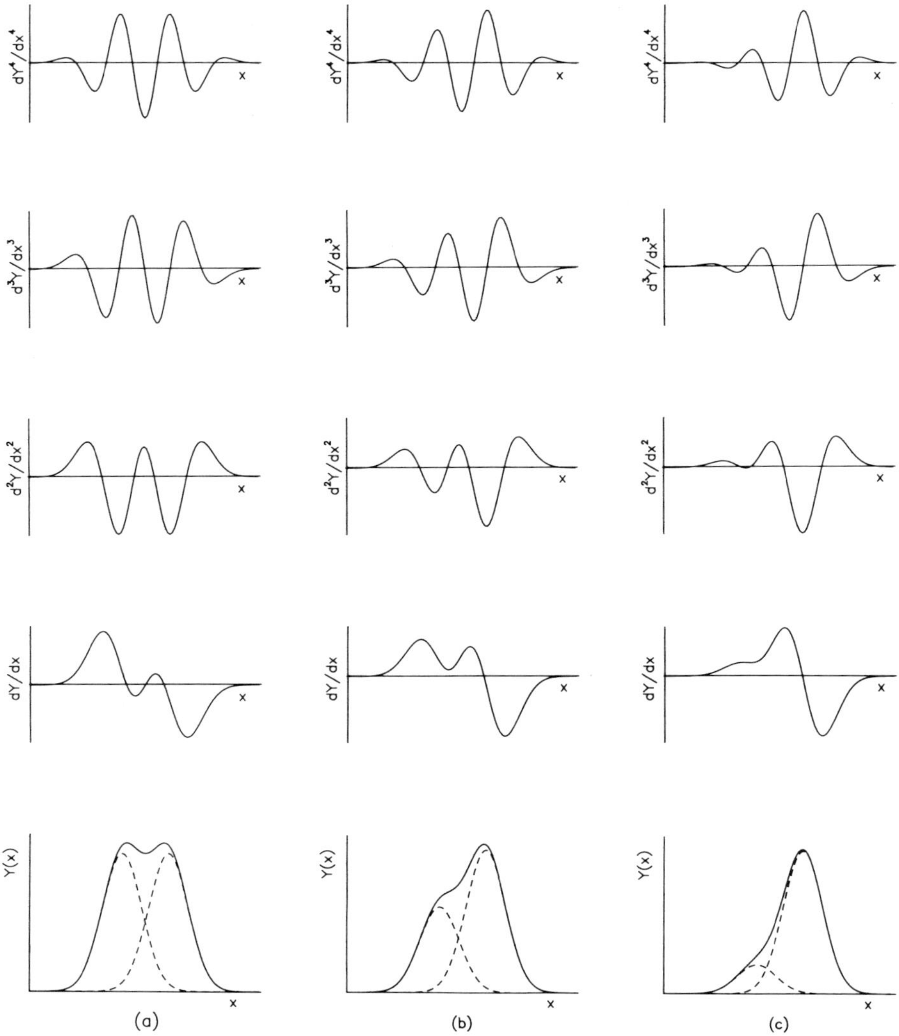

Figure 2.5 The overlap of two spectral functions of a Gaussian shape with the same halfwidths for different values of their intensity ratio: **(a)** $Y_A^0/Y_B^0 = 1.0$, **(b)** $Y_A^0/Y_B^0 = 0.6$, and **(c)** $Y_A^0/Y_B^0 = 0.2$.

Then, the following expression can be written for the Gaussian band shape:

$$Y = Y_0 \exp\left[-\frac{1}{2}\left(\frac{x - x_0}{S}\right)^2\right] \quad \textbf{(2.10)}$$

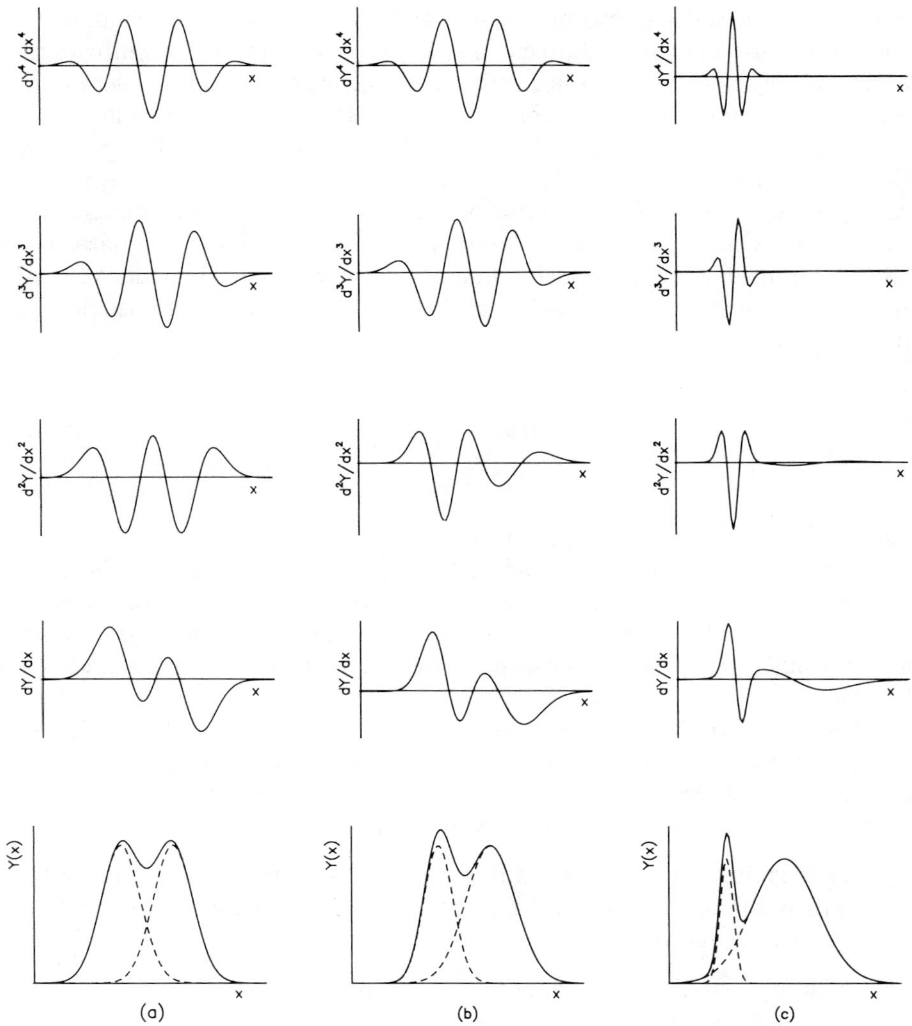

Figure 2.6 The overlap of two spectral functions of a Gaussian shape with the same intesities for different values of their halfwidth ratio: **(a)** $\Delta x_A/\Delta x_B = 1.0$, **(b)** $\Delta x_A/\Delta x_B = 0.6$, and **(c)** $\Delta x_A/\Delta x_B = 0.2$.

Let the function $\psi(x)$ be defined by the equation

$$\Psi(x) = -\frac{d(\ln Y)}{dx} = \frac{x - x_0}{S^2} \qquad \textbf{(2.11)}$$

$\psi(x)$ is the linear function with the slope $S^{-2} > 0$ and the segment $\psi(0) = -x_0/S^2$ in the area where the spectrum is described by the only Gaussian band. If the spectrum consists of more than one band, this conclusion will be sufficiently accurate only within the area in which the given band has the dominant contribution compared to the contributions of the rest of the bands.

After the transformation of the experimental spectral profile given by Equation 2.8, $\psi(x)$ can be plotted as a function of x. Several linear segments with positive slope are found on this plot. Each such linear segment specifies the area in which one Gaussian band of the original profile dominates the others. Let us assume that such a linear area was found. The values S and x_0 are subsequently determined from this line. The value Y_0 is specified by the equation

$$Y_0 = Y(x)\exp\left[-\frac{1}{2}\left(\frac{x - x_0}{S}\right)^2\right] \tag{2.12}$$

where x is chosen at the midpoint of the linear segment.

The given band can be subtracted from the total absorption profile after the determination of the given band parameters. The whole procedure is repeated for the following band. Likewise, the individual bands are subsequently subtracted from the absorption profile by the sequence method (see Figure 2.7).

The procedure is simple for spectral data recorded in equidistant increments x (frequency ν, respective to wavenumber $\tilde{\nu}$). Data with equidistant values of wavelength λ are usually available. However, the practical application of this method gives rise to two problems:

1. Performing the numerical differentiation according to Equation 2.11, which does not deform the absorption band shape and is independent of spectral noise, and
2. Finding a suitable procedure for determining the values x and S for the numerical values of $\psi(x)$.

To solve the first problem the numerical derivatives that do not distort an individual Gaussian band can be used for convenience. Moreover, an area with contributions from more than one Gaussian band can be partly deformed. However, this is not the crucial error, because the parameters of associated individual bands are calculated from the regions where those bands are dominant.

If

$$\chi(\lambda) = \Psi(\nu) \tag{2.13}$$

is defined and the equalities

$$\lambda = \nu^{-1} \tag{2.14}$$

and

$$\frac{d\lambda}{d\nu} = -\nu^{-2} = -\lambda^2 \tag{2.15}$$

are used, the following expression is obtained:

$$\Psi(x) = \frac{\lambda^2 d(\ln\chi)}{d\lambda} = \frac{d(\lambda^2 \ln\chi)}{d\lambda} - 2\lambda\ln\chi \tag{2.16}$$

If t is chosen as a constant increment λ ($t = \lambda_i - \lambda_{i-1}$), then

$$\frac{d(\lambda^2 \ln\chi)}{d\lambda} = \frac{1}{2t}\left[(\lambda+t)^2 \ln\chi(\lambda+t) - (\lambda-t)^2 \cdot \ln\chi(\lambda-t)\right] \tag{2.17}$$

This relationship is generally only an approximation. Using the separated Gaussian band it is exact, because it is exact for polynomials of the second degree, and $\lambda^2\ln\chi$ is a polynomial of the second degree in λ because

$$\lambda^2 \ln\chi = \lambda^2 \ln Y_0 - \frac{1}{2}\left(\frac{1-\lambda x_0}{S}\right)^2 \tag{2.18}$$

Then the function $\psi(x)$:

$$\Psi(x) = \frac{1}{2t}\left\{\eta(\lambda+t) + \eta(\lambda-t)\right\} - \frac{2}{\lambda}\eta(\lambda) \tag{2.19}$$

can be written, where

$$\eta(\lambda) = \lambda^2 \ln\chi - \lambda^2 \ln Y_0 \tag{2.20}$$

This relationship is sufficient for the region where a given Gaussian band is dominant.

A three-point numerical filter can be used to decrease the noise present in the derivative spectrum by

$$\xi(x) = \frac{1-\gamma}{2}\left[\eta(\lambda+t) - \eta(\lambda-t)\right] - \gamma\eta(\lambda) \tag{2.21}$$

$$\Psi(x) = \frac{1}{2t}\left[\xi(\lambda+t) - \xi(\lambda-t)\right] - \frac{2}{\lambda}\eta(\lambda) \tag{2.22}$$

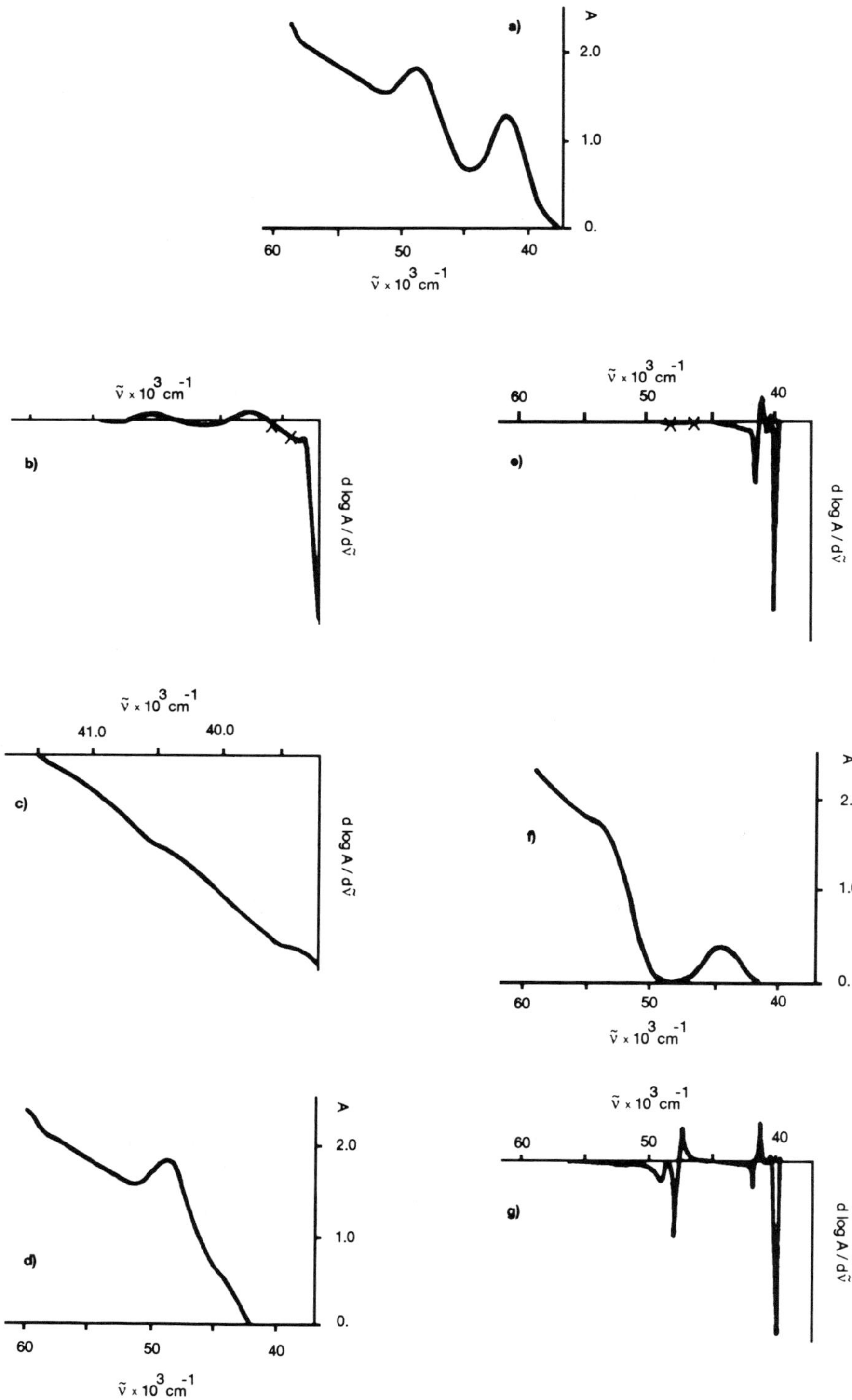
a)
A
2.0
1.0
0.
60
50
40
ṽ x 10³ cm⁻¹
b)
ṽ x 10³ cm⁻¹
d log A / dṽ
e)
ṽ x 10³ cm⁻¹
60
50
40
d log A / dṽ
c)
ṽ x 10³ cm⁻¹
41.0
40.0
d log A / dṽ
f)
A
2.0
1.0
0.
60
50
40
ṽ x 10³ cm⁻¹
d)
A
2.0
1.0
0.
60
50
40
ṽ x 10³ cm⁻¹
g)
ṽ x 10³ cm⁻¹
60
50
40
d log A / dṽ

where γ is the constant determining the filter property. For $\gamma = 1$ Equation 2.22 will change to Equation 2.19, i.e., noise reduction will not be realized. Setting $\gamma = 0.2$ seems to be most convenient for noise elimination. It can be proven that this filter does not cause deformation at any γ value for an individual Gaussian band.

The values x and S can be calculated by least squares from the numerical values $\psi(x)$. Here, it is efficient to use the method of orthogonal polynomials within the equidistant interval.[20] For this purpose the wavenumbers in the equidistant intervals of the wavelengths are indexed as follows:

$$\lambda_j = \lambda_R + jt \qquad j = 0, 1, 2, \ldots \tag{2.23}$$

where λ_R is the initial value of λ. Then

$$\tilde{\nu}_j = \left(\lambda_R + jt\right)^{-1} \tag{2.24}$$

where $\tilde{\nu}_j$ does not have equidistant increments. If j_0 is one of the j values, then

$$\lambda_0 = \lambda_R + j_0 t \tag{2.25}$$

and

$$\tilde{\nu}_j = \left[\lambda_0 + \left(j - j_0\right)t\right]^{-1} \tag{2.26}$$

If $(j - j_0)t \gg \lambda_0$, the following is approximately valid:

$$\tilde{\nu}_j = \lambda_0^{-1} - \left(j - j_0\right)t\lambda_0^{-2} \tag{2.27}$$

or

$$\tilde{\nu}_j = \tilde{\nu}_0 + \left(j - j_0\right)\delta_0 \tag{2.28}$$

Figure 2.7 The method of logarithm-differential analysis of a spectrum: **(a)** the experimental spectrum of the solution of tetra N-butylammonium iodide in triethylphosphate at 253°K, **(b)** the function $\psi(\nu) = d[\log a(\nu)]/d\nu$ of the original spectrum (a) (linear region denoted by crosswires), **(c)** expansion of $\psi(\nu)$ between the crosswires of (b), **(d)** the function $A(\nu)$ after the elimination of the first band, **(e)** the function $\psi(\nu) = d[\log a(\nu)]/d\nu$ of the original (b) (linear region denoted by crosswires), **(f)** the function $A(\nu)$ after the elimination of the second band, **(g)** the function $\psi(\nu) = d[\log a(\nu)]/d\nu$ of the spectrum (f) giving the band parameter of the small band at approximately 44,000 cm^{-1}. (From Baker, B. E., Fox, M. F., Hayon, E., and Fox, E. W., *Anal. Chem.,* 46, 1785, 1974. With permission.)

where

$$\tilde{\nu}_0 = \lambda_0^{-1} \quad \text{and} \quad \delta_0 = -t\lambda_0^{-2} \tag{2.29}$$

From Equation 2.28, it follows that $\tilde{\nu}_j$ is distributed approximately equidistantly for the points around j_0, with the distance δ_0. If $|j - j_0|$ is not too high, this approximation is sufficiently accurate.

The slope and the segment of the curve of the transformed spectrum $\psi(x)$ can be determined by fitting a polynomial of the second degree to the set of five neighboring coordinates around j_0. For ψ_j near $\tilde{\nu}_0$, the following can be written:

$$\Psi_j = C_0 + C_1\left(\tilde{\nu}_j - \tilde{\nu}_0\right) + C_2\left(\tilde{\nu}_j - \tilde{\nu}_0\right)^2 \tag{2.30}$$

where the coefficients C_i are

$$C_2 = \frac{-2T_2}{7\delta_0^2} \tag{2.31}$$

$$C_1 = \frac{T_1}{10\delta_0} \tag{2.32}$$

$$C_0 = \frac{T_0}{5} + \frac{2T_2}{7} \tag{2.33}$$

and where

$$T_0 = \Psi_{j_0-2} + \Psi_{j_0-1} + \Psi_{j_0} + \Psi_{j_0+1} + \Psi_{j_0+2} \tag{2.34}$$

$$T_1 = -2\Psi_{j_0-2} - \Psi_{j_0-1} + \Psi_{j_0+1} + 2\Psi_{j_0+2} \tag{2.35}$$

$$T_2 = -\Psi_{j_0-2} + \frac{1}{2}\Psi_{j_0-1} + \Psi_{j_0} + \frac{1}{2}\Psi_{j_0+1} - \Psi_{j_0+2} \tag{2.36}$$

The sum of the squares of the derivatives is

$$F = \sum_{m=-2}^{+2} \left[\Psi\left(\tilde{\nu}_0 + m\delta_0\right)\right]^2 - \frac{T_0^2}{5} - \frac{T_1^2}{10} - \frac{2T_2^2}{7} \tag{2.37}$$

where the parameters to be estimated are

$$S = C_1^{-\frac{1}{2}} = \left(\frac{10\delta_0}{T_1} \right)^{\frac{1}{2}} \tag{2.38}$$

$$x_0 = \tilde{\nu}_0 - \frac{C_0}{C_1} \tag{2.39}$$

The curvature is given by the value C_2. It is negligibly small if

$$e = \frac{2|C_2|\delta_0^2}{\sqrt{F}} = \frac{4|T_2|}{7\sqrt{F}} << 1 \tag{2.40}$$

where F is the sum of the squares of the deviations.

It is convenient to eliminate large, wide bands first because of their low values of ψ, but this method is not suitable if there exists an area with no dominant band.

The efficiency of this method was tested by decomposing two overlapped Gaussian bands.[19] The six independent variables can be reduced after introducing the ratios Y_A / Y_B and $\Delta x_A/\Delta x_B$. If the values Y_B, Δx_B, and x_0^B are fixed, the change in value of the other variables (Y_A, Δx_A, x_0^A) is sufficient. The area of limited resolution, in which the values x_0^A and x_0^B were estimated to within 1% of the interval $\Delta = x_0^A - x_0^B$, is shown in Figure 2.8.

An alternative method of solution arises from the fact that during the computer on-line scanning the acquired spectra at equidistant wavelengths can be recorded with a sufficient density of points. Such spectra can be simply recalculated for representation in the nonequidistant wavenumber scale by using suitable interpolation or smoothing functions for representation at equidistant frequency, with respect to the wavenumber scale.

DETERMINATION OF SPECTRAL BAND PARAMETERS FROM PROPERTIES OF SPECTRAL DERIVATIVES

Methods of this type are based on properties of the first four derivatives of the maxima of spectral bands:

$$Y' = \frac{dY}{dx} = 0 \tag{2.41}$$

$$Y'' = \frac{d^2Y}{dx^2} < 0 \tag{2.42}$$

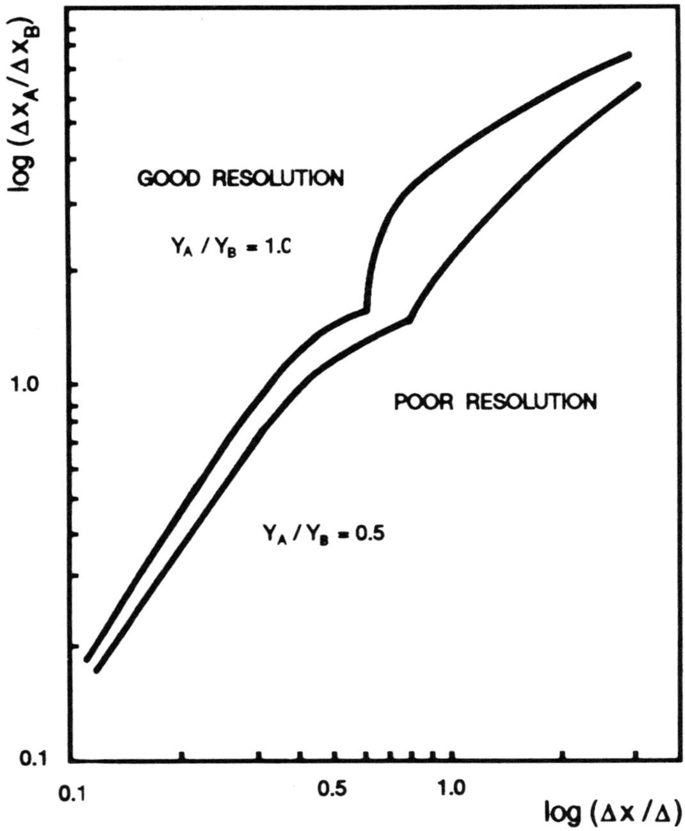

Figure 2.8 The area of limited resolution of two overlapped Gaussian functions using logarithm-differential analysis. (From Baker, B. E., Fox, M. F., Hayon, E., and Fox, E. W., *Anal. Chem.*, 46, 1785, 1974. With permission.)

$$Y''' = \frac{d^3Y}{dx^3} = 0 \tag{2.43}$$

$$Y'''' = \frac{d^4Y}{dx^4} > 0 \tag{2.44}$$

Properties of the individual derivatives of the Gaussian functions are shown in Figure 2.9. The properties of the individual derivatives of two overlapped bands of the Gaussian type are shown in Figures 2.4 through 2.6. The figures show that as the bands approach, their resolution is decreased. Smith[21] proved that the resolution of two identical Gaussian functions is only possible if their separation is greater than 0.849 of their halfwidth. For Lorentz curves this value is equal to 0.557.

Figure 2.10 shows a possible application of the derivatives for determining band positions in a complex absorption profile.

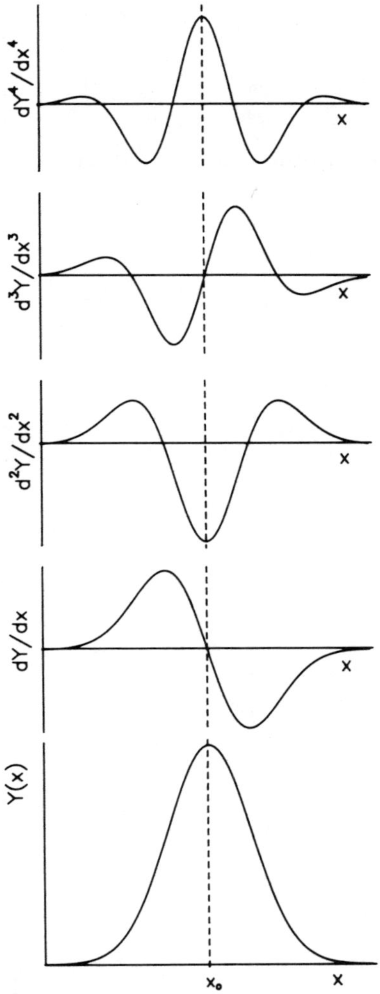

Figure 2.9 The properties of the derivations of a single Gaussian spectral curve.

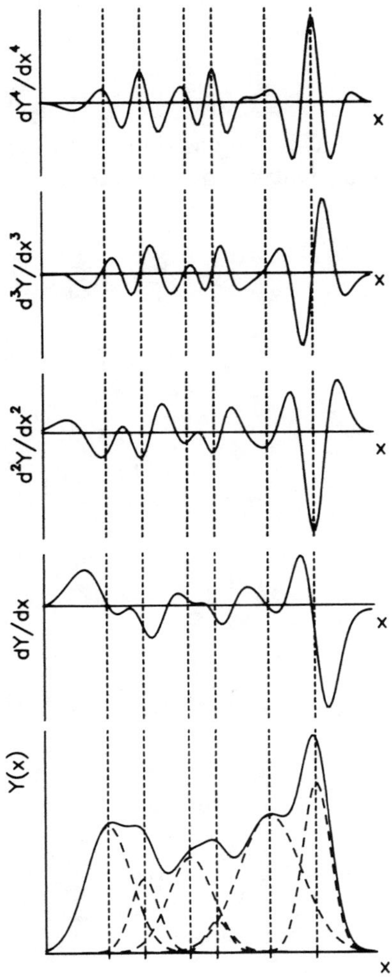

Figure 2.10 The application of derivatives of a complex spectral curve for estimating the individual curve positions.

The most frequently used methods of profiling the derivative properties of the absorption profiles were introduced by Savitzky and Golay[22] and Morrey,[23] who used a polynomial of the fourth degree to fit short segments of the absorption profile. Lischka and Derflinger[24] and Goméz-Beltrán et al.[25] developed a method using the numerical derivative of the absorption profile.

The Sequence Method

Let us approximate numerical values of a segment of five equidistant points of a spectral profile with a polynomial of the fourth degree:[22,23]

$$Y_j = \sum_{i=1}^{5} C_i \, x_j^{i-1} \tag{2.45}$$

where X_j is the position according to some initial value x_k, i.e., $X_j = x_j - x_k$. The derivatives at the point X_j become

$$Y' = \frac{dY_j}{dX_j} = \sum_i C_i(i-1)X_j^{i-2}$$

$$= C_2 + 2C_3X_j + 3C_4X_j^2 + 4C_5X_j^3 \tag{2.46}$$

$$Y'' = \frac{d^2Y_j}{dX_j^2} = \sum_i C_i(i-1)(i-2)X_j^{i-3}$$

$$= 2!C_3 + 3!C_4X_j + 4!C_5X_j^2 \tag{2.47}$$

$$Y''' = \frac{d^3Y_j}{dX_j^3} = \sum_i C_i(i-1)(i-2)(i-3)X_j^{i-4}$$

$$= 3!C_4 + 4!C_5X_j \tag{2.48}$$

$$Y'''' = \frac{d^4Y_j}{dx_j^4} = \sum_i C_i(i-1)(i-2)(i-3)(i-4)X_j^{i-5}$$

$$= 4!C_5 \tag{2.49}$$

Because Y' and Y'' are not particularly sensitive to the position of the band maximum, it is more efficient to determine this position by the ratio of Y''' and Y'''':

$$X_0 = -\frac{3!C_4}{4!C_5} \tag{2.50}$$

given that $C_5 > 0$. Therefore, the value of C_5 should be examined. If this value is positive, the value $X_0 = x_0 - x_k$, for which Equation 2.48 is valid within the given segment, will be specified. If such an X_0 value within the given segment exists and if $Y''(X_0) < 0$, then X_0 determines the position of the band maximum.

Afterward, the segment of five points is shifted about one point further, and the whole cycle is repeated until the whole absorption profile is examined.

The values of C_i can be determined by simple linear least squares:

$$\sum_{j=1}^{5}\left(Y_j - \sum_{i=1}^{5} C_i X_j^{i-1}\right)^2 = \min \tag{2.51}$$

giving the system of linear equations for the coefficients of C_i:

$$\sum_{j=1}^{5} Y_j X_j^k = \sum_{j=1}^{5} C_i X_j^{(i+k-1)} \qquad k = 0, 1, 2, 3, 4 \tag{2.52}$$

The shape must be assumed in order to determine the widths and heights of spectral bands. The width of bands from the value X_T, having the fourth derivative equal to zero, i.e., $Y''''(X_T) = 0$, can subsequently be determined. For the width of the spectral band the following expression can be derived:[23]

$$\Delta x = \left|X_0 - X_T\right|\alpha \tag{2.53}$$

where α is a coefficient depending on the type of distribution function ($\alpha = 0.3150$ for the Gaussian function, 0.2146 for the Student function, and 0.1625 for the Lorentz distribution function).

Because $Y''''(X_0)$ is proportional to the value $Y_0/\Delta x^4$, Y_0 can be determined by the equation

$$Y_0 = \beta\, C_5 \Delta x^4 \tag{2.54}$$

where the value of the β coefficient is 92.24 for the Gaussian function, 197.7 for the Student function, and 384.0 for the Lorentz function.

Two sources of errors occur when specifying the band position: the first is the consequence of inaccurately fitting the spectral segment using the polynomial, and the second is the presence of overlapping bands.

The error caused by inaccurate fitting and statistical fluctuations of data will appear during the determination of the coefficients C_4 and C_5, used for the specification of the band center position. This error is usually much lower than the error due to band overlapping. The critical area defined by two overlapping bands of the Gaussian, Student, and Lorentz type is shown in Figure 2.11.

Iterative Inclusion of Mutually Overlapping Bands

This method is based on properties of the first and second derivatives of an absorption profile. At point t of the assumed maximum of the ith band, the following is valid:[24,25]

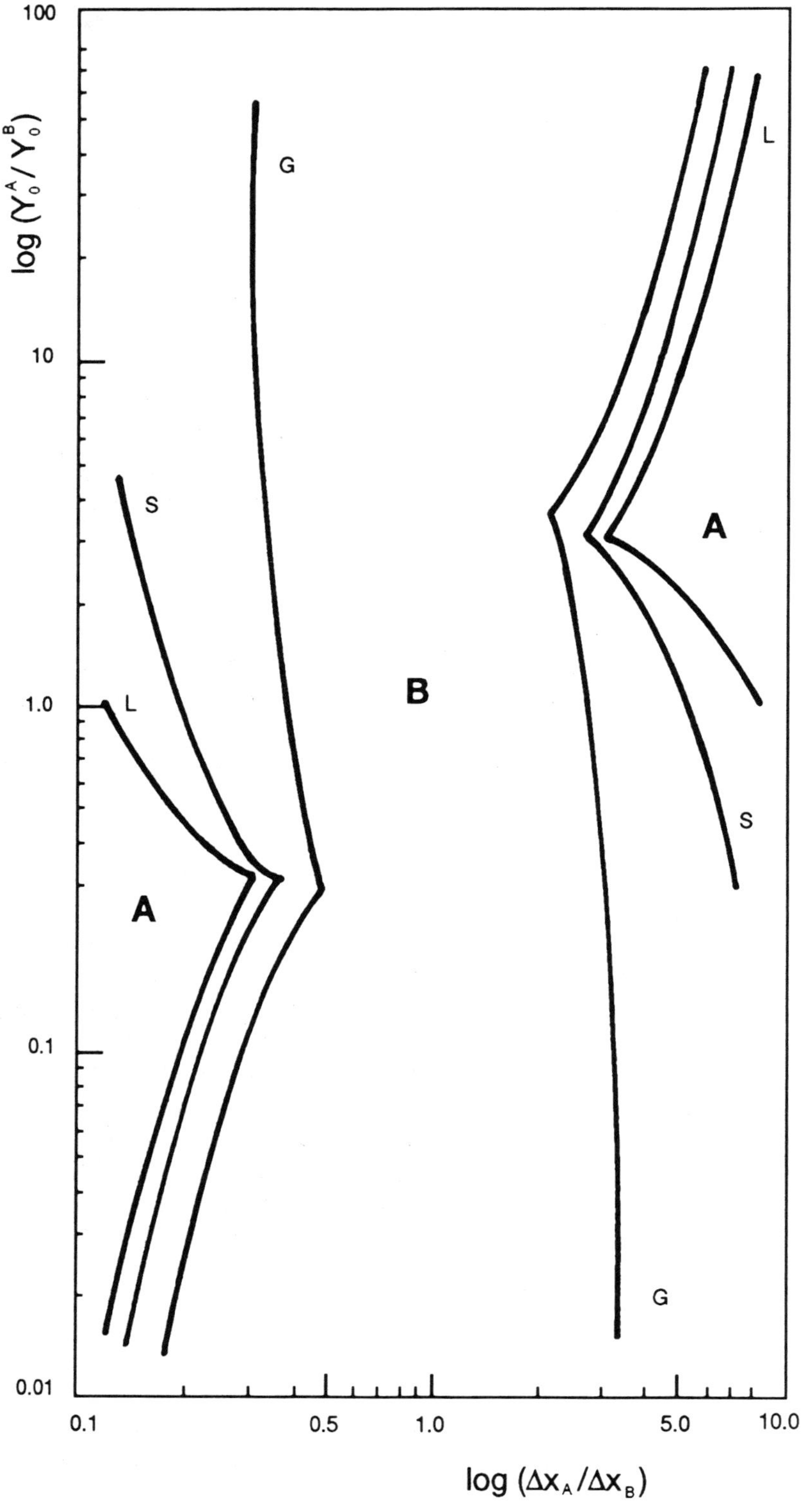
100
10
1.0
0.1
0.01
log (Y₀^A / Y₀^B)
0.1
0.5
1.0
5.0
10.0
log (ΔxA/ΔxB)
G
S
L
A
B
L
A
S
G

$$\left(\frac{d^k Y_t}{d\lambda^k}\right)_i = \left(\frac{dY_i^k}{d\lambda^k}\right)_i + \sum_{j\neq i}^{N}\left(\frac{d^k Y_j}{d\lambda^k}\right)_i \tag{2.55}$$

where N is the number of bands, $(d^kY_t/d\lambda^k)_i$ is the derivative of the kth degree of the experimental profile in the ith band maximum of $\lambda_{0,i}$, and $(d^kY_j/d\lambda^k)_i$ is the contribution of the kth derivative of the jth band at the point $\lambda_{0,i}$.

The Gaussian function describing the ith band is

$$Y_i = Y_{0,i}\exp\left[-\frac{\left(\frac{\lambda_{0,i}}{\lambda}-1\right)^2}{b_i^2}\right] \tag{2.56}$$

where

$$b_i = -\frac{\lambda_{0,i}}{2\sqrt{\ln 2}\,\Delta\lambda_{0,i}} \tag{2.57}$$

If the condition of the ith band maximum is used,

$$\left(\frac{dY_i}{d\lambda}\right)_{0,i} = 0 \tag{2.58}$$

the approximate relationship for the value $\lambda_{0,i}$ near the approximate position λ_i of the ith band maximum can be derived:

$$\lambda_{0,i} = \lambda_i + \frac{\left(\frac{dY_i}{d\lambda}\right)_i}{\left(\frac{d^2Y_i}{d\lambda^2}\right)_i} \tag{2.59}$$

Similarly, for the value b_i the approximate relationship can be obtained:

$$b_i = \frac{\left[\frac{2\,Y_{0,i}}{\left(\frac{d^2Y_i}{d\lambda^2}\right)_i}\right]^{\frac{1}{2}}}{\lambda_i} \tag{2.60}$$

Figure 2.11 The critical region of band overlap using the sequence approach to estimate the parameters: **(A)** the critical region and **(B)** the region of sufficient resolution (G, the Gaussian shape of the spectral curve, L, the Lorentz shape of the spectral curve; S, the Student shape of the spectral curve.) (From Morrey, J. R., *Anal. Chem.*, 40, 905, 1968. With permission.)

Using Equations 2.55, 2.59, and 2.60, solution of the system of simultaneous equations yields

$$Y_{0,i} = (Y_t)_i - \sum_{j\neq i}^{N} (Y_j)i \tag{2.61}$$

$$\lambda_{0,i} = \lambda_i + \left[\left(\frac{dY_t}{d\lambda} \right)_i - \sum_{j\neq i}^{N} \left(\frac{dY_j}{d\lambda} \right)_i \right] \cdot \left[\left(\frac{d^2Y_t}{d\lambda^2} \right)_i - \sum_{j\neq i}^{N} \left(\frac{d^2Y_j}{d\lambda^2} \right)_i \right]^{-1} \tag{2.62}$$

$$b_i = \frac{\left\{ 2\left[(Y_t)_i - \sum_{j\neq i}^{N} (Y_j)_i \right] \left[\left(\frac{d^2Y_t}{d\lambda^2} \right)_i - \sum_{j\neq i}^{N} \left(\frac{d^2Y_j}{d\lambda^2} \right)_i \right]^{-1} \right\}^{\frac{1}{2}}}{\lambda_i} \tag{2.63}$$

where λ_i is the original estimated value of the ith band maximum.

The system of Equations 2.61 to 2.63 can be solved iteratively in two steps. At the first step, the values of the maximal heights $Y_{0,i}$ and the b_i values for individual bands are calculated until consistency is obtained. At the second step, these values are fixed and the band positions $\lambda_{0,i}$ are calculated. Then the program is returned to the first step. This cycle is repeated until consistency of all optimized parameters is obtained.

The derivatives of an experimental absorption profile can be calculated by some of the methods introduced in Chapter 1.

The speed of convergence is shown in Figure 2.12. This method for obtaining the parameters of spectral bands can also be adapted to other types of spectral curves (see Types of Spectral Bands, above).

Method Using Linearized Gaussian Functions

This method uses the possibility of linearizing the Gaussian peak shape functions, and it enables the determination of an individual band's parameters by the linear least squares method.[26]

Let the simple Gaussian curve be assumed

$$Y = Y_0 \exp\left[-4\ln 2 \frac{(x - x_0)^2}{\Delta x^2} \right] \tag{2.64}$$

After taking the logarithm of the function Equation 2.64 the following is obtained:

$$y = a + bx + cx^2 \tag{2.65}$$

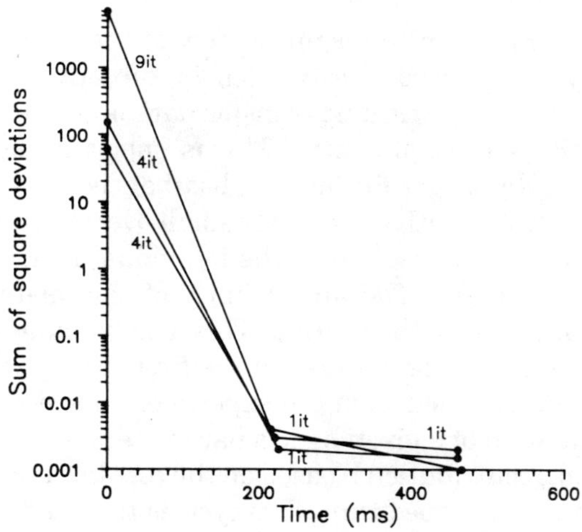

Figure 2.12 The course of the iterations resolving five overlapped Lorentzian bands. The number of iterations in each refinement cycle is indicated, and 1, 2, and 3 denote three different sets of starting parameters. (From Goméz-Beltrán, F., Salas, A., and Valero, A., *Anal. Chim. Acta,* 103, 263, 1978. With permission.)

where

$$y = \ln Y \tag{2.66}$$

$$a = \ln Y_0 - 4(\ln 2)\frac{x_0^2}{\Delta x^2} \tag{2.67}$$

$$b = 2(\ln 2)\frac{x_0}{\Delta x} \tag{2.68}$$

$$c = -4\frac{\ln 2}{\Delta x^2} \tag{2.69}$$

This transformation makes possible a fast, noniterative way of determining the desired band parameters by linear least squares method:

$$\sum_i \left(y_i^{\exp} - a - bx_i - cx_i^2\right)^2 = \min \tag{2.70}$$

In the presence of noise, or if the spectral profile is not satisfactorily described by the Gaussian functions, the method gives only a rough estimate of the spectral parameters. Parameters thus obtained are consequently not the "best parameters", from the point of view of the general least squares method, but they can be used as estimates for subsequent nonlinear minimization procedures. A logarithmic transformation guarantees the weighting, which minimizes spectral noise.[26] This is important because the noise contribution is usually larger for higher absorbancies.[2]

For a spectral profile with a relatively small overlap between individual bands, this method can be used to fit the individual bands in areas where these bands are dominant. The applicability of this method strongly decreases with the increase of the overlap. This disadvantage can be partially eliminated by an iterative procedure. In the first step, parameters of individual bands are determined from the experimental spectrum. In the next step, the determination of a given band's parameters is refined by subtracting the other band's contributions (based on the results of the previous steps) from the given area of the spectrum. This cycle is repeated until consistency of the individual parameters in two successive iterative cycles is attained.

Partitioning a spectrum into segments, in which single bands are dominant and the parameters of the individual bands are optimized, is very important for this method. The region over which a given band is optimized need not contain its maximum (although as a rule, it does), as the band maximum is often the region significantly affected by the overlap of the other bands. It is convenient to reduce the regions used to fit individual bands with the iteration process.

The value of the mean square deviation can be the measure of convergence of the iterative procedure:

$$S^2 = \sum^{N} \frac{\left(Y_i^{\mathrm{exp}} - Y_i^{\mathrm{calc}}\right)^2}{N} \tag{2.71}$$

where Y_i^{exp} are the points of the experimental spectral profile and Y_i^{calc} are the points calculated from the optimized parameters of all the bands. The dependence of the mean square deviation on the number of iterations for analysis of a 15-band chromatogram spectrum is shown in Figure 2.13.

FOURIER DECONVOLUTION OF THE SPECTRAL PROFILE TO INDIVIDUAL SPECTRAL BANDS

The Fourier transformation is used in spectroscopy for many different applications: for spectral smoothing, for the spectral derivative, and so on (see, e.g., Bracewell,[27] Horlick,[28] Stone,[29] and Chapter 1). A Fourier transformation for the resolution of overlapping bands was first reported by Kirmse and Westenberg[30] for chromatographic peaks. However, they assumed a fixed shape for individual bands (Gaussian shape with the same halfwidth), different only in their position and intensity. The general theory of Fourier self-deconvolution, which enables the recognition of strongly overlapping spectral bands, was developed by Kauppinen et al.[31]

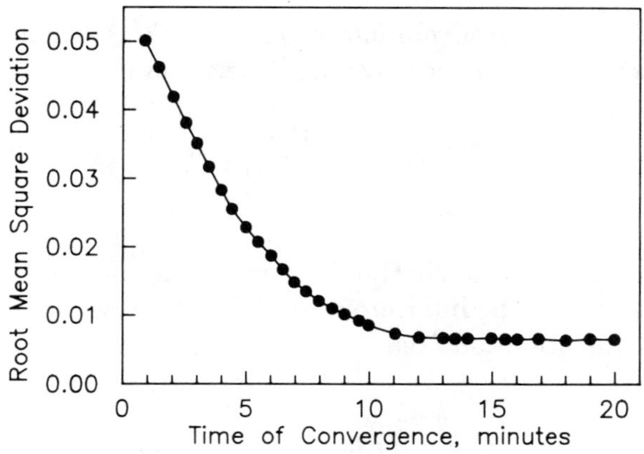

Figure 2.13 Time for convergence of an ion-exchange chromatogram fitted with 15 bands. Each square is an iteration. (From Camana, R. A., Scarle, R. B., Helleer, T., and Shupack, S. I., *Anal. Chem.*, 58, 1162, 1986. With permission.)

The relationship between a spectrum $Y(x)$ and its interferogram $I(\omega)$ is given by the expression

$$Y(x) = \int_{-\infty}^{+\infty} I(\omega) \exp(i2\pi x\omega)\, d\omega = \mathcal{F}\{I(\omega)\} \tag{2.72}$$

where

$$I(\omega) = \int_{-\infty}^{+\infty} Y(x) \exp(-i2\pi x\omega)\, dx = \mathcal{F}^{-1}\{Y(x)\} \tag{2.73}$$

where $\mathcal{F}\{\ \}$ represents the Fourier transformation and $\mathcal{F}^{-1}\{\ \}$ represents the inverse Fourier transformation (see Appendix D).

Any spectrum $Y(x)$ can be expressed as the convolution of the line shape function $G(x)$ and the spectrum $Y'(x)$

$$Y(x) = G(x) * Y'(x) = \int_{-\infty}^{+\infty} G(x')Y'(x - x')\, dx' \tag{2.74}$$

where * means the convolution operation.

The inverse Fourier transformation of both sides of Equation 2.74:

$$I(\omega) = \mathcal{F}^{-1}\{Y(x)\} = \mathcal{F}^{-1}\{G(x)\}\, I'(\omega) \tag{2.75}$$

is introduced for the deconvolution of $G(x)$ from $Y(x)$. Then, the interferogram corresponding to the deconvoluted spectrum is

$$I(\omega) = \frac{1}{\mathcal{F}^{-1}\{G(x)\}} I(\omega) \quad \textbf{(2.76)}$$

The deconvolution operation in the Fourier transformation space requires a partial apodization of the interferogram. $Y'(x)$ is subsequently obtained by the Fourier transformation $I'(\omega)$:

$$Y(x) = \int_{-\infty}^{+\infty} I'(\omega) \exp(i2\pi x\omega)\, d\omega \quad \textbf{(2.77)}$$

Generally, $Y'(x)$ is dependent on the line shape function $G(x)$. According to this, in spite of infinite resolution (assumed in Equations 2.72 and 2.73), the spectral resolution itself need not be infinite.

Let a model spectrum $Y(x)$ consist of two spectral bands in the positions $\pm x_0$ (Figure 2.14a). An even function, i.e., $Y(-x) = Y(x)$, that enables the use of the complex Fourier transformation is assumed. In the case of ideal deconvolution, $Y'(x)$ is given by

$$Y'(x) = \delta(x + x_0) + \delta(x - x_0) \quad \textbf{(2.78)}$$

where δ is the Dirac delta function. Then, from Equation 2.74:

$$\begin{aligned} Y(x) &= G(x) * Y'(x) \\ &= G_0(x) * \left[\delta(x + x_0) + \delta(x - x_0)\right] \end{aligned} \quad \textbf{(2.79)}$$

where $G_0(x)$ is the function of the line shape centered at the point $x = 0$ (Figure 2.14b).

The maximum possible improvement in spectral resolution will result from the deconvolution of $Y(x)$ with an accurate function of the line shape $G_0(x)$, i.e., from self-deconvolution.

More detailed information about the deconvolution operation can be obtained by investigating the interferograms at different stages. The inverse Fourier transformation of both sides of Equation 2.79 yields

$$\begin{aligned} I(\omega) &= \mathcal{F}^{-1}\{Y(x)\} = \mathcal{F}^{-1}\left\{G_0(x)\left[\mathcal{F}^{-1}\{\delta(x + x_0) + \delta(x - x_0)\}\right]\right\} \\ &= \mathcal{F}^{-1}\{G_0(x)\}\, 2\cos(2\pi x_0 \omega) \\ &= A(\omega)\, 2\cos(2\pi x_0 \omega) \end{aligned} \quad \textbf{(2.80)}$$

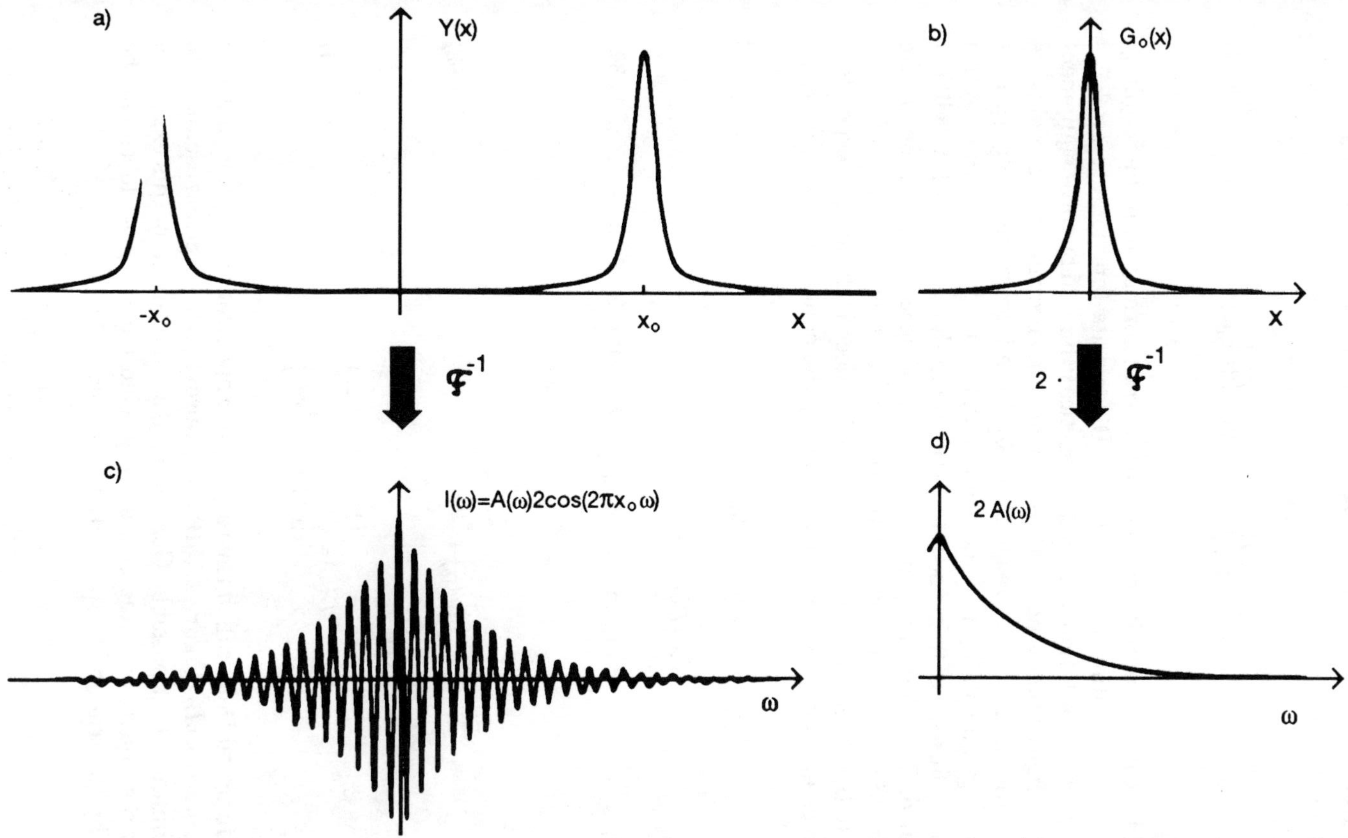

Figure 2.14 The Fourier deconvolution of a spectral band: **(a)** the model symmetrized spectrum $Y(x) = Y(-x)$; **(b)** the function of the line shape $G_0(x)$ centered in the point $x = 0$; **(c)** the inverse Fourier transformation $I(\omega) = \mathcal{F}^{-1}\{Y(x)\}$; **(d)** the inverse Fourier transformation, $A(\omega) = \mathcal{F}^{-1}\{G_0(x)\}$. (From Kauppinen, J. K., Moffatt, D. J., Mantsch, H. H., and Cameron, D. G., *Appl. Spectr.*, 35, 271, 1981. With permission.)

As is shown in Figure 2.14c, the interferogram in the form of a cosine wave modulated by $A(\omega) = \mathcal{F}^{-1}\{G_0(x)\}$ is the result (see Figure 2.14d). If the self-deconvolution is not performed, i.e., $I(\omega)$ is multiplied by $\mathcal{F}^{-1}\{G_0(x)\}$, the following is obtained:

$$I'(\omega) = \frac{I(\omega)}{\mathcal{F}^{-1}\{G(x)\}} = 2\cos(2\pi x_0 \omega) \quad \textbf{(2.81)}$$

In this way, the infinite cosine wave is transformed into the infinitely narrow lines at position $\pm x_0$, i.e., by the Dirac delta function $\delta(x \pm x_0)$.

The spectral resolution achieved in practice is limited by three equipment factors: the digitization range of the experimental spectrum, the signal/noise ratio, and the instrumental resolution. All these factors effectively restrict the interferogram validity to some point L. If, for example, Δx is the spectral resolution, information from the interferogram in the area $\Delta\omega \approx 1/\Delta x$ will not be obtained. Therefore, the upper boundary of spectral resolution in the deconvoluted spectrum is equal to the equipment resolution Δx.

Due to the deformation of the spectral self-deconvolution corresponding to two lines at points $\pm x_0$, $Y_L'(x)$ is given by the equation

$$\begin{aligned} Y_L'(x) &= \mathcal{F}\{D(\omega) \cdot I'(\omega)\} \\ &= \mathcal{F}\{D(\omega)\, 2\cos(2\pi x_0 \omega)\} \\ &= 2L\,\mathrm{sinc}[2\pi(x + x_0)] + 2L\mathrm{sinc}[2\pi(x - x_0)] \end{aligned} \quad \textbf{(2.82)}$$

where

$$\mathrm{sinc}(\alpha) = \frac{\sin\alpha}{\alpha} \quad \textbf{(2.83)}$$

$$D(\omega) = \begin{Bmatrix} 1 \text{ for } |\omega| \le L \\ 0 \text{ for } |\omega| > L \end{Bmatrix} \quad \textbf{(2.84)}$$

The function $D(\omega)$, called the rectangular apodization function, has line shape $2L\mathrm{sinc}(2\pi x L)$ (Figure 2.15B) and improves the spectral resolution to the maximum extent possible. Generally, using a function of this exact line shape is not convenient due to the large side lobes. These lobes can be reduced by the triangular apodization function:

$$D_0(\omega) = \begin{Bmatrix} 1 - \frac{|\omega|}{L} & \text{for } |\omega| \le L \\ 0 & \text{for } |\omega| > L \end{Bmatrix} \quad \textbf{(2.85)}$$

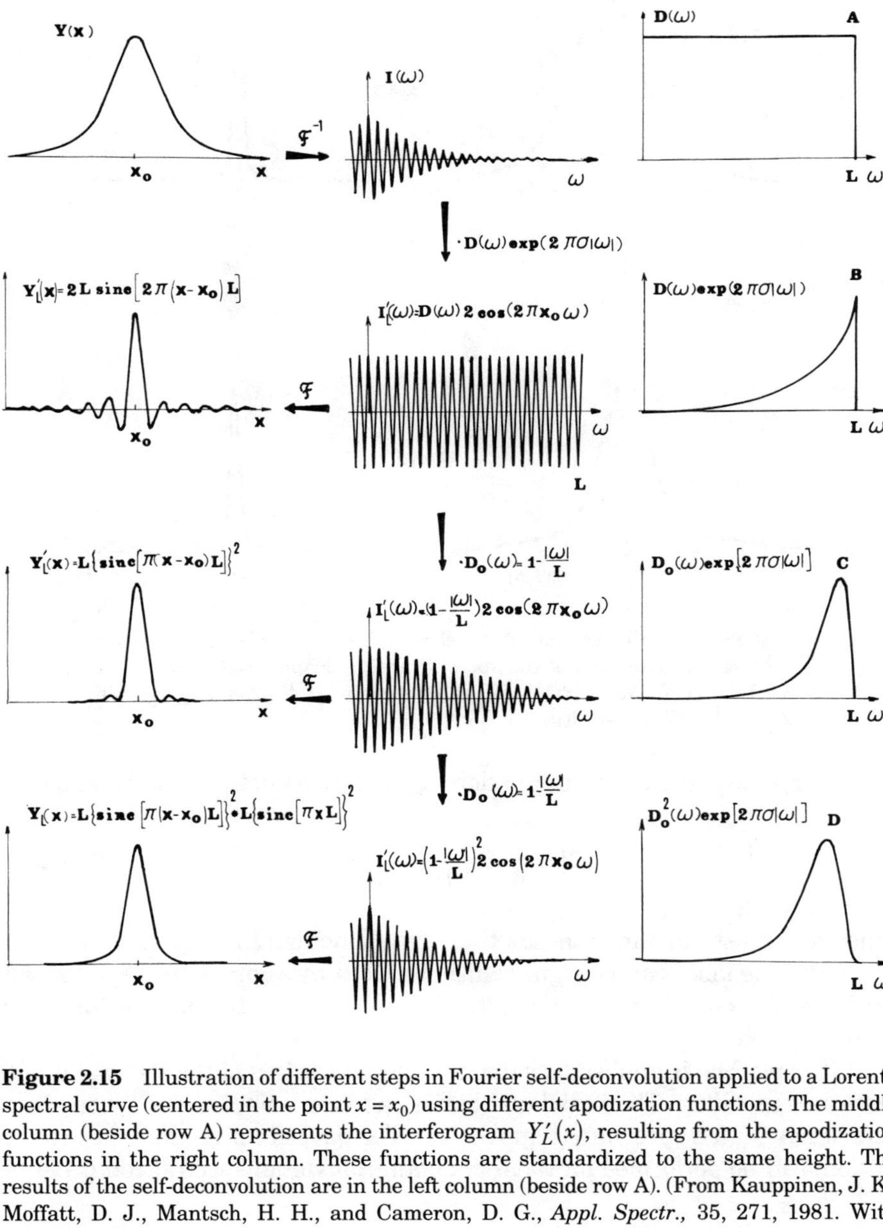

Figure 2.15 Illustration of different steps in Fourier self-deconvolution applied to a Lorentz spectral curve (centered in the point $x = x_0$) using different apodization functions. The middle column (beside row A) represents the interferogram $Y_L'(x)$, resulting from the apodization functions in the right column. These functions are standardized to the same height. The results of the self-deconvolution are in the left column (beside row A). (From Kauppinen, J. K., Moffatt, D. J., Mantsch, H. H., and Cameron, D. G., *Appl. Spectr.,* 35, 271, 1981. With permission.)

Then

$$Y'(x) = \mathscr{F}\left\{\frac{D_0(\omega)}{\mathscr{F}^{-1}\{G_0(x)\}} I(\omega)\right\} = \mathscr{F}\{I_L'(\omega)\}$$

$$= L\mathrm{sinc}^2\left[\pi(x + x_0)L\right] + L\mathrm{sinc}^2\left[\pi(x - x_0)L\right] \qquad \textbf{(2.86)}$$

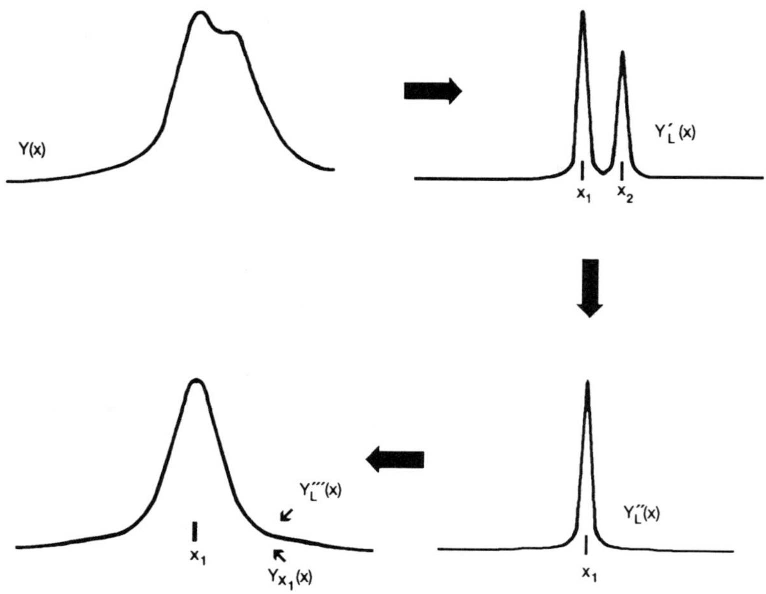

Figure 2.16 Illustration of the elimination of simple bands by successive deconvolution of the spectral profile. From Kauppinen, J. K., Moffatt, D. J., Mantsch, H. H., and Cameron, D. G., *Appl. Spectr.*, 35, 271, 1981. With permission.)

The complete deconvolution function $G_0^L(x)$ is then defined by the equation

$$G_0^L(x) = \mathcal{F}\left\{\frac{1}{D_0(\omega)}\right\} * G_0(x) \tag{2.87}$$

and the line shape function has the form $L\text{sinc}^2(\pi x L)$ (Figure 2.15C).

The side lobes can be significantly reduced by applying the apodization function $D_0^2(\omega)$,(Figure 2. 15D), although the shape of the deconvoluted line is broadened.

This method is applied by successive deconvolution of the spectral profile assumed in $G_0^L(x)$. The individual bands can be eliminated one by one from the deconvoluted spectrum (Figure 2.16). The opposite procedure, i.e., convolution, can be performed on them to obtain individual spectral components.

COMBINED FOURIER TRANSFORMATION AND LEAST SQUARES METHOD

This numerical method for spectral profile analysis[32] does not require any knowledge about the analytical form of single bands. It is based on two assumptions concerning the individual bands of a spectral profile:

1. It requires an axial symmetry of every spectral band with respect to its center.
2. It requires the same band shapes for all components.

Unlike simple Fourier analysis, this method is stable with regard to spectral noise.

Using the assumptions introduced, the spectral profile consisting of N components can be expressed as

$$Y(x) = \sum_{i=1}^{N} \left[K_i G(x - x_{0i}) \right] + U = \tilde{Y}(x) + U \tag{2.88}$$

where

$$K_i = \frac{Y_0^i}{Y_0^1} \tag{2.89}$$

is the relative intensity of the ith component ($K_1 = 1$), $G(x) = G(-x)$ is the symmetrical function of the line shape, x_{0i} is the center of mass of the ith band, and U is the constant background, independent of the position x.

At the center of mass of the spectral profile

$$\tilde{Y}(x) = Y(x) - U \tag{2.90}$$

Consequently, the following is valid:

$$x_s = \sum_{i=1}^{N} \frac{(K_i x_{0i})}{\sum_{i=1}^{N} (K_i)} \tag{2.91}$$

Using the relationship

$$x' = x - x_s \tag{2.92}$$

$\tilde{Y}(x)$ can be transformed into the form

$$\tilde{Y}(x') = \sum_{i=1}^{N} \left[K_i G(x' - x'_0) \right] \tag{2.93}$$

If the line shape function $G(x')$ is known, the spectral profile $\tilde{Y}(x')$ is characterized by $2n - 1$ parameters K_1 (l = 2, 3, 4,...N) and x_{0m}(m = 1, 2, 3,...N). By the Fourier transformation of the function $G(x')$, the following is obtained:

$$\begin{aligned} \mathcal{F}\{\tilde{Y}(x')\} &= \sum_{i=1}^{N} \mathcal{F}\{K_i G(x' - x'_{0i})\} \\ &= A'(\omega) - i\,B'(\omega) \end{aligned} \tag{2.94}$$

The application of the shift theorem (see Appendix D) leads to the equation

$$\mathscr{F}\left\{\tilde{Y}(x')\right\} = \mathscr{F}\left\{G(x')\right\} \sum_{i=1}^{N} \mathscr{F}\left\{K_i \exp\left(-i\omega x'_{0i}\right)\right\}$$

$$= \mathscr{F}\left\{G(x')\right\} \left\{ \sum_{i=1}^{N} \left\{K_i \cos\left(\omega x'_{0i}\right)\right\} - i \sum_{i=1}^{N} \left[K_i \sin\left(\omega x'_{0i}\right)\right] \right\} \quad \textbf{(2.95)}$$

Equations 2.94 and 2.95 show that the Fourier transformation of the line shape function can be written two ways:

$$\mathscr{F}\left\{G(x')\right\} = A'_0(\omega) = \frac{A'(\omega)}{\sum_{i=1}^{N} \left[K_i \cos\left(\omega x'_{0i}\right)\right]}$$

and

$$= \frac{B'(\omega)}{\sum_{i=1}^{N} \left[K_i \sin\left(\omega x'_{0i}\right)\right]} \quad \textbf{(2.96)}$$

Therefore, due to the symmetry assumed in $G(x')$, its Fourier transformation contains only the cosine part of $A'_0(\omega)$.

Moreover, the functions $A'(\omega)$ and $B'(\omega)$ can be written as

$$A'(\omega) = A(\omega) \cos\left(\omega x_s\right) + B(\omega) \sin\left(\omega x_s\right) \quad \textbf{(2.97)}$$

and

$$B'(\omega) = B(\omega) \cos\left(\omega x_s\right) - A(\omega) \sin\left(\omega x_s\right) \quad \textbf{(2.98)}$$

It is necessary to consider that $A(\omega)$ and $B(\omega)$ are dependent on the value of the background U:

$$A(\omega) = \int_{-\infty}^{+\infty} \tilde{Y}(x) \cos(\omega x) dx$$

$$= \int_{-\infty}^{+\infty} \left[Y(x) - U\right] \cos(\omega x) dx \quad \textbf{(2.99)}$$

$$B(\omega) = \int_{-\infty}^{+\infty} \tilde{Y}(x) \sin(\omega x) dx$$

$$= \int_{-\infty}^{+\infty} \left[Y(x) - U\right] \sin(\omega x) dx \quad \textbf{(2.100)}$$

Equations 2.96 to 2.100 show that besides the $2n - 1$ parameters introduced, the parameter U is also necessary for the complete description of $A_0'(\omega)$.

Let the $2n$ dimensional vector of the parameters be formed

$$\vec{h} = \left(K_2, K_3, \ldots K_N, x_{01}, x_{02}, \ldots x_{0N}, U\right) \tag{2.101}$$

Then

$$A_0'(\omega) \equiv A_0'\left(\omega, \vec{h}\right) \tag{2.102}$$

The line shape function $G(x')$ can be obtained using Equation 2.96 from the line shape of the inverse Fourier transformation:

$$G(x') = \mathcal{F}^{-1}\left\{\mathcal{F}\left\{G(x')\right\}\right\} = \mathcal{F}^{-1}\left\{A_0'\left(\omega, \vec{h}\right)\right\}$$

$$= \frac{1}{\pi}\int_{-\infty}^{+\infty} A_0'\left(\omega, \vec{h}\right)\cos(\omega x)d\omega \tag{2.103}$$

As $x' - x_0' = x - x_0$, $G(x - x_0)$ in Equation 2.88 can be expressed using Equation 2.103. Taking into consideration that the right side of Equation (2.88) is dependent on $2n$ components of the vector $\vec{h}$ parameters, this representation will be used as a model function $\tilde{Y}$ for the profile $Y(x)$:

$$\tilde{Y}\left(x, \vec{h}\right) = \frac{1}{\pi}\sum_{i=1}^{N}\left\{K_i \int_{-\infty}^{+\infty} A_0'\left(\omega, \vec{h}\right)\cos\left[\omega\left(x - x_{0i}\right)\right]d\omega\right\} + U \tag{2.104}$$

The components of the vector $\vec{h}$ can be determined by the iterative method of least squares. The sum of the squared deviations between experimental points of a spectral profile $Y(x_j)$ and the calculated functional values $\tilde{Y}(x_j, \vec{h})$ is minimized:

$$Q = \sum_{j=1}^{NB}\left[Y\left(x_j\right) - \tilde{Y}\left(x_j, \vec{h}\right)\right]^2 \tag{2.105}$$

The condition of the functional minimum Q leads to the system of $2n$ equations in the $2n$ unknown components of the vector $\vec{h}$:

$$\frac{\partial Q}{\partial K_i} = 0 \quad \text{for} \quad i = 2, 3, 4, \ldots N \tag{2.106}$$

$$\frac{\partial Q}{\partial x_{0m}} = 0 \quad \text{for} \quad m = 1, 2, 3, \ldots N \tag{2.107}$$

$$\frac{\partial Q}{\partial U} = 0 \tag{2.108}$$

These nonlinear equations can be linearized by a Taylor series expansion of the function $\tilde{Y}(x,\vec{h})$ truncated linear term. The constant term and the single partial derivatives are calculated from the estimated components of the vector $\vec{h}$. Using this approximation, Equations 2.106 to 2.108 are linear according to the increments $\Delta\vec{h}$. Solving this linear system of equations, the values of the increments $\Delta\vec{h}$ are obtained and added to the original values of the vector $\vec{h}$ parameters. This cycle is repeated until consistency of the successive iterative cycles is attained.

The required partial derivatives of the function $\tilde{Y}(x,\vec{h})$ are given by

$$\frac{\partial \tilde{Y}\left(x,\vec{h}\right)}{\partial K_j} = \frac{1}{\pi}\sum_{i=1}^{N}\left\{K_i \int_{-\infty}^{+\infty}\left[\frac{\partial}{\partial K_j} A_0'\left(\omega,\vec{h}\right)\right]\cos\left[\omega\left(x - x_{0i}\right)\right]d\omega\right\}$$
$$+\frac{1}{\pi}\int_{-\infty}^{+\infty} A_0'\left(\omega,\vec{h}\right)\cos\left[\omega\left(x - x_{0j}\right)\right]d\omega \quad \textbf{(2.109)}$$

$$\frac{\partial \tilde{Y}\left(x,\vec{h}\right)}{\partial x_{0m}} = \frac{1}{\pi}\sum_{i=1}^{N}\left\{K_i \int_{-\infty}^{+\infty}\left[\frac{\partial}{\partial x_{0m}} A_0'\left(\omega,\vec{h}\right)\right]\cos\left[\omega\left(x - x_{0i}\right)\right]d\omega\right\}$$
$$+\frac{1}{\pi}K_m\int_{-\infty}^{+\infty} \omega A_0'\left(\omega,\vec{h}\right)\sin\left[\omega\left(x - x_{0i}\right)\right]d\omega \quad \textbf{(2.110)}$$

$$\frac{\partial \tilde{Y}\left(x,\vec{h}\right)}{\partial U} = \frac{1}{\pi}\sum_{i=1}^{N}\left\{K_i \int_{-\infty}^{+\infty}\left[\frac{\partial}{\partial U} A_0'\left(\omega,\vec{h}\right)\right]\cos\left[\omega\left(x - x_{0i}\right)\right]d\omega\right\} + 1 \quad \textbf{(2.111)}$$

The standard methods of numerical integration can be used to evaluate the above integrals. Figure 2.17 shows the application of this combined method of analysis to the hyperfine structure of a Mössbauer spectral line (λ = 663.245 nm) of the isotope ^{59}Co.

RESOLUTION OF OVERLAPPED SPECTRAL BANDS USING NONLINEAR REGRESSION ANALYSIS

Recently, the application of nonlinear regression methods to resolve spectral profiles into individual bands has increased. Many programs based on the iterative solution of nonlinear least squares problems have been developed.[2,33-51] These programs differ in the type of analytical functions used to describe the individual spectral bands (see Problem of Overlap of Spectral Bands, above) and in the algorithms used for the nonlinear optimization procedure (see Appendix C.2).

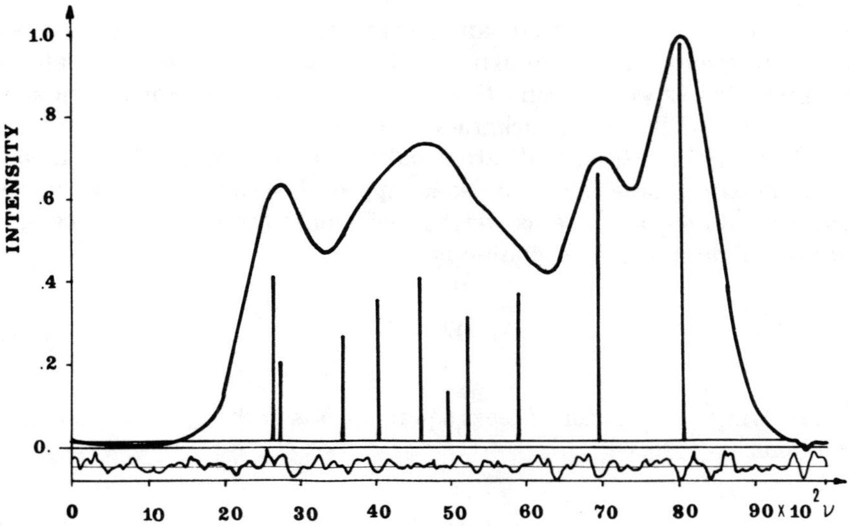

Figure 2.17 The experimental spectrum of the hyperfine structure of the spectral line λ = 663.245 nm of isotope ^{59}Co. The positions and the intensities of individual bands are represented by the lines. The difference of the experimental and simulated spectrum, multiplied by 3.4, is shown in the lower part of the figure. (From Damarowsky, M. and Guthöhrlein, G. H., *Comput. Phys. Commun.*, 52, 187, 1989.)

Nonlinear regression analysis to resolve a spectral profile into individual bands is based on minimizing the weighted sum of the squared deviations between experimental and calculated points of the spectral profile:

$$Q = \sum_{i=1}^{NB} W_i \left[Y^{\exp}(x_i) - Y^{\text{calc}}(x_i, \vec{a}) \right]^2 \quad \textbf{(2.112)}$$

where W_i are the weight coefficients, $Y^{\exp}(x_i)$ are the experimental values of the intensity at the points x_i, and $Y^{\text{calc}}(x_i)$ are the calculated values of intensity at the points x_i. The summation over i is performed for all experimentally available points of the spectral profile (N_W). The vector $\vec{a}$ consists of the parameters characterizing individual bands (height, halfwidth, position of maximum, or parameters of asymmetry) and the parameters characterizing the background.

The calculated values at the individual points x_i of a spectral profile are given by the summation of analytical band functions over all assumed bands and the value of the assumed background function:

$$Y^{\text{calc}}(x_i) = \sum_{j=1}^{NP} Y_j^{\text{calc}}(x_i, \vec{a}_j) + f(x_i, \vec{b}) \quad \textbf{(2.113)}$$

where $Y_j^{\text{calc}}(x_i, \vec{a}_j)$ is the calculated value of the contribution of the jth band at the point x_i. Y_j^{calc} is the function of the given band parameter vector $\vec{a}_j$. The summation over j is performed over all assumed bands (NP). The total number of the optimized band parameters (NPO) is given by the sum of the parameters of individual bands. The function $f(x_i, \vec{b})$ simulates the background of a spectral profile (see chapter 1). $\vec{b}$ is the vector of parameters b_i (i = 1, 2,...NPB) in the background function.

The weight coefficients W_i are usually taken as equal to 1 for all points of the spectral profile. The more exact approaches consider the variance of measured values, and use as weight coefficients the inverse values of the variance of data at corresponding points:

$$W_i = \frac{1}{\sigma_i^2} \tag{2.114}$$

Applying the criterion of least squares leads to the system of nonlinear equations:

$$\frac{\partial Q}{\partial a_i} = 0 \quad \text{for } i = 1, 2, 3, \ldots \text{NPC} \tag{2.115}$$

$$\frac{\partial Q}{\partial b_k} = 0 \quad \text{for } k = 1, 2, 3, \ldots \text{NPO} \tag{2.116}$$

This system of equations is solved by using some of the iterative procedures introduced in Appendix C.2.

The main problem in applying nonlinear regression analysis to simulation of spectral profiles is the complication caused by the minimized hypersurface of the functional Q. This surface, as a rule, consists of a series of local minima, the number of which gradually increases with the number of optimized parameters. A method of nonlinear regression analysis that would guarantee finding the global minimum as the only physically correct solution does not, as of now, exist. In order to find the global minimum, it is often necessary to perform nonlinear optimization several times, with various starting values of parameters. Suitable starting values can be obtained by some of the simple methods described in Analog Resolution of Spectra, Logarithmic Differential Analysis, and Determination of Spectral Band Parameters from Properties of Spectral Derivatives, above. The application of some common methods, which enable a high probability of yielding the absolute minimum (see Appendices C.3 and C.4), seems promising.

Sometimes an algorithm to improve convergence of the iterative procedure is used. In this algorithm, the values of some of the parameters to be optimized in the iterative steps are fixed with the aim of guaranteeing faster convergence of the other parameters. Obviously, all of the parameters have to be optimized, and the calculation must be completed by the simultaneous optimization of all parameters defining the spectral profile.

The number of points involved in regression analysis is also important. By increasing the number of experimental points, the calculation is much

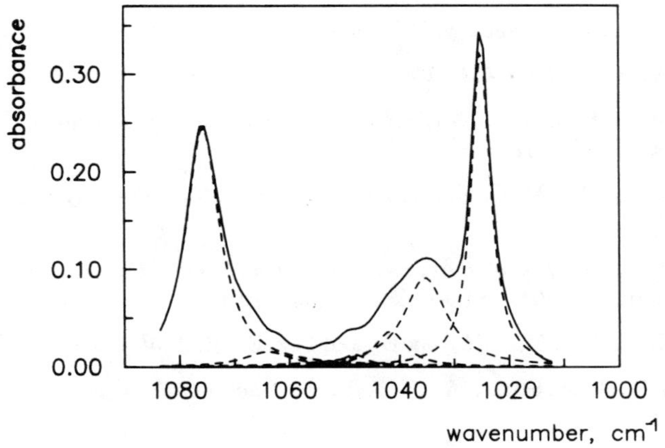

Figure 2.18 Separation of an IR spectrum of toluene into individual bands.

more time consuming, but the results obtained are more accurate. Using fewer points to determine "reasonable" values of the optimized parameters and following with minimization based on the whole set of experimentally available spectral data can be a convenient compromise. It was shown that, for example, for satisfactory description of one Gauss function, as few as 6 to 7 points are necessary.[52]

The application of the nonlinear regression procedure for resolving the absorption profile into individual bands is shown in Figure 2.18.

REFERENCES

1. Morrey, J. R., *Anal. Chem.,* 40, 905, 1968.
2. Pitha, J. and Jones, R. N., *Can. J. Chem.,* 44, 3031, 1966.
3. Fraser, R. D. B. and Suzuki, E., *Anal. Chem.,* 41, 37, 1969.
4. Schwartz, L. M., *Anal. Chem.,* 43, 1336, 1971.
5. Denisov, G. S. and Tjoruškin, B. C., *Molecular Spectroscopy,* Tonkov, M. V., Ed., University of Leningrad, Leningrad, 1981.
6. Baker, C., Cockervill, I. P., and Kelsey, J. E., *Spectrochim. Acta,* A34, 673, 1978.
7. Young, R. P., *Proc. Internat. Conf.,* Ljubljana-Zagreb, Yugoslavia, 1973.
8. Fanys, B., Rondot, D., and Mignot, J., *Spectrochim. Acta,* A34, 889, 1978.
9. Gastilovič, V. A., *Zh. Prikl. Spektr.,* 10, 878, 1969.
10. Metzler, C. M., Cahill, A. E., Petty, S., Metzler, D. E., and Lang, L., *Appl. Spectrosc.,* 39, 333, 1985.
11. Siano, D. B., *J. Chem. Educ.,* 49, 755, 1972.
12. Siano, D. B. and Metzler, D. E., *J. Chem. Phys.,* 51, 1856, 1969.
13. Klabuhn, B., Spindeer, D., and Goetz, H., *Spectrochim. Acta,* A29, 1283, 1973.

14. Du Pont Nemours Company, Delaware.

15. Baker, B. E. and Fox, M. F., *Chem. Soc. Rev.,* 9, 143, 1980.

16. Walrafen, G. E., *Water — A Comprehensive Treatise, Vol. I,* Franks, F., Ed., Plenum Press, New York, 1972.

17. Baker, B. E., Fox, M. F., and Hayon, E., *J. Chem. Soc., Faraday Trans. I.,* 74, 1776, 1978.

18. Urry, D. W., *Spectroscopic Approaches to Biomolecular Conformation,* Urry, D. W., Ed., American Medical Association, Chicago, 1970.

19. Baker, B. E., Fox, M. F, Hayon, E., and Fox, E. W., *Anal. Chem.,* 46, 1785, 1974.

20. Kendall, M. G. and Stuart, A., *The Advanced Theory of Statistics, Vol. 2.,* Hafner, New York, 1961.

21. Smith, R. C., *Rev. Sci. Instrum.,* 34, 296, 1963.

22. Sawitzky, A. and Golay, M. J. E., *Anal. Chem.,* 36, 1627, 1964.

23. Morrey, J. R., *Anal. Chem.,* 40, 905, 1968.

24. Lischka, H. and Derflinger, G., *Monatsch. Chem.,* 99, 2450, 1968.

25. Goméz-Beltrán, F., Salas, A., and Valero, A., *Anal. Chim. Acta,* 103, 263, 1978.

26. Carnana, R. A., Scarle, R. B., Helleer, T., and Shupack, S. I., *Anal. Chem.,* 58, 1162, 1986.

27. Bracewell, R., *The Fourier Transform and Its Applications,* McGraw-Hill, New York, 1965.

28. Horlick, G., *Anal. Chem.,* 44, 943, 1972.

29. Stone, H., *J. Opt. Soc. Am.,* 52, 998, 1962.

30. Kirmse, D. W. and Westenberg, A. W., *Anal. Chem.,* 43, 1035, 1971.

31. Kauppinen, J. K., Moffatt, D. J., Mantsch, H. H., and Cameron, D. G., *Appl. Spectr.,* 35, 271, 1981.

32. Damarowsky, M. and Guthöhrlein, G. H., *Comput. Phys. Commun.,* 52, 187, 1989.

33. Pitha, J. and Jones, R. N., *Can. J. Chem.,* 45, 2347, 1967.

34. Fraser, R. D. and Susuki, E., *Anal. Chem.,* 38, 1770, 1966.

35. De Tar, D. F., Ed., *Computer Programs for Chemistry, Vol.4.,* Academic Press, New York, 1972.

36. Jones, R. N., *Computer Programs for Infrared Spectroscopy,* N.R.C., Ottawa, Canada, 1976.

37. Chang, Y. S. and Shaw, J. H., *Appl. Spectrosc.,* 31, 213, 1977.

38. Hernanz, A., *Opt. P. Appl.,* 10, 231, 1977.

39. Noel, R. I. and Pires, A., *Appl. Spectrosc.,* 34, 351, 1980.

40. Baker, B. E. and Fox, M. F., *UV Spectrom. Group Bull.,* 9, 95, 1981.

41. Bartecki, A., Soltowski, J., and Kurnak, K., *Comput. Enhanced Spectrosc.,* 1, 31, 1983.

42. Scheeren, P. J. H., Barna, P., and Smit, H. C., *Anal. Chim. Acta,* 167, 65, 1985.

43. Massart, D. I., Dijkstra, A. D., and Kaufman, L., *Evaluation and Optimization of Laboratory Methods and Analytical Procedures,* Elsevier, Amsterdam, 1978.

44. Taypkin, Y. Z., *Sampling Systems Theory and its Applications, Vol.2.,* Pergamon Press, Oxford, 1964.

45. Amino, R., *Advances in Chromatography, Vol. 15,* Giddings, C. J., Grushka, E., Cazas, J., and Brown, P. H., Eds., New York, 1977.

46. Hart, R. R., J. *Molec. Spectrosc.,* 17, 368, 1965.

47. Jones, R. M., Sechadri, K. S., and Hopkins, J. W., *Can. J. Chem.,* 40, 334, 1962.

48. Marsh, R. E., Corey, R. B., and Pauling, L., *Biochim. Biophys. Acta,* 16, 1, 1955.

49. Stone, H., *J. Opt. Soc. Am.,* 52, 998, 1962.

50. Raso, M. A., Tortajada, J., Escolar, D., and Acción, F., *Comput. Chem.,* 11, 125, 1987.

51. Raso, M. A., Tortajada, J., and Acción, F., *Comput. Chem.,* 15, 29, 1991.

52. Scheeren, P. J. H., Klous, Z., and Smit, H. C., *Anal. Chim. Acta,* 171, 45, 1985.

Chapter 3

Methods of Factor Analysis in Molecular Spectroscopy

Factor analysis is a mathematical technique for interpreting a multidimensional data set[1,2]. Originally it was developed to solve problems in psychology and sociology, but it is now successfully used in various areas of chemistry[3]. Factor analysis is generally preferred for its ability, given data reflecting certain properties of the system, to determine accurately the number of components or factors (independent entities) creating the given physical or chemical system, and for its ability to classify and characterize these factors.

ABSTRACT FACTOR ANALYSIS AND THEORY OF ERRORS IN FACTOR ANALYSIS

Let N_S be the spectra of chemical mixtures composed of N_C chemical individuals — components are active in a studied spectral region (i.e., they absorb or emit radiation in this region, respectively, and they produce other signal characteristics for a given type of spectroscopy). Spectra of all mixtures are digitized to N_W points.

The intensity of a spectral signal of a mixture at every point is given by the sum of the increments of individual compounds. According to the Lambert-Beer law for absorption spectroscopy, and according to analogous relations for other types of spectra, the increment in unit thickness is given by the product of the molar absorption coefficient (generally of the pure component spectrum) and the concentration of the components:

$$a_{ij} = \sum_{k=1}^{N_C} \varepsilon_{ik}\, c_{kj} \qquad \begin{aligned} i &= 1 \rightarrow N_W \\ j &= 1 \rightarrow N_S \end{aligned} \tag{3.1}$$

where a_{ij} is the ith element of the spectral data matrix of the jth mixture in the unit thickness of the layer, ε_{ij} is the ith point of the spectrum of the kth component, and c_{kj} is the concentration of the kth component in the jth mixture. In a matrix form, Equation 3.1 can be rewritten as

$$\bar{\bar{A}} = \bar{\bar{S}}\,\bar{\bar{C}} \tag{3.2}$$

where $\bar{\bar{A}}$ is the matrix of spectral data (dimension $N_W \times N_S$), $\bar{\bar{S}}$ is the matrix of component spectra (dimension $N_W \times N_S$), and $\bar{\bar{C}}$ is the concentration matrix (dimension $N_C \times N_S$).

Spectra of individual mixtures are arranged in column in matrix $\bar{\bar{A}}$, and each spectrum is represented by an N_W-dimensional vector $\vec{A}_j$. Similarly, N_W-dimensional vectors of the component spectra $\vec{S}_K$ are placed into columns of the matrix $\bar{\bar{S}}$.

The first step of all factor analysis methods is abstract factor analysis (AFA), also called principal component analysis (PCA).

The matrix of the spectral data is not directly factor analyzable. Instead, a square, symmetrical matrix $\bar{\bar{Z}}$ (also called the scatter matrix) is constructed:

$$\bar{\bar{Z}} = \bar{\bar{A}}\,\bar{\bar{A}}^T \tag{3.3}$$

The scatter matrix $\bar{\bar{Z}}$ is decomposed into a set of "abstract" factors. Decomposition is performed by diagonalization (by solving the characteristic equation) of the matrix $\bar{\bar{Z}}$ (see Appendix A):

$$\bar{\bar{Q}}^{-1}\,\bar{\bar{Z}}\,\bar{\bar{Q}} = \bar{\bar{\lambda}} \tag{3.4}$$

The diagonal matrix $\bar{\bar{\lambda}}$ contains eigenvalues λ_i of the matrix $\bar{\bar{Z}}$, arranged by decreasing magnitude. Corresponding eigenvectors $\vec{Q}$, defining an orthonormal set, are placed into the columns of matrix $\bar{\bar{Q}}$.

With regard to the orthonormality of eigenvectors, the following is given:

$$\bar{\bar{Q}}^{-1} = \bar{\bar{Q}}^T \tag{3.5}$$

and the left side of Equation 3.4 can be written as

$$\bar{\bar{Q}}^T\,\bar{\bar{A}}\,\bar{\bar{A}}^T\,\bar{\bar{Q}} = \bar{\bar{U}}\,\bar{\bar{U}}^T \tag{3.6}$$

where

$$\bar{\bar{U}} = \bar{\bar{Q}}^T\,\bar{\bar{A}} \tag{3.7}$$

The rows of matrix $\bar{\bar{U}}$ are orthogonal to one other:

$$\bar{\bar{U}}\,\bar{\bar{U}}^T = \bar{\bar{\lambda}} \tag{3.8}$$

$$\lambda_i = \sum_{k=1}^{N_S} u_{ik}^2 \tag{3.9}$$

Following Equation 3.7:

$$\bar{\bar{A}} = \bar{\bar{Q}}\,\bar{\bar{U}} \tag{3.10}$$

Decomposition of the matrix $\bar{\bar{A}}$ according to Equation 3.10 is formally analogous to Equation 3.2. Matrices $\bar{\bar{Q}}$ and $\bar{\bar{U}}$ produce a matrix of spectral data without any physical meaning. They represent only the first step in the solution of the problem, called the abstract solution.

A set of eigenvectors $\bar{\bar{Q}}$ forms the orthonormal basis of a space of spectral data matrix $\bar{\bar{A}}$. Each spectrum of a mixture represented by a corresponding spectral vector $\vec{A}_j$ (with elements a_{ij}) can be defined in terms of

$$a_{ij} = \sum_{k=1}^{N_B} q_{ik}\, u_{kj} \tag{3.11}$$

The intrinsic dimension of the space of the spectral data N_B is identical with the ranks of the $\bar{\bar{A}}$ and $\bar{\bar{Z}}$ matrices, which are equal to the number of nonzero eigenvalues λ_i. If the spectral data were not influenced by experimental errors (if the pure data were studied), the intrinsic dimension N_B would be equal to the number of components N_C. This situation is obviously hypothetical.

Each element of the real matrix $\bar{\bar{A}}$ (i.e., the matrix of raw data) consists of two parts:

$$a_{ij} = a_{ij}^x + e_{ij} \tag{3.12}$$

where a_{ij}^x is the "pure" part and e_{ij} is the experimental error of the measurement. $\bar{\bar{A}}$, the matrix of raw spectral data, is consequently the sum of the pure data matrix $\bar{\bar{A}}^x$ and the error matrix $\bar{\bar{E}}$:

$$\bar{\bar{A}} = \bar{\bar{A}}^x + \bar{\bar{E}} \tag{3.13}$$

Although all the matrices in Equation 3.13 have dimensions of $N_W \times N_S$, their ranks are different. $\bar{\bar{A}}^x$, the matrix of pure data, has rank N_C, while the ranks of the matrices $\bar{\bar{A}}$ and $\bar{\bar{E}}$ are equal to the smaller of N_W and N_S (for simplicity, $N_W > N_S$ will be assumed later).

Having included experimental errors in the matrix $\bar{\bar{A}}$, the diagonalization of the matrix $\bar{\bar{Z}}$ produces a larger number of nonzero eigenvalues than would be produced by the diagonalization of the matrix of pure spectral data. This means that, for specification of the raw data space, a larger set of eigenvectors is necessary than for specification of the pure data space. However, the application of this complete basis leads to total reproduction of raw data in the reproduction step (Equation 3.10), including the reproduction of experimental errors. The choice of such a subset of the eigenvectors $\vec{Q}_k^{\ddagger}$, which describes the pure data space, can lead to the elimination of a certain portion of errors in the reproduced data.

Malinowski[3,4] developed the theory of errors in abstract factor analysis and showed how experimental errors are introduced into the scheme of factor analysis. According to this theory, one part of experimental error is always associated with the reproduced data. The total inaccuracy e_{ij} of the given element of the matrix $\bar{\bar{A}}$ is given by the sum of two terms:

$$e_{ij} = e_{ij}^{\ddagger} + e_{ij}^{0} \tag{3.14}$$

where $e_{ij}^{\ddagger}$ is the imbedded error and e_{ij}^{0} is the pure error (i.e., the error that can be eliminated by the "correct" choice of basis). After replacing Equations 3.12 to 3,14 the following expression is obtained:

$$a_{ij} = a_{ij}^{\ddagger} + e_{ij}^{0} \tag{3.15}$$

where $a_{ij}^{\ddagger}$ is the element of the spectral data matrix reproduced by the "correct" subset of the eigenvectors $\bar{\bar{Q}}^{\ddagger}$:

$$a_{ij}^{\ddagger} = a_{ij}^{x} + e_{ij}^{\ddagger} \tag{3.16}$$

Malinowski[3,4] divided a set of eigenvalues λ_k (arranged in descending order) and their eigenvectors $\vec{Q}_k$ into two groups:

1. The primary eigenvalues $\lambda_k^{\ddagger}$, k = 1 → N_C
2. The secondary eigenvalues $\lambda_k^{\ddagger}$, k = N_C + 1 → N_S

The primary eigenvalues $\lambda_k^{\ddagger}$ are the largest, and directly related to the chemical information contained in a spectral data matrix, but they also reflect a part of spectral data errors (called imbedded errors). The secondary eigenvalues λ_k^{0} are smaller and related exclusively to the experimental errors in spectral data. The eigenvectors corresponding to the primary eigenvalues form the set of primary eigenvectors. The eigenvectors belonging to the secondary eigenvalues are labeled secondary eigenvectors. The set of primary eigenvectors (the primary axes) defines the factor space basis, with the dimension N_C.

From Equation 3.14, the matrix of the experimental errors is the sum of the matrix of the imbedded errors $\bar{\bar{E}}^{\ddagger}$ and the matrix of the pure errors $\bar{\bar{E}}^{0}$:

$$\bar{\bar{E}} = \bar{\bar{E}}^{\ddagger} + \bar{\bar{E}}^{0} \quad \textbf{(3.17)}$$

From matrix $\bar{\bar{E}}$ a scatter matrix can be formed:

$$\bar{\bar{Z}}_e = \bar{\bar{E}}\,\bar{\bar{E}}^{T} \quad \textbf{(3.18)}$$

and can be diagonalized. The same orthonormal basis made from the eigenvectors $\bar{\bar{Q}}$, describing the space of raw spectral data, can be used to describe the experimental error space[4]:

$$\bar{\bar{Q}}^{T}\,\bar{\bar{Z}}_e\,\bar{\bar{Q}} = \bar{\bar{\lambda}}_e \quad \textbf{(3.19)}$$

According to Equation 3.14 an element of the error matrix is expressed in the base $\bar{\bar{Q}}$ by the relationship

$$e_{ij} = \sum_{k=1}^{N_C} q_{ik}\,\sigma^{\ddagger}_{kj} + \sum_{k=1}^{N_C} q_{ik}\,\sigma^{0}_{kj} \quad \textbf{(3.20)}$$

where q_{ik} is the ith element of the eigenvector $\vec{Q}_k$, $\sigma^{\ddagger}_{ij}$ is the projection of the imbedded error $e^{\ddagger}_{ij}$ to the corresponding primary axis, and σ^{0}_{ij} is the analogous projection of the pure error e^{0}_{ij} to the corresponding secondary axis. In matrix form, Equation 3.20 is changed into

$$\bar{\bar{E}} = \bar{\bar{Q}}\,\bar{\bar{\sigma}}^{\ddagger} + \bar{\bar{Q}}\,\bar{\bar{\sigma}}^{0} \quad \textbf{(3.21)}$$

$\bar{\bar{\sigma}}^{\ddagger}$ is the matrix of dimension $N_S \times N_S$, in which the first N_C rows form the projections $\sigma^{\ddagger}_{jk}$ and the rest contain zero values. $\bar{\bar{\sigma}}^{0}$ is the matrix of dimension $N_S \times N_S$, in which the first N_C rows contain zero values and the projections σ^{0}_{kj} are in the rows from N_C+1 to N_S. The first product of the right side of Equation 3.21 corresponds to the matrix of imbedded errors and the second part to the matrix of pure errors:

$$\bar{\bar{E}}^{\ddagger} = \bar{\bar{Q}}\,\bar{\bar{\sigma}}^{\ddagger} \quad \textbf{(3.22)}$$

$$\bar{\bar{E}}^{0} = \bar{\bar{Q}}\,\bar{\bar{\sigma}}^{0} \quad \textbf{(3.23)}$$

According to orthogonality

$$\bar{\bar{\sigma}}^{\ddagger}\,\bar{\bar{\sigma}}^{0T} = \bar{\bar{Q}} \quad \textbf{(3.24)}$$

then

$$\bar{\bar{E}}\,\bar{\bar{E}}^{T} = \bar{\bar{E}}^{\ddagger}\,\bar{\bar{E}}^{\ddagger T} + \bar{\bar{E}}^{0}\,\bar{\bar{E}}^{0T} \tag{3.25}$$

Subsequently, the scatter matrix of the experimental errors is the sum

$$\bar{\bar{Z}}_{e} = \bar{\bar{Z}}_{e}^{\ddagger} + \bar{\bar{Z}}_{e}^{0} \tag{3.26}$$

where

$$\bar{\bar{Z}}_{e}^{\ddagger} = \bar{\bar{E}}^{\ddagger}\,\bar{\bar{E}}^{\ddagger T} \tag{3.27}$$

$$\bar{\bar{Z}}_{e}^{0} = \bar{\bar{E}}^{0}\,\bar{\bar{E}}^{0T} \tag{3.28}$$

Then the matrix $\bar{\bar{Z}}_{e}$ trace is the sum of the traces of the matrices $\bar{\bar{Z}}_{e}^{\ddagger}$ and $\bar{\bar{Z}}_{e}^{0}$:

$$tr\left(\bar{\bar{Z}}_{e}\right) = tr\left(\bar{\bar{Z}}_{e}^{\ddagger}\right) + tr\left(\bar{\bar{Z}}_{e}^{0}\right)$$

namely

$$\sum_{i=1}^{N_W}\sum_{j=1}^{N_S}\left(e_{ij}\right)^2 = \sum_{i=1}^{N_W}\sum_{j=1}^{N_S}\left(e_{ij}^{\ddagger}\right)^2 + \sum_{i=1}^{N_W}\sum_{j=1}^{N_S}\left(e_{ij}^{0}\right)^2 \tag{3.29}$$

The left side of Equation 3.29 represents the sum of the squares of experimental errors, i.e., the total error of the spectral matrix $\bar{\bar{A}}$, it is connected with the residual standard deviation (RSD), which has $N_W \cdot N_S$ degrees of freedom:

$$N_W\, N_S (\mathrm{RSD})^2 = \sum_{i=1}^{N_W}\sum_{j=1}^{N_S}\left(e_{ij}\right)^2 \tag{3.30}$$

The right side of Equation 3.29 represents the decomposition of the total experimental error into the imbedded error with $N_W \cdot N_C$ degrees of freedom and into the pure error with $N_W \cdot (N_S - N_C)$ degrees of freedom:

$$N_W\, N_C (\mathrm{RSD})^2 = \sum_{i=1}^{N_W}\sum_{j=1}^{N_S}\left(e_{ij}^{\ddagger}\right)^2 \tag{3.31}$$

$$N_W\left(N_S - N_C\right)(\mathrm{RSD})^2 = \sum_{i=1}^{N_W}\sum_{j=1}^{N_S}\left(e_{ij}^{0}\right)^2 \tag{3.32}$$

From the invariability of the matrix $\bar{\bar{Z}}$ trace to the similarity transformation it follows that:

$$\sum_{i=1}^{N_W} \sum_{j=1}^{N_S} a_{ij}^2 = \sum_{k=1}^{N_S} \lambda_k = \sum_{k=1}^{N_C} \lambda_k^{\ddagger} + \sum_{k=N_C+1}^{N_S} \lambda_k^0 \tag{3.33}$$

An analogous relationship can be developed for the spectral data $a_{ij}^{\ddagger}$, i.e., for the data obtained by reproducing of Equation 3.10 using the set of primary eigenvectors:

$$\sum_{i=1}^{N_W} \sum_{j=1}^{N_S} \left(a_{ij}^{\ddagger}\right)^2 = \sum_{k=1}^{N_C} \lambda_k^{\ddagger} \tag{3.34}$$

By subtracting Equation 3.34 from Equation 3.33 the following is obtained:

$$\sum_{i=1}^{N_W} \sum_{j=1}^{N_S} \left[a_{ij}^2 - \left(a_{ij}^{\ddagger}\right)^2\right] = \sum_{k=N_C+1}^{N_S} \lambda_k^0 \tag{3.35}$$

Equation 3.35 describes the relationship between raw and reproduced spectral data and secondary eigenvalues. Because the matrix of reproduced data is orthogonal to the matrix of pure errors,[4]

$$\bar{\bar{A}}^{\ddagger T} \bar{\bar{E}}^{0} = \bar{\bar{0}} \tag{3.36}$$

according to Equation 3.35, the following results:

$$\sum_{i=1}^{N_W} \sum_{j=1}^{N_S} \left(e_{ij}^0\right)^2 = \sum_{i=1}^{N_W} \sum_{i=1}^{N_S} \left[a_{ij}^2 - \left(a_{ij}^{\ddagger}\right)^2\right] = \sum_{k=N_C+1}^{N_S} \lambda_k^0 \tag{3.37}$$

Now, from the Equations 3.32 and 3.37 the residual standard deviation (RSD) can be expressed as

$$\mathrm{RSD} = \left[\frac{\sum_{k=N_C+1}^{N_S} \lambda_k^0}{N_W\left(N_S - N_C\right)}\right]^{\frac{1}{2}} \tag{3.38}$$

Placing Equations 3.30, 3.31, and 3.32 into Equation 3.29 gives

$$(\mathrm{RSD})^2 = \frac{N_C}{N_S}(\mathrm{RSD})^2 + \frac{N_S - N_C}{N_S}(\mathrm{RSD})^2 \tag{3.39}$$

Based on Equation 3.39, Malinowski[3,4] defined three types of errors emerging in the scheme of abstract factor analysis:

1. The real error (RE), equal to the residual standard deviation, is the measure of the difference between pure and raw data:

$$\mathrm{RE} = \mathrm{RSD} = \left[\frac{\sum_{k=N_C+1}^{N_S} \lambda_k^0}{N_W \left(N_S - N_C\right)} \right]^{\frac{1}{2}} \tag{3.40}$$

2. The extracted error (XE) is the measure of the difference between raw and reproduced data, i.e., it is a measure of pure error:

$$\mathrm{XE} = \left(\frac{N_S - N_C}{N_S} \right)^{\frac{1}{2}} \cdot \mathrm{RSD}$$

$$= \left(\frac{\sum_{k=N_C+1}^{N_S} \lambda_k^0}{N_W\, N_S} \right)^{\frac{1}{2}} \tag{3.41}$$

3. The imbedded error (IE) is the measure of the difference between reproduced and pure data, i.e., it characterizes an experimental error remaining in reproduced data:

$$\mathrm{IE} = \left(\frac{N_C}{N_S} \right)^{\frac{1}{2}} \cdot \mathrm{RSD}$$

$$= \left(\frac{N_C \sum_{k=N_C+1}^{N_S} \lambda_k^0}{N_W\, N_S \left(N_S - N_C\right)} \right)^{\frac{1}{2}} \tag{3.42}$$

Substituting Equations 3.40, 3.41 and 3.42 into Equation 3.39 gives the expression of the relationship between these errors:

$$(\mathrm{RE})^2 = (\mathrm{XE})^2 + (\mathrm{IE})^2 \tag{3.43}$$

The significance of correctly determining the basic dimension of the factor space follows from the theory of error of abstract factor analysis. According to Equation 3.41, the portion of experimental errors eliminated in the factor analysis scheme, and not transferred for the following calculation, is directly connected with the used number of the primary axes, N_C. Overestimation of the number of significant factors (i.e., $N_C > N_C$ [real])

means decreasing the extracted error and increasing the imbedded error, which influences the final analysis results. The opposite situation leads to a physically incorrect description of a spectral data matrix. The number of varying components in the spectral set studied, as indicated by the number of significant eigenvalues, also represents the first information about the chemical nature of the system under study.

DETERMINATION OF THE NUMBER OF ACTIVE COMPONENTS IN THE SPECTRAL SET

Instead of using the methods of factor analysis, a rank of a spectral data matrix $\bar{\bar{A}}$ can be searched directly to find the common methods of matrix calculus (used in the initial study dealing with determination of the number of active components in the spectral set of mixtures).[5-9] The existence of experimental errors is taken into consideration either by calculating the standard deviations of the chosen criteria (determinants of the nonsingular matrix)[5,7] or by performing all the matrix operations for the error estimates of matrix $\bar{\bar{A}}$ elements.[6,8]

Because real spectral data are affected by experimental errors, determining the number of active components in mixture spectra is not a trivial problem.[1,3,10,11] Various methods defining the primary and secondary axes of the spectral data space in a set of eigenvectors $\bar{\bar{Q}}$ have been developed. However, no method can be accepted as universal. In practice, the simultaneous application of several criteria increases the reliability of results.[3,10]

The ability of any factor analysis method to obtain the correct number of active components in spectral mixtures is determined by the degree to which it meets two assumptions:[1,12]

1. Any spectrum of pure components is not a linear combination of the rest of the component spectra.
2. The concentration of any component cannot be expressed by a constant ratio of the concentrations of other components, i.e., the relative ratio of components in mixtures is not constant.

In the reaction

$$A \rightarrow 2B + C \tag{3.44}$$

the concentration ratio of the components B and C is always 2:1:

$$c_c = 0.5c_b \tag{3.45}$$

in all mixtures. Abstract factor analysis indicates only two components, even in the case where components A, B, and C have different characteristic spectra.

Reproduction of the Experimental Error[13]

The measure of the reproduction of experimental error in factor analysis is the real error (RE). According to Equation 3.40 RE values can be evaluated as a function of the dimension of the factor space:

$$\mathrm{RE} = f\left(N_C\right) \tag{3.46}$$

Values of RE significantly greater than the true experimental error indicate that some primary eigenvalues $\lambda_k^{\ddagger}$ were also used for the calculation, and the real dimension of the factor space is higher. If the experimental error is reproduced well by RE, the N_C value used to generate RE can be regarded as the number of significant factors, i.e., as the number of active components.

Reproduction of the Spectral Data Matrix

Equation 3.11 defines the reproduction of the spectral data matrix elements and their dependence on the dimension of the factor space N_C (i.e., on the number of eigenvectors used). In this way a set of reproduced matrices $\bar{\bar{A}}(N_C)$, which are certain approximations of a real matrix $\bar{\bar{A}}$, can be constructed. The closeness of approach is improved as N_C increases. The first matrix $\bar{\bar{A}}(N_C)$ reproducing $\bar{\bar{A}}$ within the experimental precision of its elements can be considered the best approximation. The corresponding N_C value then coincides with the dimension of the factor space, i.e., with the number of active components. Two tests were proposed to determine if the matrix $\bar{\bar{A}}(N_C)$ reproduces the matrix $\bar{\bar{A}}$ within the experimental error:

a. Chi-square test[14-16] — the value of the criterion $\chi^2(N_C)$:

$$\chi^2\left(N_C\right) = \sum_{i=1}^{N_W} \sum_{j=1}^{N_S} \left[\frac{a_{ij}^{\ddagger}\left(N_C\right) - a_{ij}}{\sigma\left(a_{ij}\right)} \right]^2 \tag{3.47}$$

where $\sigma(a_{ij})$ is the estimation of experimental error of the spectral data matrix element a_{ij} and is compared with the critical value

$$\left[\chi^2\left(N_C\right)\right]_{\text{crit}} = \left(N_W - N_C\right)\left(N_S - N_C\right) \tag{3.48}$$

For $N_C < N_C$ (real), the value $[\chi^2(N_C)]_{\text{crit}}$ will be substantially greater than the $\chi^2(N_C)$ value. For $N_C = N_C$(real), the value of $\chi^2(N_C)$ will approach the value of $[\chi^2(N_C)]_{\text{crit.}}$

b. Analysis of the difference distribution $|a_{ij}^{\ddagger}(N_C) - a_{ij}|$[14,16,17] — each difference $|a_{ij}^{\ddagger}(N_C) - a_{ij}|$ can be classified into one of the following intervals:

$$\left[(\alpha - 1)\ \sigma\left(a_{ij}\right),\ \alpha\,\sigma\left(a_{ij}\right)\right]$$

where α is a positive integer. The approximation degree of the matrix $\bar{\bar{A}}$ by the matrix $\bar{\bar{A}}(N_C)$ is characterized by distributing their element differences into the integer intervals of the multiples of the experimental errors. For example, the 3σ rule can be used as a criterion for satisfactory reproduction.

Reproduction of the Covariance Matrix[18]

The expression for reproducing the covariance matrix using N_C eigenvalues $\lambda_k^{\ddagger}$ and eigenvectors $\vec{Q}_k^{\ddagger}$ can be derived from the equation for matrix decomposition (Equation 3.4):

$$\bar{\bar{Z}}^{\ddagger}(N_C) = \sum_{k=1}^{N_C} \lambda_k^{\ddagger} \vec{Q}_k^{\ddagger} \vec{Q}_k^{\ddagger T} \tag{3.49}$$

or

$$z_{ij}^{\ddagger}(N_C) = \sum_{k=1}^{N_C} \lambda_k^{\ddagger} q_{ik}^{\ddagger} q_{jk}^{\ddagger} \tag{3.50}$$

The standard deviation of the elements of the covariance matrix is given by

$$\sigma(z_{ij}) = \sum_{k=1}^{N_S} \left[a_{ik}^2 \sigma^2(a_{jk}) + a_{jk}^2 \sigma^2(a_{ik}) \right]^{\frac{1}{2}} \tag{3.51}$$

The value N_C, at which satisfactory reproduction of the covariance matrix is attained, corresponds to the dimension of the factor space N_C (real). It is possible to use analogous tests to determine the adequacy of reproduction, as shown in Reproduction of the Spectral Data Matrix, above.

Cumulative and Residual Variance[3,19]

The cumulative variance is the measure of the portion of total variability that arises when the spectral data matrix, using a basis set consisting of primary eigenvectors, is reproduced:

$$V_C(N_C) = \frac{\sum_{i=1}^{N_W} \sum_{j=1}^{N_S} \left[a_{ij}^{\ddagger}(N_C) \right]^2}{\sum_{i=1}^{N_W} \sum_{j=1}^{N_S} \left(a_{ij} \right)^2} \tag{3.52}$$

From Equations 3.33 and 3.34, V_C can be expressed by the eigenvalues of the matrix $\bar{\bar{Z}}$:

$$V_C(N_C) = \frac{\sum_{k=1}^{N_C} \lambda_k^{\ddagger}}{\sum_{k=1}^{N_S} \lambda_k} \quad \textbf{(3.53)}$$

The cumulative variance criterion accepts the largest eigenvalues as the primary set, which explains the variance to the degree desired. The difficulty in using this test is in the selection of this degree. The specifications of 90, 95, 99%, and so on may not be reliable. Until the real value of the variance in spectral data is known, this method can give only a preliminary estimate of the number of components N_C.

The residual variance is a more convenient criterion for the determination of N_C:

$$V_r(N_C) = \frac{\sum_{i=1}^{N_W} \sum_{j=1}^{N_S} \left\{ \left[a_{ij}^{\ddagger}(N_C) \right]^2 - a_{ij}^2 \right\}}{\sum_{i=1}^{N_W} \sum_{j=1}^{N_S} \left(a_{ij}^2 \right)}$$

$$V_r(N_C) = \frac{\sum_{k=N_C+1}^{N_C} \lambda_k^0}{\sum_{k=1}^{N_S} \lambda_k} \quad \textbf{(3.54)}$$

Ideally, the slope of a function $V_r = f(N_C)$ steeply approaches zero before the point $N_C = N_C(\text{real})$.[19,20]

The IE Function[3,21]

If the experimental errors are randomly and uniformly distributed over the whole matrix of spectral data, the sum of squares of their projections to any secondary axis is approximately constant:

$$\lambda_N^0 \simeq \lambda_{N+1}^0 \simeq \lambda_{N+2}^0 \simeq \ldots \simeq \lambda_N^0 = \lambda^0 \quad \textbf{(3.55)}$$

Placing this assumption into Equation 3.42 the following is obtained:

$$\text{IE} = \left(\frac{N_C\, \lambda^0}{N_W\, N_S} \right)^{\frac{1}{2}} = (N_C)^{\frac{1}{2}} K \qquad \text{for } N_C > N_C(\text{real}) \quad \textbf{(3.56)}$$

where

$$K = \left(\frac{\lambda^0}{N_W \, N_S} \right)^{\frac{1}{2}} \quad \textbf{(3.57)}$$

As the value of N_C increases the value of the function IE(N_C) decreases until $N_C \leq N_C$(real). According to Equation 3.56, the IE(N_C) for $N_C > N_C$(real) is increasing. The point at which IE(N_C) attains the minimum corresponds to the dimension of the factor space.

In real cases IE $= f(N_C)$ often possesses a shallow minimum, which does not allow determination of the correct N_C (real) value. The IE(N_C) function is very sensitive to inhomogeneities of the error distribution and puts emphasis on raw errors. This situation is especially likely when:

1. The error distribution is not random and uniform
2. Data are influenced by systematic errors
3. Systematic and rough errors occur in data

The Factor Indicator Function (IND)[3,21]

This function is empirical, intuitively formed. Its properties can be used to determine dimension of the factor space more effectively than the properties of the IE(N_C) function can. The IND(N_C) function is expressed by RE(N_C):

$$\text{IND}(N_C) = \frac{\text{RE}(N_C)}{(N_S - N_C)^2} \quad \textbf{(3.58)}$$

From experience in studying an extended set of factor analysis problems,[21] the following properties of this function were determined:

1. IND(N_C) attains a minimal value at $N_C = N_C$(real). This minimum is also easily observable even when IE(N_C) does not reach the minimal value.
2. Each element of a matrix $\bar{\bar{A}}$ influenced by raw error can produce an additional factor indicated by a secondary local minimum of the function IND(N_C) or by the extension of the minimum at $N_C = N_C$ (real) to a shallow plateau.
3. For a raw data matrix, where elements are random numbers, the function IND(N_C) increases monotonically.

The function IND(N_C) is especially effective in situations when there are no available estimations of errors in spectral data $\sigma(a_{ij})$, necessary for other methods.

The Average Eigenvalue[20,22]

According to this criterion (also known as the Kaiser criterion), those eigenvalues greater than an arithmetic average of the total set of eigenvalues are primary:

$$\bar{\lambda} = \frac{1}{N_S} \sum_{k=1}^{N_S} \lambda_k \tag{3.59}$$

Estimation of the Standard Errors of Eigenvalues[14]

Hugus and El-Awady derived an expression for the standard error of an eigenvalue λ_k

$$\sigma(\lambda_k) = \left[\sum_{l=1}^{N_S} \sum_{m=1}^{N_S} q_{lk}^2 \, q_{mk}^2 \, \sigma^2(z_{ml}) \right]^{\frac{1}{2}} \tag{3.60}$$

where q_{lk} and q_{mk}, respectively, are the lth and mth elements of the eigenvector $\vec{Q}_k$. The standard errors of the covariance matrix elements are given by

$$\sigma^2(z_{lm}) = \sum_{i=1}^{N_W} \left[a_{il}^2 \, \sigma^2(a_{im}) + a_{im}^2 \, \sigma^2(a_{il}) \right] \qquad \text{for } l \neq m \tag{3.61}$$

$$\sigma^2(z_{ll}) = 4 \sum_{i=1}^{N_W} a_{il}^2 \, \sigma^2(a_{il}) \qquad \text{for } l = m \tag{3.62}$$

where $\sigma(a_{im})$ and $\sigma(a_{il})$ are estimations of the experimental errors of the corresponding elements of the spectral matrix.

According to this test, the number of components (the dimension of the factor space) N_C is equal to the number of eigenvalues λ_k that are larger than their standard errors $\sigma(\lambda_k)$.

The Function log(λ)[23,24]

The magnitudes of the primary eigenvalues $\lambda_k^{\ddagger}$ are significantly different from those of the secondary eigenvalues λ_k. The magnitude of all secondary eigenvalues is approximately the same (Equation 3.53), therefore, it is possible to recognize them as the special group in the total eigenvalue set. This test is carried out by plotting the discrete function

$$\log(\lambda_i) = f(i) \tag{3.63}$$

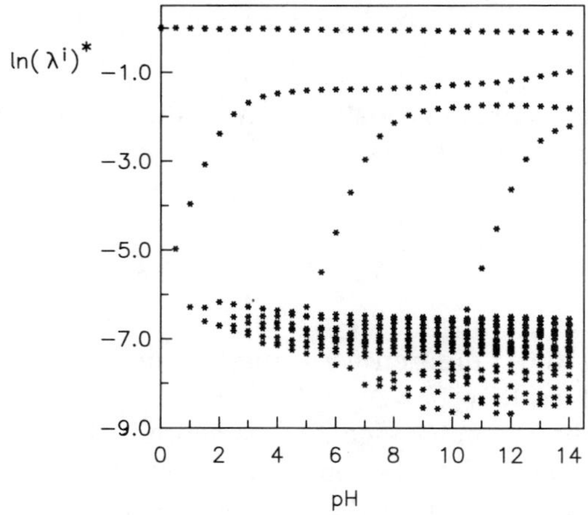

Figure 3.1 Eigenvalues λ_i for various noise levels of the spectral data: the original spectral data (○), the spectral data with superimposed uniformly distributed random noise with maximum value 0.005 (□), and the spectral data with superimposed uniformly distributed random noise with maximum value 0.01 (Δ).

where i is the number of the eigenvalue in the set, arranged in descending order.

As was shown during the study of model spectral sets,[24] this test is not sufficiently sensitive to the presence of relatively low amounts of components with respect to other, minority components. Recognizing the primary and secondary eigenvalues can not be unambiguous in such cases.

Antoon et al.[24] concluded that while the magnitude of primary eigenvalues is substantially independent of the noise in the spectral data, the magnitude of secondary eigenvalues increases as noise level is increased. Therefore, if the functions (Equation 3.63) are plotted for the eigenvalue sets obtained by diagonalizing covariance matrices belonging to spectral data sets with a different noise level, the separation of primary and secondary eigenvalues on the diagram (Figure 3.1) will be more observable. Sets of spectral data with various levels of noise can be obtained, for example, by repeating measurements with different spectrophotometer ranges, by accumulating various numbers of single measurements, or by superpositioning synthetic noise of various levels onto the only set of spectral data.

Reduced Eigenvalues[25]

Malinowski[25] showed that the probability distribution function of secondary eigenvalues belonging to data influenced by a uniformly distributed error has the following form:

$$P\left(\lambda_k^0\right) = N\left(N_W - k + 1\right)\left(N_S - k + 1\right) \tag{3.64}$$

where N is the normalizing constant. The reduced eigenvalues (REV) are defined by the following expression:

$$\mathrm{REV}_k = \frac{\lambda_k}{\left(N_W - k + 1\right)\left(N_S - k + 1\right)} \tag{3.65}$$

According to (3.64) differences in magnitudes of individual secondary reduced eigenvalues $\mathrm{REV}_k^0 (k = N_C + 1, N_C + 2, \ldots N_S)$ will not be statistically significant. Because primary eigenvalues contain the contributions of experimental errors, as well as the contributions of real factors, their distribution is not in agreement with the distribution (Equation 3.64), and the corresponding reduced eigenvalues, $\mathrm{REV}_k^{\ddagger} (k = 1, 2, \ldots N_C)$, are significantly higher than REV_k^0.

For recognizing primary and secondary reduced eigenvalues, Malinowski[26] proposed the null hypothesis:

$$H_0 : \mathrm{REV}_k = \overline{\mathrm{REV}}_k^0 \tag{3.66}$$

against the alternative hypothesis:

$$H_a : \mathrm{REV}_k > \overline{\mathrm{REV}}_k^0 \tag{3.67}$$

where $\overline{\mathrm{REV}}_k^0$ is the weighted average of secondary reduced eigenvalues. Because the statistical weights are given by the distribution function (3.64), for $\overline{\mathrm{REV}}_k^0$ the following is given:

$$\overline{\mathrm{REV}}_k^0 = \frac{\sum_{j=k+1}^{N_S} \lambda_j^0}{\sum_{j=k+1}^{N_S} \left[\left(N_W - j + 1\right)\left(N_S - j + 1\right)\right]} \tag{3.68}$$

For the validity of H_0, the test criterion

$$F_k = \frac{\mathrm{REV}_k}{\overline{\mathrm{REV}}_k^0} \tag{3.69}$$

has the F distribution with 1 and $(N_S - k)$ degrees of freedom.

The determination of the number of real factors (active components) starts from the lowest eigenvalue, i.e., k is changed so that $k = N_S - 1$, $N_S - 2, \ldots 1$. If the level of significance corresponding to F_k is high enough (higher than the reference value), it is not possible to reject the null hypothesis (Equation 3.66), and the kth eigenvalue is considered secondary. If the level of significance falls under the chosen value, the null hypothesis is rejected and the kth eigenvalue is regarded as the lowest primary eigenvalue, i.e., the number of real factors in the set of spectral data is k.

Frequency Analysis of Eigenvectors[27]

The rank estimation by frequency analysis of eigenvectors (REFAE) is based on the assumption that spectral information contained in primary eigenvectors is Fourier transformed into low frequency Fourier coefficients, while secondary eigenvectors contain a greater portion of high frequency coefficients after the transformation (see Appendix D).

The discrete Fourier transformation has the following form for $\vec{Q}_k$ eigenvectors in the REFAE method:

$$V_k(u) = \frac{1}{N_W} \sum_{x=0}^{N_W - 1} v_k(x) \exp\left(-i2\pi \frac{x\,u}{N_W} \right) \tag{3.70}$$

for the direct transformation and

$$v_k(x) = \frac{1}{N_W} \sum_{u=0}^{N_W - 1} V_k(u) \exp\left(i2\pi \frac{x\,u}{N_W} \right) \tag{3.71}$$

for the inverse transformation. The representation in the time domain $v_k(x)$ is a discrete function, regularly sampled on N_W points with an arbitrary step of 1s. The components of the vectors $\vec{Q}_k$ correspond to the amplitude of the function $v_k(x)$. The representation of $V_k(u)$ in the frequency domain is phase shifted so that the coefficient of the null frequency corresponds to the value $u = (N_W/2) + 1$. The complete function $V_k(u)$ is represented by the Fourier spectrum:

$$\left| V_k(u) \right| = \left\{ \left[V_k^{\mathrm{real}}(u) \right]^2 + \left[V_k^{\mathrm{imag}}(u) \right]^2 \right\}^{\frac{1}{2}} \tag{3.72}$$

where $v_k^{\mathrm{real}}(u)$ and $v_k^{\mathrm{imag}}(u)$ are the real and imaginary coefficients of the function $v_k(u)$.

The relative importance of the low frequency region of the spectrum $|v_k(u)|$, restricted by the points $\pm u_{\mathrm{lim}}$ for reproduction in the time domain, is expressed by the relation

$$R_k(u_{\mathrm{lim}}) = \frac{A_k(u_{\mathrm{lim}})}{T}\ 100\% \tag{3.73}$$

where

$$T = \sum_{u=0}^{N_{W-1}} \left| V_k(u) \right| \tag{3.74}$$

is the total area below the function $v_k(u)$ and

$$A_k(u_{\lim}) = \sum_{u=-u_{\lim}}^{u_{\lim}} \left| V_k(u) \right| \tag{3.75}$$

is the area below the investigated low frequency area. In accordance with the basic assumption of this method, it is expected that the values of R_k for primary eigenvectors will be substantially higher than the R_k values for secondary eigenvectors.

Rossi and Warner[27] applied the REFAE method to a large number of test sets of spectral data, and based on the results they formulated the rules for the application of this method. Recommended intervals of low frequency regions with corresponding threshold values of R_k are shown in Table 3.1. Eigenvectors for which the obtained value $R_k(u_{\lim})$ is higher than that of the boundary value are considered primary. It is possible to extend this criterion by an investigation of the plots:

$$R_k(u_{\lim}) = f(k) \tag{3.76}$$

called the frequency distribution of eigenvectors by Rossi and Warner. Primary and secondary eigenvectors form separate sets on this graph.

Evolving Factor Analysis[28] *and Submatrix Analysis*[29,30]

Many sets of spectra have a certain (known beforehand) internal structure. It is possible to add a concrete value describing this structure to each spectrum in the set. Such sets can be, for example, spectra of reaction mixtures measured over certain intervals of the reaction course; spectra of

Table 3.1 Threshold Values for the REFAE Method

Frequency range (Hz)	A (%)
±0.258	80
±0.196	65
±0.133	46
±0.102	36
±0.055	25

From Rossi, T. M. and Warner, I. M., *Anal. Chem.*, 58, 810, 1986. With permission.

equilibrium mixtures (spectra whose changes depend on the changes in an equilibrium concentration of components — the parameter of the change can be pH, temperature, pressure, or the analytical concentration of any component); a mixture spectra for various spectrophotometric titrations; and so on.

According to the concept of evolving factor analysis (EFA),[28] the structure of spectral sets correlates with the structure of eigenvalues of the scatter matrix $\bar{\bar{Z}}$. Cartwight called this method "submatrix analysis" and generalized its application using absorption spectra of phosphoric acid.[29] In the pH range 0 to 14, four components appear in the spectra of aqueous solutions of phosphoric acid: the acid and the three conjugated bases.

Spectra of individual samples are arranged according to increasing value of the pH parameter in the columns of a matrix $\bar{\bar{A}}$, as the vectors $\vec{A}_j$. From the first n columns of the matrix $\bar{\bar{A}}$ (i.e., from the first n spectra) the submatrices $\bar{\bar{A}}^n$ are subsequently created for $n = 1 \rightarrow$ NS. The scatter matrices $\bar{\bar{Z}}^n$ are formed from each submatrix of $\bar{\bar{A}}^n$, and the eigenvalues $\lambda_k^n (k = 1 \rightarrow n)$ obtained by their diagonalization are normalized:

$$\lambda_k^{*n} = \frac{\lambda_k^n}{\sum_{k=1}^{n} \lambda_k^n} \tag{3.77}$$

The set of normalized eigenvalues $\lambda_k^{*n} (k = 1 \rightarrow n)$ is the result of each value of the pH parameter. Functions

$$\ln\left(\lambda_k^{*n}\right) = f(pH) \tag{3.78}$$

are plotted (Figure 3.2). The secondary eigenvalues are grouped in the lower part of the diagram along the whole axis of the pH parameter, while primary eigenvalues increase to the top as the system evolves (i.e., as the pH parameters change). This phenomenon is so universal that it can be used for the recognition of primary and secondary eigenvalues in spectral sets with relatively high levels of experimental error.[29] Submatrix analysis in this form can be used in situations where the number of components appearing in spectra is dependent on the value of a certain parameter. The actual nature of this parameter is unimportant from the point of view of analysis. Its values are not included in the calculations and they are necessary only for the correct ordering of single spectra into the columns of the matrix $\bar{\bar{A}}$.

Cartwright and Farley extended the submatrix analysis method to universal disordered sets of spectra without any defined structure.[30] The modified algorithm consists of four steps:

1. The scatter matrix $\bar{\bar{Z}}$ belonging to a complete set of N_S spectra is diagonalized and a set of eigenvalues $\lambda_k^{N_S} (k = 1 \rightarrow N_S)$ is obtained. A variable N'_S with an initial value $N'_S = N_S$ is defined.

2. The following procedure is performed for each index $j = 1 \rightarrow N'_S$. From the spectra $\vec{A}_1 \ldots \vec{A}_{j-1}$, $\vec{A}_{j+1} \ldots \vec{A}_{N_S}$, ($\vec{A}_2 \ldots \vec{A}_{N_S}$, for $j = 1$ and $\vec{A}_1 \ldots \vec{A}_{N'_S-1}$, for $j = N'_S$) i.e., from all spectra except $\vec{A}_j$, a scatter matrix $\bar{\bar{Z}}$ is created and diagonalized. The $N'_S - 1$ eigenvalues obtained are summed into $T_j^{N'_S-1}$.

3. An eigenvalue set, to which the highest value $T_j^{N'_S-1}$ belongs, is kept as the set $\lambda_k^{N'_S-1}$ for step 4. A spectrum, to which the maximal value $T_j^{N'_S-1}$ belongs, is eliminated from the set of spectra, and the value of a variable $N'_S = N'_S - 1$ is redefined. If $N'_S \geq 1$, i.e., if there is at least one spectrum in a set of spectra, the method is repeated for step 2.

4. Sets of eigenvalues $\lambda_k^i (i = N_S \rightarrow 1, k = 1 \rightarrow i)$ are normalized to the unit sum:

$$\lambda_k^{*i} = \frac{\lambda_k^i}{\sum_{k=1}^{i} \lambda_k^i} \quad \textbf{(3.79)}$$

and the functions

$$\ln\left[\lambda_k^{*i}\right] = f(i) \quad \textbf{(3.80)}$$

are plotted. The following procedure is analogous to the submatrix analysis of spectra in ordered sets.

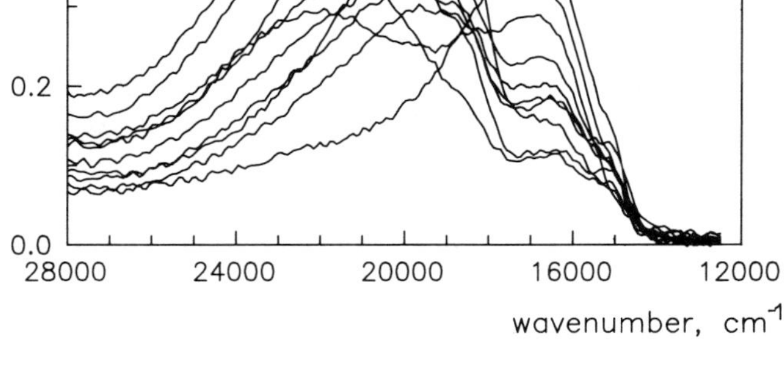

Figure 3.2 Submatrix analysis of the set of spectra of phosphoric acid for the pH range 0 to 14. (From Cartwright, H., *J. Chemometrics,* 1, 111, 1987. With permission.)

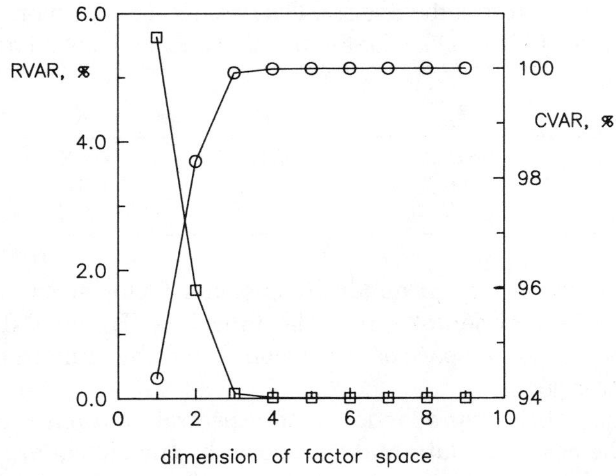

Figure 3.3 UV-VIS set of spectra of the mixture of four dyes.

Application of AFA for Determining the Number of Active Components in Mixture Spectra

Figure 3.3 shows the set of UV-VIS spectra of 10 mixtures of four dyes (bromcresol green, methylorange, alkali blue, and methylene blue). The choice of mixture composition ensures that the spectral signal of one component (methylene blue) will be significantly lower than the spectral signals of other components in all spectra. An "additional experimental error" uniformly and randomly distributed with a maximal value of ±0.005 was superimposed onto the spectra of the set. Various tests of abstract factor analysis to determine the number of active (absorbing) components in this set of spectra were performed.

The values of functions RE (Equations 3.40 and 3.44), IE (Equations 3.42 and 3.56), and IND (Equation 3.58) are shown in Table 3.2. The reproduction of a real error (RE) for a factor space of dimensions 1, 2, and 3 is higher than the estimation of an experimental error of the spectral data

Table 3.2 AFA for the Set of Mixture Spectra — RE, IE, and IND

Factor space dimension	RE	IE × 10^3	IND × 10^4
1	0.056	17.93	7.00
2	0.033	14.75	5.15
3	0.007	4.10	1.53
4	0.003	1.78	0.78
5	0.003	1.93	1.09
6	0.003	2.06	1.66
7	0.003	2.19	2.90
8	0.002	2.28	6.36
9	0.002	2.36	24.86

Table 3.3 χ^2 — Test of the Spectral Data Matrix Reproduction and Standard Errors of Eigenvalues for Estimating the Experimental Error of 0.005

Factor space dimension	χ^2_{crit}	χ^2	λ	$\sigma(\lambda)$
3	819	173914	0.996	0.058
4	696	228	0.041	0.022
5	575	179	0.001	0.058

(0.005), while the RE for a quadridimensional factor space satisfactorily reproduces an experimental error. The functions IE and IND reached a minimum for the factor space of dimension 4, and the minimum of the IND function is sharper.

Evaluation of the reproduction of the spectral data matrix (see Reproduction of the Spectral Data Matrix, above) by the chi-square test (Equations 3.47, and 3.48) and estimates of standard errors of eigenvalues (Equations 3.60–3.62) are shown in Table 3.3. Comparison of χ^2 and χ^2_{crit} indicates the presence of four components in the studied set of spectra. The eigenvalues λ_3 and λ_4 are higher than the estimates of their standard errors, while the standard error of the eigenvalue λ_5 is significantly higher than the eigenvalue itself. This test also suggests four active components.

Residual (Equation 3.54) and cumulative (Equation 3.53) variances (Figure 3.4) do not provide, in this case, the correct estimate of the dimension of a factor space. On the other hand, the gradient of the function $V_R = f(N_C)$ (see Cumulative and Residual Variance, above) in the vicinity of the point $N_C = 4$ is indeed steeply approaching zero (Table 3.4). The slope of the discrete function (Equation 3.63) indicates four primary eigenvalues whose magnitudes are significantly different from the secondary eigenvalues with relatively the same value (Figure 3.1). This distribution of the eigenvalues

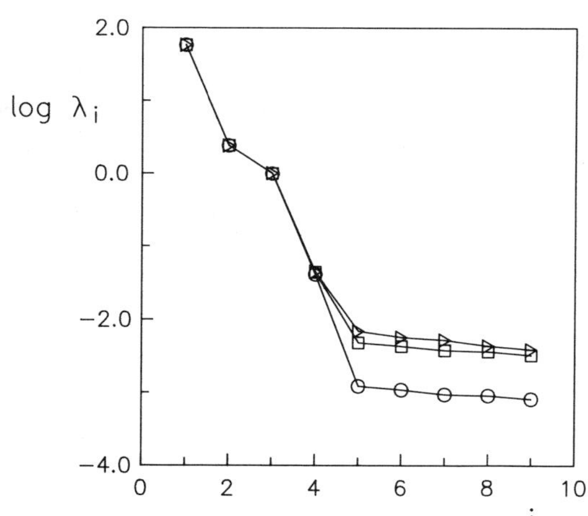

Figure 3.4 Residual (□) and cumulative (○) variances as a function of the dimension of the factor space.

Table 3.4 Slope of the Function $V_R = f(N_C)$

Factor space dimension	Slope $V_R = f(N_C)$
1	–3.94
2	–1.61
3	–0.07
4	–0.0002
5	–0.0002
6	–0.0002
7	–0.0001

into primary and secondary sets is confirmed also by the fact that the first four highest eigenvalues are practically independent from the level of noise in the spectral data, whereas the magnitude of the secondary eigenvalues increases as the noise level increases.

The average eigenvalue (Equation 3.59) of the set (Table 3.5) is 6.18, which, in harmony with the Kaiser criterion (see The Average Eigenvalue, above) shows incorrectly that there is only one active component in the set of spectra. The character of this criterion is not clearly defined, though, and it is known that it underestimates the real dimension of the factor space in many sets of spectral data.[31]

It is possible to clearly distinguish four primary eigenvalues from other secondary eigenvalues (Table 3.5) in the set of the reduced eigenvalues (see Reduced Eigenvalues, above). The significance level of rejecting the hypothesis (Equation 3.66) in favor of the alternative hypothesis (Equation 3.67) is very high for $k = 9 \rightarrow 5$, while the hypothesis

$$H_0{:}\mathrm{REV}_4 = \overline{\mathrm{REV}}_4^0 \tag{3.81}$$

Table 3.5 Eigenvalues, Reduced Eigenvalues, and F-Test of Reduced Eigenvalues

k	λ_k	REV_k	F_k	$\alpha(1, N_S - k)$[a] (%)
1	58.273	4.86×10^{-2}	73.17	0.0001
2	2.4304	2.27×10^{-3}	9.05	1.7
3	0.9967	1.06×10^{-3}	72.38	0.006
4	0.0414	5.05×10^{-5}	21.25	0.4
5	0.0012	1.75×10^{-6}	0.66	46
6	0.0011	1.89×10^{-6}	0.63	48
7	0.0009	2.05×10^{-6}	0.56	51
8	0.0009	2.67×10^{-6}	0.58	53
9	0.0008	3.63×10^{-6}	0.54	60
10	0.0007	6.67×10^{-6}		

[a] $\alpha(1, N_S - k)$ is the level of significance for rejecting the hypothesis; $H_0{:}\mathrm{REV}_k = \overline{\mathrm{REV}}_k^0$ against and $H_a{:}\mathrm{REV}_k > \overline{\mathrm{REV}}_k^0$.

can be rejected in favor of the alternative hypothesis:

$$H\text{:REV} > \overline{\text{REV}}^0 \tag{3.82}$$

at the very low significance level 0.4%. Thus, the F-test (see Reduced Eigenvalues, above) also shows four active components in the studied set of spectra.

The Fourier spectra (Equation 3.72) of the scatter matrix eigenvectors of the set of spectra studied are shown in Figure 3.5. In the sense of the REFAE method (see Frequency Analysis of Eigenvectors, above), the first four eigenspectra can be classified as low frequency spectra, while the Fourier spectra of the other eigenvectors contain a significant portion of high

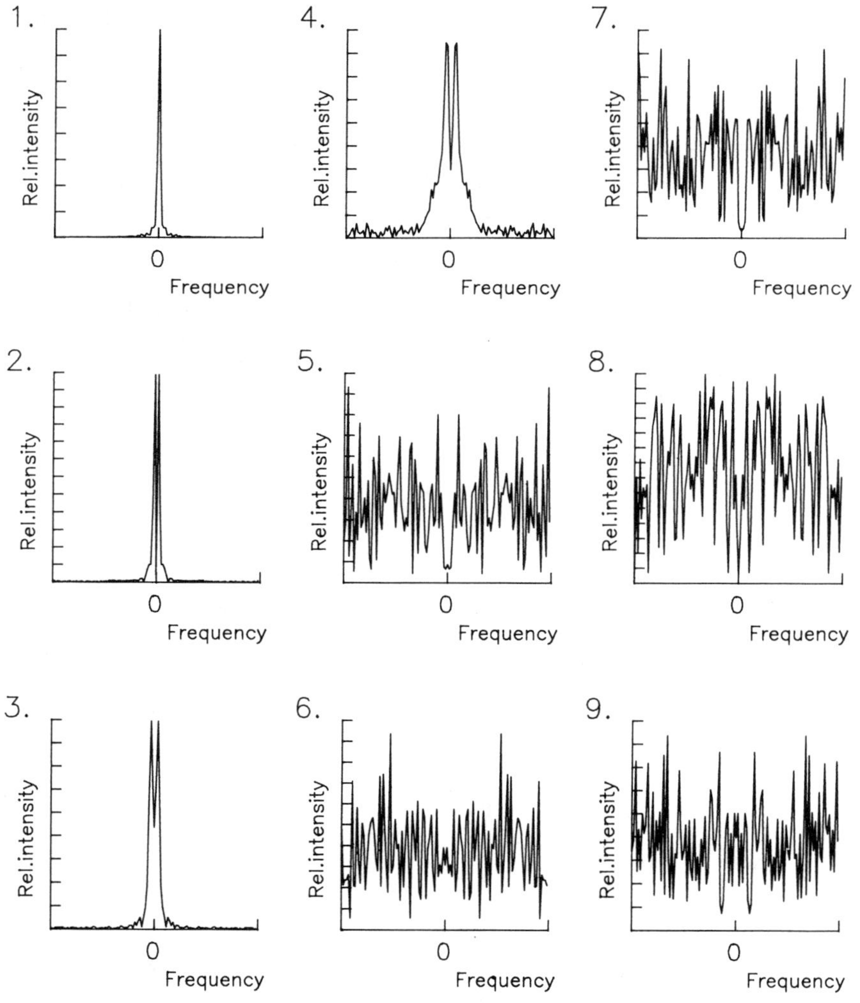

Figure 3.5 Fourier spectra of the eigenvectors.

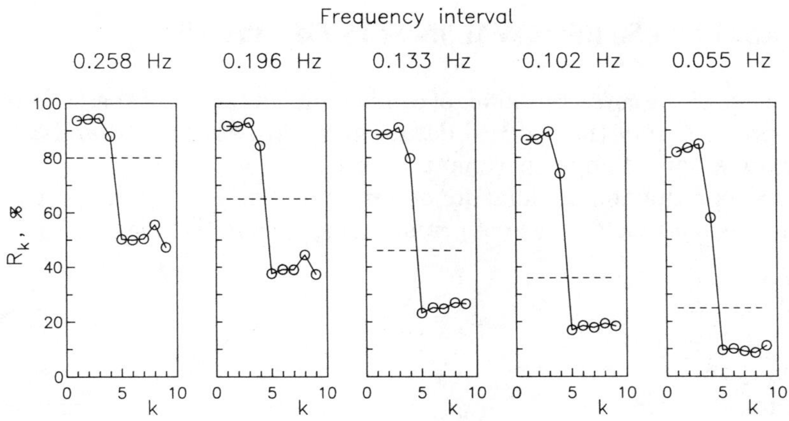

Figure 3.6 Frequency distribution of the eigenvectors: the boundary values for RA according to Table 3.1 (– – – –).

frequency coefficients. It is possible to recognize clearly the set of the four primary eigenvectors within all the recommended frequency intervals on the diagrams of the eigenvector frequency distribution (Figure 3.6). It is also possible to see that the R_k values (Equation 3.73) of these vectors are higher than the threshold values (Table 3.1).

The diagram of the submatrix analysis of the disordered set of spectra (see Evolving Factor Analysis and Submatrix Analysis, above) shows that four primary eigenvalues rise up from the level of secondary eigenvalues as the number of eigenvalues is increased (Figure 3.7).

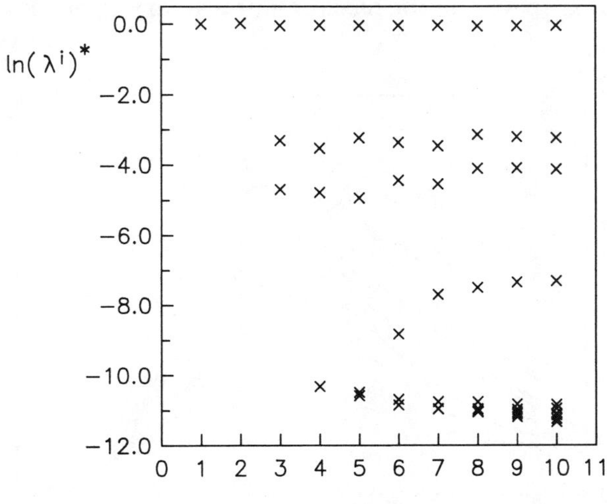

Figure 3.7 Submatrix analysis of the spectra set.

AFA AND NOISE REMOVAL IN SETS OF SPECTRA

Knowledge of the correct number of components in the set of spectra enables the reproduction of the spectral data matrix (Equations 3.10 and 3.11) to eliminate a part of any experimental error.

The reproduction of elements of a data matrix of a set of spectra in a N_C-dimensional factor space of primary eigenvectors is given by

$$a_{ij}^{r} = \sum_{k=1}^{N_C} q_{ik}\, u_{kj} \qquad \textbf{(3.83)}$$

or, in matrix form,

$$\bar{\bar{A}}^{r} = \bar{\bar{Q}}^{\ddagger}\, \bar{\bar{U}}^{\ddagger} \qquad \textbf{(3.84)}$$

where $\bar{\bar{Q}}^{\ddagger}$ is the matrix of primary eigenvectors and $\bar{\bar{Q}}^{\ddagger}$ are the corresponding rows of the matrix $\bar{\bar{U}}$ (Equation 3.7).

According to the error theory of factor analysis (see Abstract Factor Analysis and Theory of Errors in Factor Analysis, above), the relationship (Equation 3.84) reproduces both pure spectral data and a part of the experimental error, the imbedded error (Equation 3.42), while the second part of the experimental error, extracted error (Equation 3.41), is not transferred to reproduced data. Therefore, the reproduction (Equation 3.84) improves the signal/noise ratio and increases the information content of the whole spectral set.

The result of reproducing the set of spectra of the mixtures (see The Application of AFA for Determining the Number of Active Components in Mixture Spectra, above), on the basis of four primary eigenvectors, is shown in Figure 3.8.

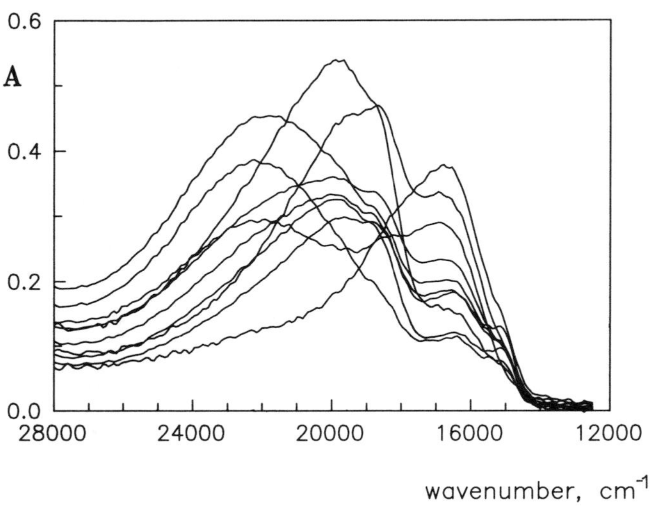

Figure 3.8 Reproduction of the spectral data set.

Analogous methods can eliminate various effects from spectral data, those interfering with or undesirable for a given study,[32] e.g., spectral signals of small amounts of components in mixtures, the influence of minor or parasitic chemical reactions, and so on.

TRANSFORMATION OF THE ABSTRACT SOLUTION — THE ISOLATION OF COMPONENTS OF MIXTURE SPECTRA

Identifying the components of a mixture from their spectra and obtaining the composition of the mixture, i.e., determining the component concentrations in individual mixtures, is the aim of spectroscopic analysis of an unknown mixture set. From the mathematical point of view, this problem resides in the determination of matrices $\bar{\bar{S}}$ and $\bar{\bar{C}}$ based on a known matrix $\bar{\bar{A}}$ (Equation 3.2).

The abstract solution (Equation 3.10) has no physical meaning. Eigenvectors $\vec{Q}_k^{\ddagger}$ represent the formal mathematical solution. However, transformation relationship

$$\bar{\bar{S}} = \bar{\bar{Q}}^{\ddagger}\, \bar{\bar{T}} \qquad \textbf{(3.85)}$$

exists between these "mathematical spectra" and the real — but unknown — spectra of the components. The transformation relationship for the concentration matrix is obtained from Equations 3.85 and 3.10:

$$\bar{\bar{Q}}^{\ddagger} = \bar{\bar{S}}\, \bar{\bar{T}}^{-1} \qquad \textbf{(3.86)}$$

$$\bar{\bar{A}}^{\ddagger} = \bar{\bar{Q}}^{\ddagger}\, \bar{\bar{U}}^{\ddagger} = \bar{\bar{S}}\, \bar{\bar{T}}^{-1}\, \bar{\bar{U}}^{\ddagger} \qquad \textbf{(3.87)}$$

By comparing Equations 3.2 and 3.87 the following results:

$$\bar{\bar{C}} = \bar{\bar{T}}^{-1}\, \bar{\bar{U}}^{\ddagger} \qquad \textbf{(3.88)}$$

Equation 3.85 is the expression of the component spectra $\bar{\bar{S}}$ on the basis of the primary eigenvectors $\bar{\bar{Q}}^{\ddagger}$. For the spectrum of the component "k" (i.e., the *k*th column of the matrix $\bar{\bar{S}}$), the following is valid:

$$\vec{S}_k = \sum_{l=1}^{N_C} t_{lk}\, \vec{Q}_k^{\ddagger} \quad \left(k = 1 \rightarrow N_C\right) \qquad \textbf{(3.89)}$$

Self Modeling Curve Resolution Methods

In 1971, Lawton and Sylvestre[33] proposed a method for isolating the component spectra from the spectra of two-component mixtures. They named it self modeling curve resolution (SMCR), emphasizing that their method, contrary to the methods of spectra decomposition into spectral bands (Chapter 2), does not assume any particular form of spectral functions.

For the construction of the matrix $\bar{\bar{T}}$ (Equation 3.85), SMCR methods utilize two restrictions resulting from the physical nature of the problem:

1. The matrix $\bar{\bar{S}}$ elements are non-negative, i.e., the component spectral amplitudes can acquire only non-negative values.
2. The matrix $\bar{\bar{C}}$ elements are non-negative, i.e., the component concentrations in mixtures cannot have negative values.

These restrictions are usually supplemented by certain normalization conditions.

For the two-component system ($N_C = 2$), Equation 3.89 has the following form:

$$\vec{S}_k = t_{1k}\,\vec{Q}_l^{\ddagger} + t_{2k}\,\vec{Q}_2^{\ddagger} \quad (k = 1, 2) \tag{3.90}$$

According to the first restriction, the following is given:

$$s_{ik} = t_{1k}\,q_{1l}^{\ddagger} + t_{2k}\,q_{i2}^{\ddagger} \geq 0 \tag{3.91}$$

for $i = 1 \rightarrow N_W$ and $k = 1, 2$.

The solution of the inequality (Equation 3.91) is

$$t_{2k} \geq -t_{1k}\,\min\left(\left|\frac{q_{11}^{\ddagger}}{q_{12}^{\ddagger}}\right|, \left|\frac{q_{21}^{\ddagger}}{q_{22}^{\ddagger}}\right|, \left|\frac{q_{N_W,1}^{\ddagger}}{q_{N_W,2}^{\ddagger}}\right|\right) \qquad \text{for } q_{i2}^{\ddagger} \geq 0 \tag{3.92}$$

and

$$t_{2k} \leq -t_{1k}\,\min\left(\left|\frac{q_{11}^{\ddagger}}{q_{12}^{\ddagger}}\right|, \left|\frac{q_{21}^{\ddagger}}{q_{22}^{\ddagger}}\right|, \left|\frac{q_{N_W,1}^{\ddagger}}{q_{N_W,2}^{\ddagger}}\right|\right) \qquad \text{for } q_{i2}^{\ddagger} < 0 \tag{3.93}$$

According to Equation 3.2 a spectral vector $\vec{A}_j$ is a linear combination of the spectral vectors of components:

$$\vec{A}_j = c_{1j}\,\vec{S}_1 + c_{2j}\,\vec{S}_2 \tag{3.94}$$

Keeping in mind Equation 3.10 and 3.11, it is also possible to express the vector $\vec{A}_j$ as a linear combination of the primary eigenvectors:

$$\vec{A}_j = u_{1j}\,\vec{Q}_1^{\ddagger} + u_{2j}\,\vec{Q}_2^{\ddagger} \tag{3.95}$$

If vectors $\vec{S}_1$ and $\vec{S}_2$ of Equation 3.94 are replaced with Equations 3.90 and 3.95, the following is obtained:

$$\left(\vec{Q}_1^{\ddagger}, \vec{Q}_2^{\ddagger}\right)\begin{pmatrix} u_{1j} \\ u_{2j} \end{pmatrix} = \left(\vec{Q}_1^{\ddagger}, \vec{Q}_2^{\ddagger}\right)\begin{pmatrix} t_{11} & t_{12} \\ t_{21} & t_{22} \end{pmatrix}\begin{pmatrix} c_{1j} \\ c_{2j} \end{pmatrix} \qquad \textbf{(3.96)}$$

The solution of Equation 3.96 for elements of a concentration matrix is

$$\begin{pmatrix} c_{1j} \\ c_{2j} \end{pmatrix} = \begin{pmatrix} t_{11} & t_{12} \\ t_{21} & t_{22} \end{pmatrix}^{-1}\begin{pmatrix} u_{1j} \\ u_{2j} \end{pmatrix} \qquad \textbf{(3.97)}$$

Due to the second restriction, all elements of a matrix $\bar{\bar{C}}$ are non-negative:

$$\begin{pmatrix} c_{1j} \\ c_{2j} \end{pmatrix} \geq \begin{pmatrix} 0 \\ 0 \end{pmatrix} \qquad \textbf{(3.98)}$$

The solutions of the system (Equations 3.97 and 3.98) are the two inequalities

$$t_{21} \geq t_{11} \max\left(\frac{u_{21}}{u_{11}}, \frac{u_{22}}{u_{12}}, \ldots \frac{u_2, N_S}{u_1, N_S}\right) \qquad \textbf{(3.99)}$$

and

$$t_{22} \geq t_{12} \min\left(\frac{u_{21}}{u_{11}}, \frac{u_{22}}{u_{12}}, \ldots \frac{u_2, N_S}{u_1, N_S}\right) \qquad \textbf{(3.100)}$$

The set of UV spectra of five mixtures of benzene-benzaldehyde are shown in Figure 3.9. The SMCR analysis of this system is illustrated in Figure 3.10. The inequalities (Equations 3.92, 3.93, 3.99, and 3.100) define two segments in a factor space, in which the spectral vectors fulfilling the restrictions are located. Spectra of components are usually normalized to unit area:

$$\sum_{i=1}^{N_W} s_{ik} = 1 \qquad \textbf{(3.101)}$$

The equation of a line in a factor space results from Equation 3.101:

$$t_{2k} = \frac{1 - \sum_{i=1}^{N_W} q_{i1}^{\ddagger}}{\sum_{i=1}^{N_W} q_{i2}^{\ddagger}} \qquad \textbf{(3.102)}$$

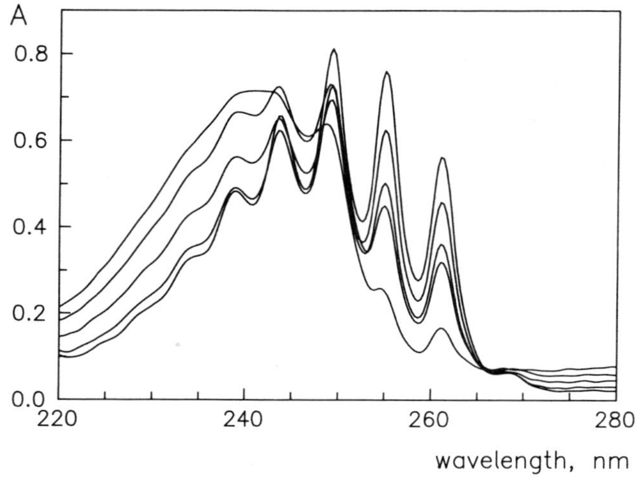

Figure 3.9 UV-VIS spectra of the benzene-benzaldehyde mixtures.

The line (Equation 3.102) intercepts the boundaries of the "allowed area" at the points A_1, A_2 and B_1, B_2. A coordinate origin of a factor space, together with these points, define vectors ${}^1(\vec{S}_1)$, ${}^2(\vec{S}_1)$ and ${}^1(\vec{S}_2)$, ${}^2(\vec{S}_2)$.

The pairs of spectral curves belonging to the vectors ${}^1(\vec{S}_1)$, ${}^2(\vec{S}_1)$ and ${}^1(\vec{S}_2)$, ${}^2(\vec{S}_2)$ are called the spectral bands of the components (Figure 3.11). The component spectra are placed inside these bands and can be expressed by the relationships

$$\vec{S}_k = a_k \; {}^1\left(\vec{S}_k\right) + \left(1 - a_k\right) \; {}^2\left(\vec{S}_k\right) \tag{3.103}$$

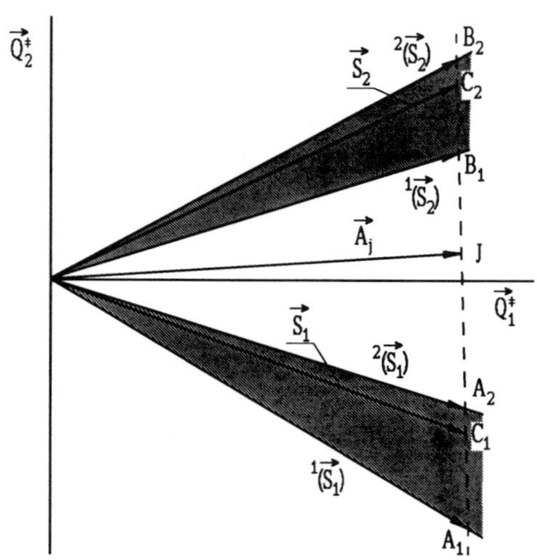

Figure 3.10 Illustration of SMCR analysis of the spectral set of the benzene-benzaldehyde mixture: the areas satisfying the inequalities (Equations 3.92, 3.93, 3.99, and 3.100) (▨) and the line (Equation 3.102) (– – –).

where, for $k = 1, 2$:

$$0 \le a_k < 1 \qquad \textbf{(3.104)}$$

Sharaf and Kowalski[34,35] analyzed the problem of quantitative characterization of mixtures in a two-dimensional factor space. If a spectrum of mixtures is normalized to fit the unit area, its spectral vector $\vec{A}_j$ lies between the bands of "allowed" regions and intercepts the line (Equation 3.102) at the

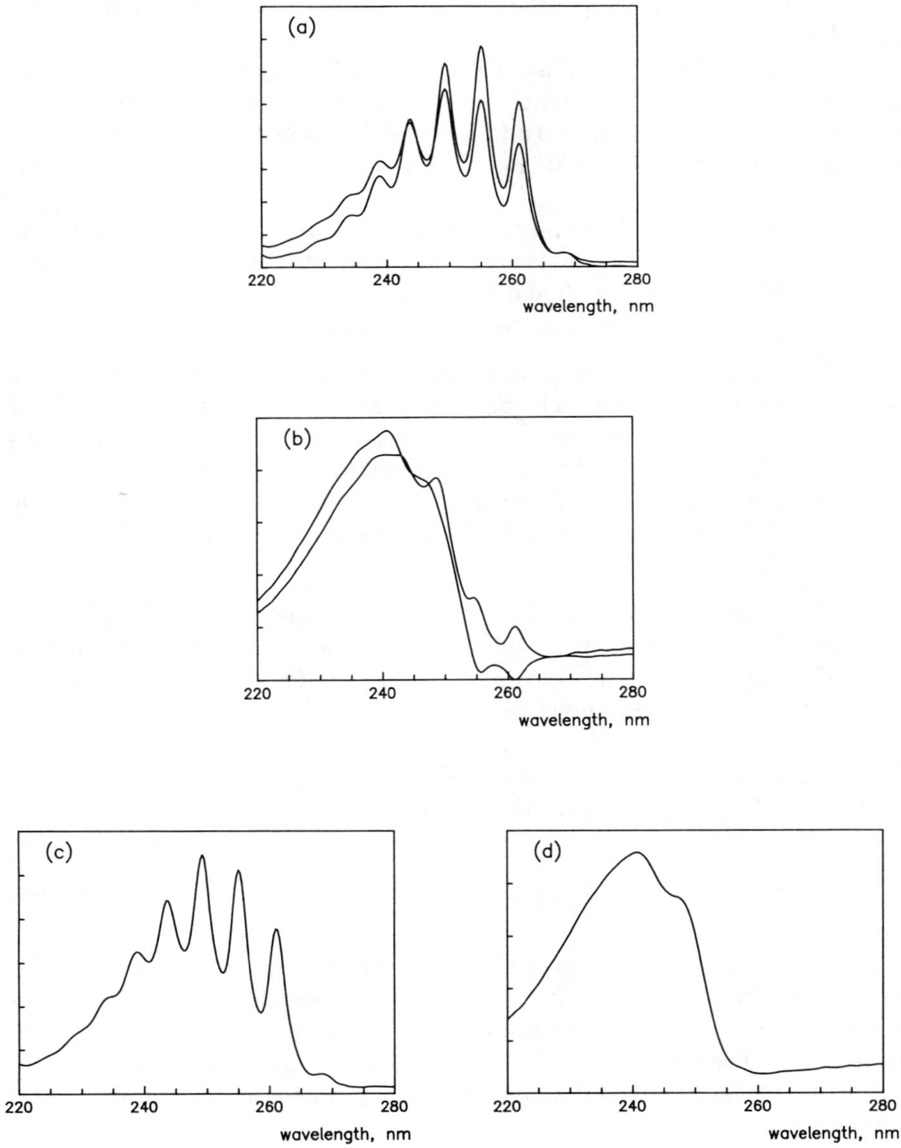

Figure 3.11 Spectral bands of the components of the benzene-benzaldehyde mixtures set of spectra **(a, b)**; estimation of the component spectra is based on the minimization of Equation 3.109 **(c, d)**.

point J_j (Figure 3.10). C_1 and C_2 are the intercepts of spectral vectors of components ($\vec{S}_1$ and $\vec{S}_2$) with the line (Equation 3.102). Then, the relative abundance of the component "*k*" in the mixture "*j*" is given by the quotient:

$$f_{kj} = \frac{d(C_k - J_j)}{d(C_1 - C_2)} \tag{3.105}$$

where $d(C_k - J_j)$ and $d(C_1 - C_2)$ are the Euclidean distances of the corresponding points.

The principal problem of using Equation 3.105 to quantitatively evaluate mixtures of two unknown components is the estimation of spectra of pure components $\vec{S}_1$ and $\vec{S}_2$. Sharaf and Kowalski[35] discussed three such general situations occurring during SMCR analysis of two-component mixture spectra.

1. The band width of the allowed regions is zero. This occurs when pure spectra of both components are in the in the set. The vectors ${}^1(\vec{S}_1)$, ${}^2(\vec{S}_1)$ or ${}^1(\vec{S}_2)$, ${}^2(\vec{S}_2)$ are confluent and identical with the spectral components $\vec{S}_1$ and $\vec{S}_2$; and they can be used for analysis.
2. The band width of the allowed region is nonzero, however, in spectra the pure signals for each component exist (a pure signal is a row of a matrix of spectral data $\bar{\bar{A}}$, in which the only component contributes by the spectral signal to all its elements). In this case the spectra of the components are identical with the external boundaries of the allowed regions (in Figure 3.10 these would be the vectors ${}^1[\vec{S}_1] \equiv \vec{S}_1$ and ${}^2[\vec{S}_2] \equiv \vec{S}_2$).
3. The band widths of the allowed regions are nonzero, and the presence of the pure component signals are not known. In this case the spectra of the components are estimated using certain assumptions, simplifications, and approximations.

Sharaf and Kowalski[34] recommended estimating component spectra as midpoints of spectral bands of the components:

$$\vec{S}_k = 0\ 5\ {}^1\left(\vec{S}_k\right) + 0\ 5\ {}^2\left(\vec{S}_k\right) \tag{3.106}$$

Sasaki et al.[36,46] proposed applying the criterion according to which real component spectra are the simplest in the set of possible candidates, i.e., in the set of spectra suitable for Equations 3.103 and 3.104. In this case, the simplicity is directly related to the smoothness of the corresponding spectral curve. To measure spectral smoothness, they proposed using the entropy function

$$H_k = -\sum_{i=1}^{N_W} p_{ik} \ln\left(p_{ik}\right) \tag{3.107}$$

where

$$p_{ik} = \frac{\left| s_{ik}^{(d)} \right|}{\sum_{i=1}^{N_W} \left| s_{ik}^{(d)} \right|} \qquad \textbf{(3.108)}$$

is the normalized first ($d = 1$)[36] or second ($d = 2$)[46] derivative of the spectrum of the component "k" in the absolute value ($s_{ik}^{(d)}$ are the corresponding dth derivatives of the point i). According to Equations 3.103 and 3.108 H_k is consequently a function of the parameter a_k:

$$H_k = f(a_k) \qquad \textbf{(3.109)}$$

Sasaki et al.[36,46] recognized that the function (Equation 3.107) is analogous to the entropy, and they compare its minimization to reducing the uncertainty of the stochastic process. According to this, the component spectra are those that correspond to the minimum of the function Equation 3.109.

The measure of the smoothness of the spectral line, and the measure of its simplicity, can also be taken as the sum of squares of the second derivatives:

$$Q_k = \sum_{i=1}^{N_W} \left[s_{ik}^{(2)} \right]^2 \to \min \qquad \textbf{(3.110)}$$

The component spectra illustrated in Figure 3.3 were estimated by minimizing the function of Equation 3.110.

Ohta[38] applied the SMCR method to analyze spectra of the mixtures of three dyes. He defined the region of the solution by mapping the three-dimensional factor space as constrained by the conditions of the restriction that neither $\bar{\bar{S}}$ nor $\bar{\bar{C}}$ can contain any negative elements. By Monte Carlo selection he partly reduced the number of candidates for the component spectra in the region of the solution.

Chen and Hwang[39] showed that by normalization it is possible to describe a spectrum of a three-component set in a two-dimensional space, using polar coordinates:

$$\vec{S}_k = \cos\theta\, \vec{Q}_1^{\ddagger} + \sin\theta \cos\phi\, \vec{Q}_2^{\ddagger} + \sin\theta \sin\phi\, \vec{Q}_3^{\ddagger} \qquad \textbf{(3.111)}$$

The satisfying of the first restriction is obtained by mapping the θ-ϕ space (θ-ϕ plot). In the p-q coordinates

$$p = \theta \cos\phi$$

$$q = \theta \sin\phi \qquad \textbf{(3.112)}$$

the allowed area is created by a polygon with several peaks (Figure 3.12). If pure signals are present in mixture spectra, the three sharpest vertices of this polygon (I, II, and III in Figure 3.12) represent spectra of the components.

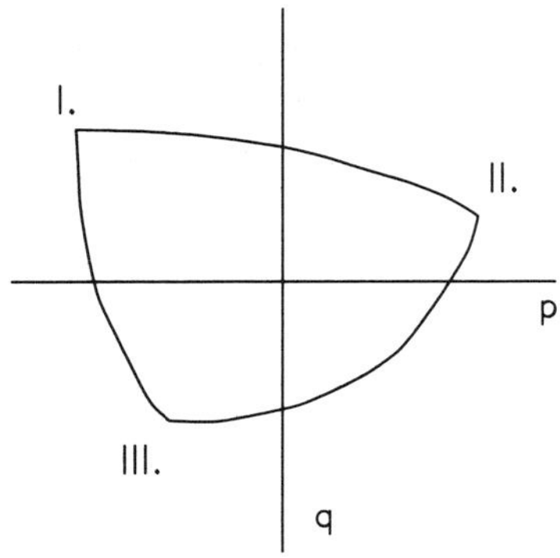

Figure 3.12 Θ-ϕ plot for the three-component set of mass spectra.

Vandeginste et al.[40] added the condition of maximal simplicity of the component spectra to the method of mapping the polar coordinates. As a measure of simplicity, they proposed using the ratio between the area of a spectrum and its Euclidean norm. Component spectra are obtained as minima in this ratio calculated for all spectra corresponding to the edge of the solution polygon. The method includes a correction for the second restriction as well.[40]

Meister[41] proposed seeking spectra of pure components by extracting the most widely different spectra satisfying the first restriction. This criterion is realized by constructing an N_C-dimensional simplex with vertices placed on the edge of the polygon bordering the area of validity of the first restriction and with the largest volume in the set of all these simplexes. Meister reported the algorithm for constructing an optimal simplex by the iterative method. The vertices of this optimal simplex correspond to the component spectra.

Mauro and Delaney[42,43] connected SMCR analysis with the information contained in libraries of spectra. In this method, named SMCR/LS (Library Search), N_C points whose spectra correspond to those of the N_C library spectra are found in the solution area.

Borgen and Kowalski[37] generalized the restrictions on the spectra of N_C-component mixtures and developed another method for three-component mixtures. Regions satisfying the restrictions are narrowed through a detailed analysis of their geometric proportions near the boundary conditions established in this algorithm.

Sharaf[44] developed an SMCR method for quantitative characterization of the spectra of three-component mixtures.

The universal SMCR method for spectral set analysis of multicomponent mixtures based on nonlinear constrained optimization was proposed by Sasaki et al.[45,46] They constructed an objective function as a sum of two terms:

$$U = H + P \qquad \textbf{(3.113)}$$

The penalty term P, which ensures that the first restriction (the non-negativity of elements of component spectral vectors) and second restriction (the non-negativity of component concentration in mixtures) are fulfilled, is given by the relation

$$P = \gamma \left[\sum_{i=1}^{N_W} \sum_{k=1}^{N_C} F(s_{ik})\, s_{ik}^2 + \sum_{k=1}^{N_C} \sum_{j=1}^{N_S} F(c_{kj})\, c_{kj}^2 \right] \quad \textbf{(3.114)}$$

where γ is a scale factor and

$$F(y) = \begin{Bmatrix} 0, & \text{if } y \geq 0 \\ 1, & \text{if } y < 0 \end{Bmatrix} \quad \textbf{(3.115)}$$

The term H ensures, in the objective function, that the choice of "simplest" spectra from the spectral set satisfies the two restrictions:

$$H = \sum_{k=1}^{N_C} H_k \quad \textbf{(3.116)}$$

where H_k, a measure of "simplicity" of the spectrum of component "k", is given by Equations 3.107 and 3.108.

The terms H and P are the functions of the elements of the matrix $\bar{\bar{S}}$ of the component spectra and of the concentration matrix $\bar{\bar{C}}$. According to Equations 3.85 and 3.88 the elements of the matrices $\bar{\bar{S}}$ and $\bar{\bar{C}}$ are functions of the transformation matrix $\bar{\bar{T}}$ for the given matrix of spectral data $\bar{\bar{A}}$.

If the normalization condition (Equation 3.101) is valid, it is possible to express one element as a linear combination of the rest of the row elements in each row of the matrix $\bar{\bar{T}}$:

$$t_{N_C,k} = \frac{1}{\sum_{i=1}^{N_W} q_{i,N_C}} \left(1 - \sum_{i=1}^{N_W} \sum_{l=1}^{N_C-1} q_{il}^{\ddagger}\, t_{lk} \right) \quad \textbf{(3.117)}$$

Then, H and P, as well as the objective function, are consequently functions of $N_C(N_C - 1)$ elements of the matrix $\bar{\bar{T}}$. By minimization of the objective function

$$U = f\left(t_{11}, t_{12}, \ldots t_{1,N_C-1}, t_{21}, t_{22}, \ldots t_{2,N_C-1}, \ldots \right.$$
$$\left. t_{N_C,1}, t_{N_C,2}, \ldots t_{N_C,N_C-1}\right) \quad \textbf{(3.118)}$$

the elements of the transformation matrix $\bar{\bar{T}}$ corresponding to spectra of components are obtained. Sasaki et al.[46] proposed optimizing the function

(Equation 3.118) by the simplex method (Appendix C), with a change of the scaling factor value in every optimization cycle.

Key Set Factor Analysis Method

The basic strategy of Key Set Factor Analysis (KSFA) is based on an idea by Knorr and Futrell.[47] They isolated mass spectra of components from mixture spectra by searching for pure masses, i.e., signals in the spectra of mixtures that are produced by one component exclusively.

The method proposed by Knorr and Futrell works reliably in two- and three-dimensional factor space, if pure masses (or other responses) can be found.

The general procedure for the KSFA was developed by Malinowski[48] and by Malinowski et al.[49] This method leads to identification of a key set of N_C typical rows of an eigenvector matrix $\bar{\bar{Q}}^{\ddagger}$. The key set of typical rows is that set of N_C rows of the matrix $\bar{\bar{Q}}^{\ddagger}$, which has the "most unique" properties and best expresses the properties of the components. The uniqueness of the properties of this set corresponds to the most orthogonal vector dimensions in its rows.

The KSFA procedure treats the normalized row vectors of the matrix $\bar{\bar{Q}}^{\ddagger}$. The elements of the normalized row vectors $\vec{q}_i^{\,*}$ are given by the relation

$$q_{ik}^{*} = \frac{q_{ik}}{\left(\sum_{l=1}^{N_C} q_{i,1} \right)^{\frac{1}{2}}} \tag{3.119}$$

The N_C points defined by the complete set of row vectors $\vec{q}_i^{\,*}$ $(i = 1 \rightarrow N_W)$ are placed, in this manner, in the factor space on the surface of an N_C-dimensional hypersphere. The normalization (Equation 3.119) enables comparison of the vectors $\vec{q}_i^{\,*}$ according to their directions.

The various sets of the rows $\vec{q}_i^{\,*}$ in the KSFA are arranged into square matrices and their determinants are evaluated. The absolute value of the determinant is the measure of the orthogonality of the vectors $\vec{q}_i^{\,*}$, and the maximal absolute value corresponds to the set that is the closest approximation to an orthogonal set. The determinant also expresses the N_C-dimensional volume. Because the vectors $\vec{q}_i^{\,*}$ of the key set subtend all other row vectors, the volume corresponding to the optimal set will be the largest among the volumes of all possible sets.

The first eigenvector $\vec{Q}_1$ describes the maximal part of the spectral data matrix, and in the direction of the first eigenvector $\vec{Q}_1$ the density of the points defined by the row vectors $\vec{q}_i^{\,*}$ is the largest. The first typical row $\vec{q}$ (key 1) will be that vector whose direction is the most different from the direction of the vector $\vec{Q}_1$, i.e., the vector $\vec{q}_1^{\,*}$ with the smallest element $|q_{i1}^{*}|$.

The second typical row is found in the two-dimensional factor space defined by the vectors $\vec{Q}_1^{\ddagger}$ and $\vec{Q}_2^{\ddagger}$. The first two elements of the row vectors $\vec{q}_i^{\,*}$ (q_{i1}^{*} and q_{i2}^{*}) are used during the identification of this row. If the first typical vector $\vec{q}$ (key 1) is the vector $\vec{q}_j^*$, then the second typical vector will be the one that is the most orthogonal to the vector $\vec{q}_j^*$ in the two-dimensional space ($\vec{Q}_1^{\ddagger}$, $\vec{Q}_2^{\ddagger}$). Mathematically, it means finding among the determinants D_i

$$D_i^{(2)} = \begin{vmatrix} q_{j1}^* & q_{j2}^* \\ q_{i1}^* & q_{i2}^* \end{vmatrix} \tag{3.120}$$

for $i = 1 \rightarrow N_W$ and $i \neq j$, that determinant $D_k^{(2)}$ with the highest absolute value. The corresponding vector $\vec{q}_k^*$ is consequently the second typical row $\vec{q}$ (key 2).

The first three components of the vectors $\vec{q}_i^*$ are considered during identification of the third typical row. The third typical row $\vec{q}$ (key 3) is the row in which the vector $\vec{q}_l^*$ is the most orthogonal to the plane defined by the vectors of the first two typical rows in three-dimensional factor space. The vector $\vec{q}_l^*$ is found to be that vector $\vec{q}_i^*$ $(i = 1 \rightarrow N_W, i \neq j, i \neq k)$ with the highest value of the determinant

$$D_i^{(3)} = \begin{vmatrix} q_{j1}^* & q_{j2}^* & q_{j3}^* \\ q_{k1}^* & q_{k2}^* & q_{k3}^* \\ q_{i1}^* & q_{i2}^* & q_{i3}^* \end{vmatrix} \tag{3.121}$$

Generally, the determinants with the dimension of $m \times m$ are examined by searching the mth typical row $\vec{q}$ (key m):

$$D_i^{(m)} = \begin{vmatrix} q_{j1}^*, & q_{j2}^*, & \cdots & q_{j,m}^* \\ q_{k1}^*, & q_{k2}^*, & \cdots & q_{k,m}^* \\ \cdot & \cdot & \cdots & \cdot \\ \cdot & \cdot & \cdots & \cdot \\ q_{i,1}^*, & q_{i,2}^*, & \cdots & q_{i,m}^* \end{vmatrix} \tag{3.122}$$

The elements of those vectors $\vec{q}$ (key l), $l = 1 \rightarrow m - 1$ already identified as typical rows, are placed in the first $m - 1$ rows of the determinants $D_i^{(m)}$. The elements of the rest of the vectors $\vec{q}_i^*$ are changed, one after the other, in the mth row. The mth typical row and its vector $\vec{q}_m^*$ correspond to the determinant with the highest absolute value $|D_m^{(m)}|$. The selection of the typical rows continues until the N_C-member key set is fully identified.

An analysis done with KSFA cannot differentiate the signals of pure noise in a base line from the real signals; instead, they are identified only as typical rows. To suppress this artifact, it is necessary to eliminate those vectors $\vec{q}_i^{\,*}$ from analysis whose corresponding rows in the matrix $\bar{\bar{A}}$ contain mostly data with weak intensities. Analyzing only those rows in which the average signal is at least five times higher than the experimental error is recommended.[49]

Because the selection of typical rows in KSFA is sequential, not all of the N_C-dimensional vectors $\vec{q}_i^{\,*}$ are used in individual steps during the calculation of determinants. The most orthogonal set of rows does not have to be the result of this process. This situation occurs especially with the KSFA spectra of mixtures, which are composed of a large number of components. The iterative method of working in total N_C-dimensional factor space was proposed to obtain the most orthogonal selection of the key set of typical rows (Iterative Key Set Factor Analysis, IKSFA).

The IKSFA procedure uses the results of KSFA as a first approximation of the real key set. All N_C-dimensional vectors $\vec{q}_i^{\,*}$ not originally placed in the key set are subsequently tested in place of every row $\vec{q}\,(\text{key } m)$, $m = 1 \rightarrow N_C$. The vector $\vec{q}_i^{\,*}$, which minimizes the corresponding determinant, is assigned to the key set as a typical row $\vec{q}\,(\text{key } m)$. This process is repeated through all typical rows until convergence is obtained.

It is possible to append the value of the spectral coordinate (wavenumber, wavelength, mass coordinate, and so on) to every row of the matrix $\bar{\bar{Q}}$; as a result, the typical rows of the key set correspond to those spectral coordinates containing pure signals in the spectral set of mixtures, i.e., signals that are the most typical for the given component in the given spectral set. Typical rows and their spectral coordinates can be used for various specific applications,[50] for example, for analytical multicomponent spectrophotometric calibrations (Chapter 4), for isolating component spectra and for modeling.

The KSFA method always identifies the most orthogonal set of rows $\vec{q}_i^{\,*}$, even when the pure signals of components do not exist in the spectral set. It is necessary to evaluate the results with a certain vigilance and interpret the key set of the typical rows in harmony with chemical experiences and intuition; for example, negative amplitudes in the component spectra or negative values of concentrations indicate a false solution.

If the typical rows of the key set correspond to spectral coordinates with pure signals of components, the spectral vectors of the components $\vec{S}_k$ will cross the points defined in the factor space by the vectors $\vec{q}\,(\text{key } k)$, $k = 1 \rightarrow N_C$. The matrix transforming the abstract solution into the real spectra of components (Equation 3.85) is subsequently obtained by replacing the row vectors $\vec{q}\,(\text{key } k)$ with their columns:

$$\bar{\bar{T}} = \left[\vec{q}^{\,T}(\text{key } 1), \vec{q}^{\,T}(\text{key } 2), \ldots \vec{q}^{\,T}(\text{key } N_C)\right] \qquad \textbf{(3.123)}$$

Target Factor Analysis

Target factor analysis (TFA)[3,51,52] is a procedure projecting a test vector to the factor space defined by a matrix of primary eigenvectors $\bar{\bar{Q}}^{\ddagger}$. It enables one to test individually the presence of a component with a defined spectrum in spectral mixtures without any knowledge of the spectra of other components.

A transformation of the abstract solution (Equations 3.85 or 3.89) to the general spectral vector $\vec{P}_k$ can be obtained by

$$\vec{P}_k = \bar{\bar{Q}}^{\ddagger} \vec{T}_k \qquad \textbf{(3.124)}$$

where $\vec{P}_k$, called the predicted vector, lies in a factor space. With the correct choice of the transformation vector $\vec{T}_k$ (the kth column of the transformation matrix $\bar{\bar{T}}$), $\vec{P}_k$ approximates the real factor describing the component spectrum $\vec{S}_k$. The construction of the transformation vector $\vec{T}_k$ in target factor analysis is based on the test vector $\vec{R}_k$. The test vector is regarded as certain, often an incomplete hypothesis about the real spectrum of the component $\vec{S}_k$. It is a "target" toward which the abstract solution is transformed.

From the assumption that the spectral vector $\vec{P}_k$ lies in the factor space as close as possible to the target vector $\vec{R}_k$ and based on the method of least squares, Malinowski derived[3,51,53] the relation for the vector $\vec{T}_k$:

$$\vec{T}_k = \bar{\bar{Q}}^{\ddagger T} \vec{R}_k \qquad \textbf{(3.125)}$$

Substituting Equation 3.125 into Equation 3.124, the transformation equation of target factor analysis is obtained:

$$\vec{P}_k = \bar{\bar{Q}}^{\ddagger} \bar{\bar{Q}}^{\ddagger T} \vec{R}_k \qquad \textbf{(3.126)}$$

For practical applications of target factor analysis it may not always be possible to define all of the elements of the target vector $\vec{R}_k$. For these cases, Malinowski[3] modified Equations 3.125 and 3.126 by the Free Floating procedure:

$$\vec{T}_k = Q^{*T} \vec{R}_k^{*} \qquad \textbf{(3.127)}$$

$$\vec{P}_k = \bar{\bar{Q}}^{\ddagger} Q^{*T} R_k^{*} \qquad \textbf{(3.128)}$$

where $\vec{R}_k^{*}$ is an incomplete target vector with N_W^{*} elements ($N_W^{*} \leq N_W$) corresponding to the spectrum in which the intensities for N_W^{*} spectral coordinates only are known and $\bar{\bar{Q}}^{*}$ contains only those rows of the matrix $\bar{\bar{Q}}^{\ddagger}$ that correspond to the elements of the vector $\vec{R}_k^{*}$. Equation 3.128 produces the total spectral vector $\vec{P}_k$ with N_W elements from the incomplete target vector $\vec{R}_k$.

Experimental errors in target spectral vectors $\vec{R}_k$ and a matrix of noisy spectral data $\bar{\bar{A}}$ can cause, by mutual effects, difficulties in evaluating the success of the projection of the target vector $\vec{R}_k$ to the factor space, and the decision regarding whether $\vec{R}_k$ is or is not a real factor can be problematic.

Malinowski[53] investigated the influence of experimental errors on TFA transformation and defined three types of errors connected to each other by the relation

$$(\mathrm{AET}_k)^2 = (\mathrm{RET}_k)^2 + (\mathrm{REP}_k)^2 \tag{3.129}$$

The apparent error in the target vector (AET_k) is a measure of the difference between the raw target vector $\vec{R}_k$ and its projection $\vec{P}_k$:

$$\mathrm{AET}_k = \left[\frac{N_W - N_C}{N_W^* - N_C} \frac{\sum_{i=1}^{N_W^*} (r_{ik} - p_{ik})^2}{N_W} \right]^{\frac{1}{2}} \tag{3.130}$$

The real error of the predicted vector (REP_k) is a measure of the difference between the projection $\vec{P}_k$ and the pure hypothetical target vector $\vec{R}_k^x$, which does not contain any experimental error:

$$\mathrm{REP}_k = (\mathrm{RE}) \left(\frac{\vec{T}_k^T \vec{T}_k}{\lambda_k} \right)^{\frac{1}{2}} \tag{3.131}$$

REP has its origin mainly due to errors in the matrix of spectral data $\overline{\overline{A}}$.

The real error in the target vector (RET_k) is a measure of the difference of the real raw target vector $\vec{R}_k$, and the pure target vector $\vec{R}_k^x$.

To measure the quality of the projection of the target vector $\vec{R}_k$ into the factor space of spectral data, Malinowski introduced the SPOIL_k function:[53]

$$\mathrm{SPOIL}_k = \frac{\mathrm{RET}_k}{\mathrm{REP}_k} \tag{3.132}$$

Based on a study of model sets of data, Malinowski[53] classified target vectors with the SPOIL value in the interval (0, 3) as acceptable. Values of SPOIL > 6 indicate unacceptable factors. The target vectors with a SPOIL value in the interval (3, 6) are classified as marginally acceptable. Such target vectors are probably real factors, but their RETs are relatively high (the target vector is influenced by a significant error).

Malinowski also developed another method, which evaluates the acceptability of the target vector $\vec{R}_k$ as the real factor by the statistical F test.[26] The difference between the expected value of the vector $\vec{P}_k$ and the expected value of the vector $\vec{R}_k$ represents the variance $V(T_k)$ with ($N_W^* - N_C$) degrees of freedom:

$$V(T_k) = \frac{N_w \left| \vec{R}_k - \vec{P}_k \right|^2}{\left(N_W^* - N_C \right) \sum_{j=1}^{N_C} \frac{t_{jk}^2}{\lambda_j}} \tag{3.133}$$

The value of the testing criterion F_k is obtained as a ratio of the reduced variance $V(T)$, which has a statistical weight of $[(N_W - N_C + 1) \cdot (N_S - N_C + 1)]$ and weighted average of the secondary reduced eigenvalues REV_{NC}^0 (Equation 3.68):

$$F_k\left(N_W^* - N_C, N_S - N_C\right) = \frac{\sum_{l=N_C+1}^{N_S} \left(N_W - l + 1\right)\left(N_S - l + 1\right)}{\left(N_W - N_C + 1\right)\left(N_S - N_C + 1\right)} \cdot \frac{N_w \left|\vec{R}_k - \vec{P}_k\right|^2}{\left(N_W^* - N_C\right) \sum_{l=N_C+1}^{N_S} \lambda_l^0 \sum_{j=1}^{N_S} \frac{t_{jk}^2}{\lambda_j}} \tag{3.134}$$

If the significance level $\alpha(F_k)$ for the obtained value F_k is higher than the chosen reference value, it is not possible to reject the null hypothesis

$$H_0 : \vec{R}_k \equiv \vec{P}_k \tag{3.135}$$

and the target vector $\vec{R}_k$ is accepted as a real factor at significance level $\alpha(F_k)$. If the value $\alpha(F_k)$ is lower than the reference value, the null hypothesis will be rejected in favor of the alternative hypothesis:

$$H_a : \vec{R}_k \not\equiv \vec{P}_k \tag{3.136}$$

Acceptance of the null hypothesis (Equation 3.135) means that the target spectrum $\vec{R}_k$ lies inside the factor space of a spectral data matrix $\bar{\bar{A}}$ and so component "k" is an active component of this spectral set.

Tests for the presence of components, candidates in spectra of chemical mixtures, are carried out so that the spectra of these components are submitted as the TFA targets.[26,54-56] Analysis of the infrared spectra of 11 mixtures of o-xylene, m-xylene, p-xylene, ethylbenzene, and chloroform is used as an example (Table 3.6).[26,55] Abstract factor analysis confirmed the presence of five components in this spectral set. From the seven target spectra tested, the SPOIL function and the F-test reliably eliminated the presence of toluene and 1,2,4-trimethylbenzene in the mixture spectra. On the other hand, both tests indicate the presence of four components (m-xylene, p-xylene, ethylbenzene, and chloroform) without any doubts. The value of the SPOIL function for o-xylene (6.71) just exceeds the threshold of acceptability. The F-test allows rejection of the hypothesis concerning the presence of o-xylene in the mixtures because the significance level is greater that 1%. The presence of o-xylene is not confirmed by this result, but neither is it reliably excluded. It is estimated that the nonuniqueness of the tests is caused by impurities in the sample of o-xylene, which give rise to inaccuracies in its spectrum.

Table 3.6 TFA Analysis of the Infrared Spectra of 11 Mixtures of o-Xylene, m-Xylene, p-Xylene, Ethylbenzene, and Chloroform

Target	% α[a]	SPOIL
o-Xylene	1.2	6.71
m-Xylene	65.2	0.94
p-Xylene	45.4	1.26
Ethylbenzene	10.9	2.21
Chloroform	96.3	0.0
Toluene	0.0	104.96
1,2,4-Trimethylbenzene	0.0	318.18

[a] α is the significance level of the F-test.
From Malinowski, E. R., *J. Chemometrics,* 3, 49, 1989. With permission.

After the identification of all of the components of the mixtures, the transformation vectors $\vec{T}_k$ (Equations 3.125 and 3.127) can be arranged columnwise into the transformation matrix $\bar{\bar{T}}$ and used to estimate the composition of individual mixtures by transforming the concentration matrix (Equation 3.88). The average error transformed into the concentration of the component "k" is calculated from the formula[57]

$$\left(E_C\right)_k = \mathrm{RE}\left(\sum_{l=1}^{N_C} \tilde{t}_{jk}\right)^{\frac{1}{2}} \tag{3.137}$$

where $\tilde{t}_{jk}$ are elements of the matrix $\bar{\bar{T}}^{-1}$.

Not only can target spectral vectors be constructed directly from the experimental data as in the previous example, but they can be based on certain assumptions and hypotheses as well. In the automated spectral isolation method (ASI),[58,59] target vectors (called the vectors of prototype spectra in this method) are defined as certain mathematical functions, the form of which is specific for the given type of spectra. For ultraviolet and visible spectra the Gaussian function was used (see Chapter 2, Types of Spectral Bands), and for infrared spectra the single needle function (a simple peak with a height of 1.0 at the baseline 0.08 in the absorbance scale) was employed. The target spectrum of every component $\vec{R}_k$, characterized by the set of parameters defining the given function, can be projected by the TFA method to the factor space of the mixture spectra. Based on the results of such projections, the parameter set is optimized so that the relation

$$\vec{R}_k \leftrightarrow \vec{P}_k \tag{3.138}$$

is the closest. In the ASI method it is assumed that the best prediction of the component spectrum $\vec{P}_k$ (opt) is given by the projection of the optimal target vector $\vec{R}_k$ (opt). In the ASI method the RISK function is used as the objective function for the optimization:

$$\mathrm{RISK}_k = \frac{1}{N_W} \sum_{i=1}^{N_W} \left(p_{ik} - r_{ik}\right)^2 \sum_{i=1}^{N_W} \left(p_{ik}\right)^2 \quad \mathbf{(3.139)}$$

The results of ASI analysis of the nine infrared spectral sets measured on three-component mixtures are shown in Figure 3.13.

The TFA projection is also used to determine the number of active components in the spectra by the cross validation method (CV).[60] The matrix of spectral data is divided into a submatrix $\bar{\bar{A}}_a$ containing l columns of the original matrix $\bar{\bar{A}}$ and a submatrix $\bar{\bar{A}}_b$ with NS $- l$ remaining columns of $\bar{\bar{A}}$. As a rule

$$1 \le l \le \mathrm{NS} - l \quad \mathbf{(3.140)}$$

The submatrix $\bar{\bar{A}}_b$ is submitted for abstract factor analysis, and the eigenvector set $(\vec{Q}_m)_b$, $m = 1 \to \mathrm{NS} - l$ is obtained. For each spectral vector of a mixture, a column of submatrix $\bar{\bar{A}}_a$, a target, is subjected to TFA projections in the factor subspaces defined successively by m vectors $(\vec{Q}_m)_b$, $m = 1 \to \mathrm{NS} - l$.

The above procedure is repeated until all column vectors of the $\bar{\bar{A}}$ matrix have been used as targets in l member groups of the submatrix $\bar{\bar{A}}_a$. The standard deviation of target spectra from their TFA projections is evaluated for every dimension of the factor subspaces m. The standard deviation decreases with the increasing "m" value, until the primary eigenvectors are added to the $(\vec{Q}_m)_b$ subspaces. The increase in the standard deviations for $m = N_C$ and $m = N_C + 1 \to \mathrm{NS} - l$ will be insignificant.

Table 3.7 contains results of CV analysis of the spectral set in Figure 3.2. All mixture spectra were placed into submatrix $\bar{\bar{A}}_a$ in randomly selected couples. In Table 3.7 the standard deviations of TFA projections decrease from factors 1 to 4 with increasing factor subspace dimension. The difference in values of standard deviation for factor subspace dimensions four through eight is insignificant. This is confirmed by high levels of significance for the rejection of the null hypothesis in the F-test of the standard deviations (40 to 50%).

Evolving Factor Analysis

The concept of the evolving factor analysis (EFA) method (see Evolving Factor Analysis and Submatrix Analysis, above) is also used to determine the mixture composition and to isolate the component spectra from the set of mixture spectra.[28,61-65] N_S mixture spectra are arranged according to an evolutionary variable. Abstract factor analysis is successively performed with subsets of the first $N_C, N_C + 1, \ldots, N_S - N_C + 1$ spectra, and corresponding sets of eigenvalues $(\lambda_k)_F^j$, $k = 1 \to N_C$ are obtained for every such spectral subset $j = N_C \to N_S - N_C + 1$. The primary eigenvalues of this "forward" EFA increase over the evolutionary variable, unlike the secondary eigenvalues,

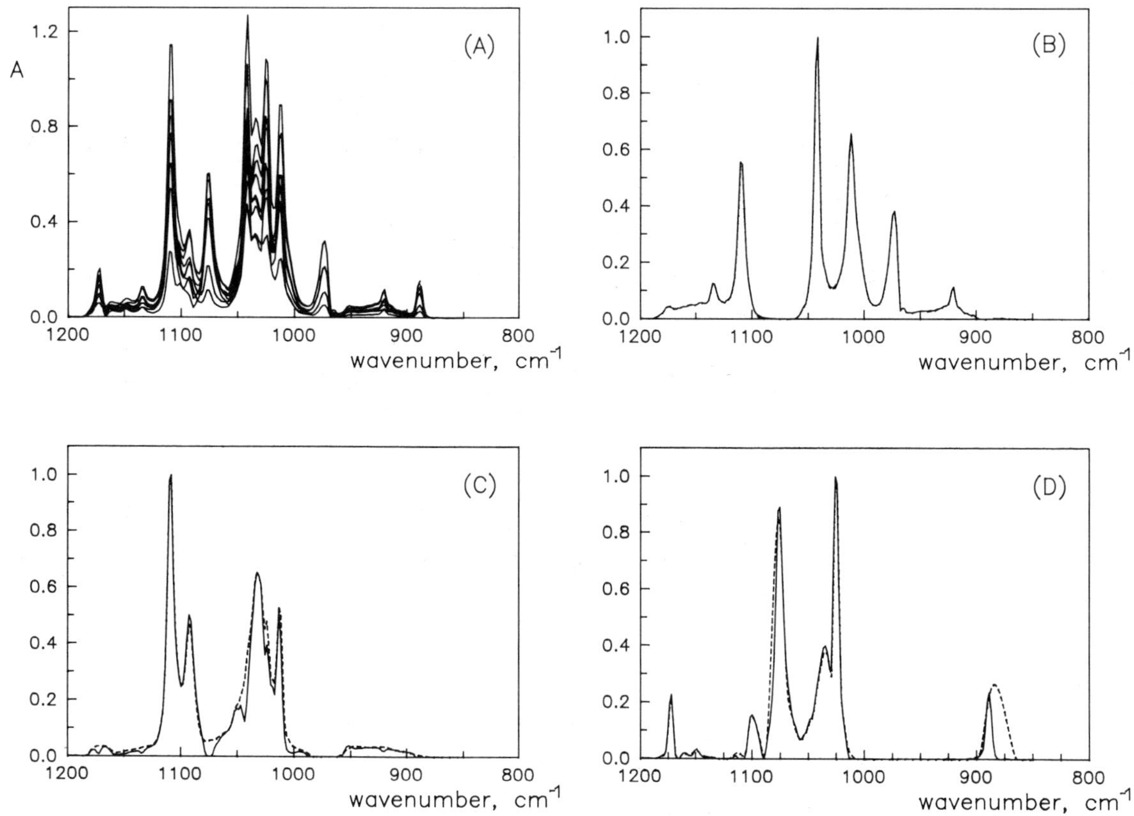

Figure 3.13 ASI analysis of the IR spectra of the nine mixtures of o-xylene, p-xylene, and toluene: the mixture spectra **(A)**, the component spectra **(B–D)**, the true component spectra (– – –), the ASI results (——).

Table 3.7 Cross Validation Analysis of the Spectral Set of Figure 3.2

Factor subspace dimension	Standard deviation of TFA projections	α[a] (%)
1	0.0191	0.01
2	0.0133	0.04
3	0.0035	0.06
4	0.0013	46.46
5	0.0013	47.19
6	0.0013	48.53
7	0.0012	49.09
8	0.0012	

[a] α is the level of significance for rejecting the hypothesis: H_0:(standard deviation)$_k$ = (standard deviation)$_{k+1}$ and H_a:(standard deviation)$_k$ > (standard deviation)$_{k+1}$.

as new active components in the spectra appear during the evolution of the set of spectra. The backward EFA uses an analogous procedure starting with the last N_C spectra (i.e., spectra N_S, $N_S - 1,\ldots N_S - N_C + 1$) and successively adds the spectra $N_S - N_C$, $N_S - N_C - 1$ up to the spectrum N_C into the AFA analyzed subsets. The primary eigenvalues of the backward EFA $(\lambda_k)_B^j$ $k = 1 \rightarrow N_C, j = N_C \rightarrow N_S - N_C + 1$ correspond to the disappearance of the components. The eigenvalue sets for both the forward $(\lambda_k)_F^j$ and backward $(\lambda_k)_B^j$ EFA are plotted together as a function of the evolutionary variable. The concentration profile of the kth component is found as the boundaries of the region under the curves of the ith forward and the $(N_C + 1 - k)$th backward eigenvalues:

$$c_{kj} = \min\Big((\lambda_k)_F^j,\ (\lambda_k)_B^j \Big) \tag{3.141}$$

Primary concentration profiles of EFA analysis of model spectral data for the consecutive reaction (Figure 3.14A)

$$B \rightarrow C \rightarrow D \tag{3.142}$$

are shown in Figure 3.15A.

The concentration profiles obtained by this method are usually relatively rough. Gampp et al.[61,63] proposed the following iterative procedure for their improvement.

The concentration profiles of the individual components are normalized and arranged into rows of the matrix $\overline{\overline{C}}^*$, and the spectra of the components are obtained by solving Equation 3.2 for matrix $\overline{\overline{S}}$:

$$\overline{\overline{S}} = \overline{\overline{A}}\,\overline{\overline{C}}^{*T}\left(\overline{\overline{C}}^*\,\overline{\overline{C}}^{*T}\right)^{-1} \tag{3.143}$$

Physically incorrect negative values in the matrix $\overline{\overline{S}}$ are set equal to zero, and matrix $\overline{\overline{S}}$ thus corrected is used to calculate the new concentration matrix:

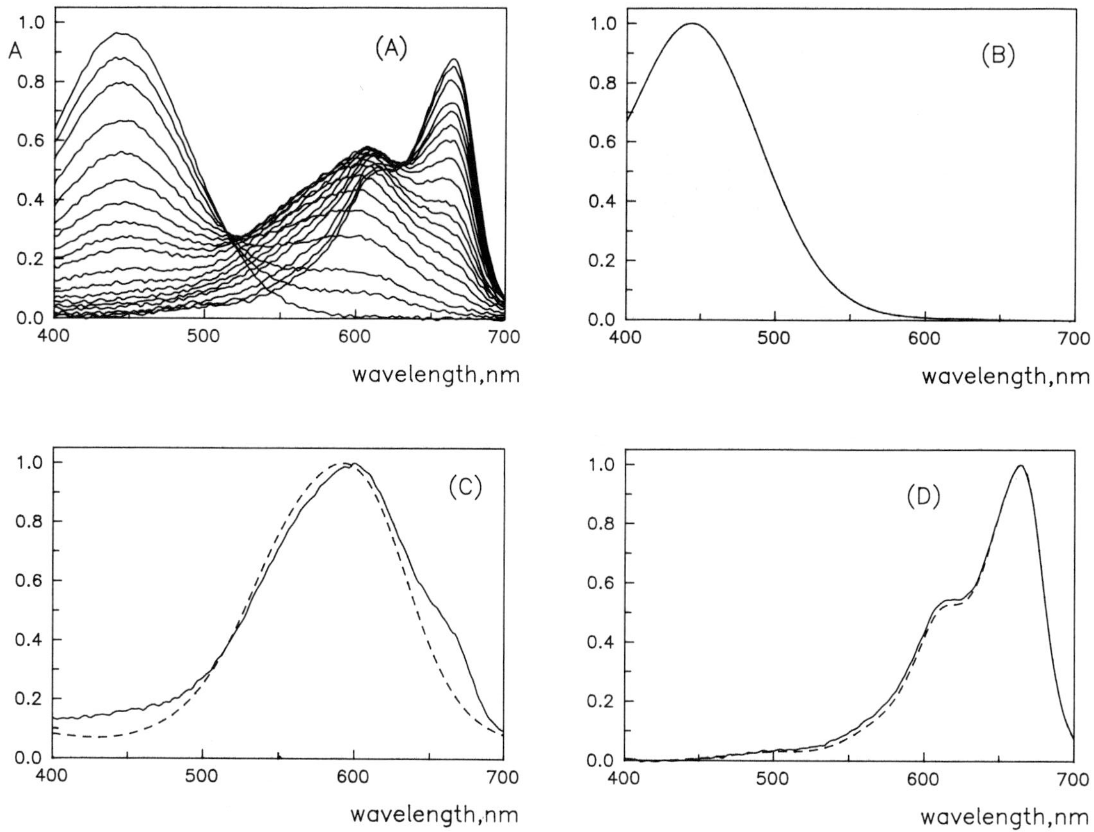

Figure 3.14 Spectra of EFA analysis of the model spectral data of consecutive reactions B → C → D: the spectra of 19 reaction mixtures **(A)**, the spectra of the reaction components **(B–D)**, the true component spectra (– – –), the EFA results (——) — spectrum B was fixed as a "known" spectrum.

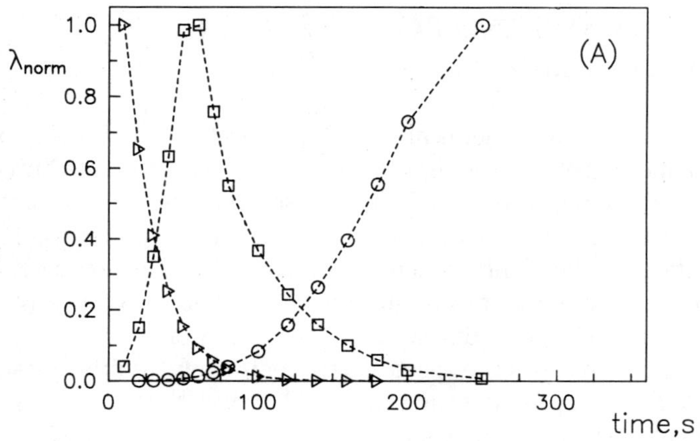

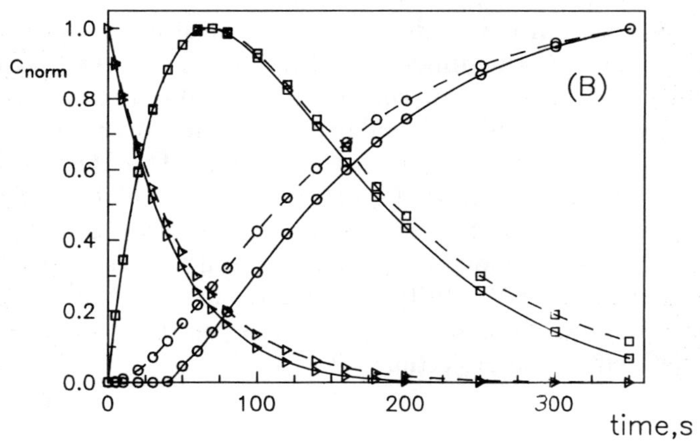

Figure 3.15 Normalized concentration curves of EFA analysis of the model spectral data of consecutive reactions B → C → D: primary concentration profiles **(A)**, the EFA results (——), and the true component concentrations (– – –) **(B)**.

$$\bar{\bar{C}} = \left(\bar{\bar{S}}^T \bar{\bar{S}}\right)^{-1} \bar{\bar{S}}^T \bar{\bar{A}} \qquad \textbf{(3.144)}$$

Any negative values in the elements of matrix $\bar{\bar{C}}$ are set equal to zero, the rows of matrix $\bar{\bar{C}}$ are normalized to the matrix $\bar{\bar{C}}^*$ and iteration is repeated from Equation 3.143 until convergence is obtained.

The construction of concentration profiles and the isolation of component spectra by evolving factor analysis can be problematic if the curves describing the real component concentrations are complex (e.g., two maxima and so on).

FACTOR ANALYSIS AND COMPRESSION OF SPECTRAL LIBRARIES

Comparing an unknown chemical compound spectrum with standard spectra is the basic method of spectroscopic identification. An automated procedure for such analysis assumes the availability of a standard set of spectra in libraries saved on certain storage media. The more items placed in the spectral library, the more effective the identification of unknown chemical compounds. On the other hand, as the library's contents increase, often dramatically, so do the requirements for storage media capacity. The ability of abstract factor analysis to reduce the dimension of the spectral data set is useful in filtering superfluous information and compressing the spectral library.[31,66,67]

The first step in compressing the spectral library is abstract factor analysis (see Abstract Factor Analysis and Theory of Errors in Factor Analysis, above) arranging the library spectra into columns of a matrix $\bar{\bar{A}}$ (Equation 3.2). The result of this analysis is the set of eigenvectors $\vec{Q}_j, j = 1 \rightarrow N_S$ describing the total library space of the spectral data. It is necessary to define in this set the subset of primary eigenvectors $\vec{Q}_k^{\ddagger}, k = 1 \rightarrow N_F$ (N_F is the dimension of the primary library space of the spectral data). Factors, in this sense, correspond to the typical spectral characteristics of various basic subgroups of the library items. Correct determination of the primary factor space (see The Determination of the Number of Active Components in the Spectral Set, above) is a critical moment in the compression of spectral libraries. Underestimating N_F leads to the loss of relevant information and can cause false identification. Overestimating N_F limits the effective extent of the compression.

The compression of the library spectrum $\vec{A}_j$ to a basis of primary eigenvectors is defined by the projection[31,66]

$$\vec{W}_j = \bar{\bar{Q}}^{\ddagger T} \vec{A}_j \tag{3.145}$$

Archiving the original spectra library, which contains N_S spectra, each of N_W points, requires saving $N_W \times N_S$ data. To archive the compressed library, it is necessary to save the matrix $\bar{\bar{Q}}^{\ddagger}$ of $N_W \times N_F$ dimension and N_S vectors containing the projections $\vec{W}_j$ of dimension N_F. The degree of compression consequently gives the ratio

$$\frac{N_F(N_W + N_S)}{N_W\, N_S} \tag{3.146}$$

For a library of 2300 spectra, where each has 185 points, and $N_F = 37$, the memory capacity required is about 21% of that required for the original data set.[66]

REFERENCES

1. Rummel, R. J., *Applied Factor Analysis,* Northwestern University Press, Evanston, IL, 1970.
2. Harman, H. H., *Modern Factor Analysis,* The University of Chicago Press, Chicago, 1976.
3. Malinowski, E. R., *Factor Analysis in Chemistry,* Wiley-Interscience, New York, 1991.
4. Malinowski, E. R., *Anal. Chem.,* 49, 606, 1977.
5. Wallace, R. M., *J. Phys. Chem.,* 64, 899, 1960.
6. Wallace, R. M. and Katz, S. M., *J. Phys. Chem.,* 68, 3890, 1964.
7. Ainsworth, S., *J. Phys. Chem.,* 65, 1968, 1961.
8. Katakis, D., *Anal. Chem.,* 37, 876, 1965.
9. Coleman, J. S., Varga, L. P., and Mastin, S. H., *Inorg. Chem.,* 9, 1015, 1970.
10. Duewer, D. L., Kowalski, B. R., and Fasching, L. J., *Anal. Chem.,* 48, 2002, 1976.
11. Gillette, P. C., Lando, J. B., and Koenig, J. L., *Fourier Transform Infrared Spectrosc.,* 4, 1, 1985.
12. Koenig, J. L., *Adv. Polym. Sci.,* 54, 87, 1984.
13. Kankare, J. J., *Anal. Chem.,* 42, 1322, 1970.
14. Hugus, Z. Z., Jr. and El-Awady, A. A., *J. Phys. Chem.,* 75, 2954, 1973.
15. Bulmer, J. T. and Shurvell, H. F., *J. Phys. Chem.,* 77, 257, 1973.
16. Shurvell, H. F. and Bulmer, J. T., *Vibrational Spectra and Structure,* Vol. 6, During, J. R., Ed., Elsevier, New York, 1977.
17. Shurvell, H. F., Chitumbo, K., Dunham, A., and Chappel, L., *Can. J. Spectrosc.,* 21, 53, 1976.
18. Ritter, G. L., Lowry, S. R., Isenhour, T. L., and Wilkins, C. L., *Anal. Chem.,* 36, 40, 1982.
19. Cattell, R. B., *Multivariate Behav. Res.,* 1, 245, 1966.
20. Rozet, R. W. and Petersen, E. M., *Anal. Chem.,* 47, 1301, 1975.
21. Malinowski, E. R., *Anal. Chem.,* 49, 612, 1977.
22. Kaiser, H. F., *Educ. Psych. Meas.,* 20, 141, 1960.
23. Yamaoka, K. and Takatsuki, M., *Bull. Chem. Soc. Japan.,* 51, 3182, 1978.
24. Antoon, M. K., D'Esposito, L., and Koenig, J. K., *Appl. Spectrosc.,* 33, 351, 1979.
25. Malinowski, E. R., *J. Chemometrics,* 1, 33 1987.
26. Malinowski, E. R., *J. Chemometrics,* 3, 49, 1989.
27. Rossi, T. M. and Warner, I. M., Anal. Chem., 58, 810, 1986.
28. Gamp, H., Maeder, M., Meyer, C. J., and Zuberbühler, A. D., *Talanta,* 32, 810, 1986.
29. Cartwright, H., *J. Chemometrics,* 1, 111, 1987.

30. Cartwright, H. and Farley, H. A., *J. Chemometrics,* 2, 111, 1988.

31. Harrington, P. B. and Isenhour, T. L., *Appl. Spectrosc.,* 41, 449, 1987.

32. Edward, J. T. and Wong, S. C., *J. Am. Chem. Soc.,* 99, 4229, 1977.

33. Lawton, W. H. and Sylvestre, E. A., *Technometrics,* 13, 617, 1971.

34. Sharaf, M. A. and Kowalski, B. R., *Anal. Chem.,* 53, 518, 1981.

35. Sharaf, M. A. and Kowalski, B. R., *Anal. Chem.,* 54, 1291, 1981.

36. Kawata, S., Komeda, H., and Minami, S., *Appl. Spectrosc.,* 39, 610, 1985.

37. Borgen, O. S. and Kowalski R., *Anal. Chim. Acta,* 174, 1, 1985.

38. Ohta, N., *Anal. Chem.,* 45, 553, 1973.

39. Chen, J.-H. and Hwang, L.-P., *Anal. Chim. Acta,* 133, 271, 1981.

40. Vandeginste, B., Essers, R., Bosman, T., Reijnen, J., and Kateman, G., *Anal. Chem.,* 57, 971, 1985.

41. Meister, A., *Anal. Chim. Acta,* 161, 149, 1984.

42. Mauro, D. M. and Delaney, M. F., *Anal. Chem.,* 58, 2622, 1986.

43. Delaney, M. F. and Mauro, D. M., *Anal. Chim. Acta,* 172, 193, 1985.

44. Sharaf, M. A., *Anal. Chem.,* 58, 3084, 1986.

45. Sasaki, K., Kawata, S., and Minami, S., *Appl. Optics,* 22, 3599, 1983.

46. Sasaki, K., Kawata, S., and Minami, S., *Appl. Optics,* 23, 1955, 1984.

47. Knorr, J. F. and Futrell, J. H., *Anal. Chem.,* 51, 1236, 1979.

48. Malinowski, E. R., *Anal. Chim. Acta,* 134, 129, 1982.

49. Malinowski, E. R., Cox, R. A., and Haldna, U. L., *Anal. Chem.,* 56, 778, 1984.

50. Schostack, K. J. and Malinowski, E. R., *Chemometrics Intell. Lab. Systems,* 6, 21, 1989.

51. Weiner, P. H., Malinowski, E. R., and Levinstone, A. R., *J. Phys. Chem.,* 74, 4537, 1970.

52. Hopke, P. K., *Chemometrics Intell. Lab. Systems,* 6, 7, 1989.

53. Malinowski, E. R., *Anal. Chim. Acta,* 103, 339, 1978.

54. Malinowski, E. R. and McCue, M., *Anal. Chem.,* 49, 284, 1977.

55. McCue, M. and Malinowski, E. R., *Anal. Chim. Acta,* 133, 125, 1981.

56. McCue, M. and Malinowski, E. R., *Appl. Spectrosc.,* 37, 463, 1983.

57. Malinowski, E. R., *Anal. Chim. Acta,* 122, 327, 1980.

58. Lin, C.-H. and Liu, S.-C., *J. Chinese Chem. Soc.,* 25, 167, 1978.

59. Lin, C.-H. and Lin, L.-C., *Proc. Natl. Sci. Counc. ROC,* 3, 1, 1979.

60. Malinowski, E. R., Application of target factor analysis to quantitative absorption spectroscopy, in *Computerized Quantitative Infrared Analysis,* McClure, G. L., Ed., ASTM Special Technical Publ., Philadelphia, 1984.

61. Gamp, H., Maeder, M., Meyer, C. J., and Zuberbühler, A. D., *Chimia,* 39, 315, 1985.

62. Maeder, M. and Zuberbühler, A. D., *Anal. Chim. Acta,* 181, 287, 1986.

63. Gamp, H., Maeder, M., Meyer, C. Y., and Zuberbühler, A. D., *Talanta,* 33, 943, 1986.

64. Gamp, H., Maeder, M., Meyer, C. Y., and Zuberbühler, A. D., *Comments Inorg. Chem.,* 6, 41, 1987.

65. Malinowski, E. R., Evolutionary factor analysis in analytical spectroscopy, in *Computer-Enhanced Analytical Spectroscopy,* Meuzelaar, H. L. C. and Isenhour, T. L., Ed., Plenum Press, New York, 1987.

66. Hagnac, C. P., Wiebolt, R. C., Lam, R. M., Isenhour, T. L., *Appl. Spectrosc.,* 36, 40, 1982.

67. Wang, C. P., Isenhour, T. L., *Appl. Spectrosc.,* 41, 185, 1987.

Chapter 4

Quantitative Analysis in Molecular Spectroscopy

Spectroscopic prediction of quantitative mixture composition is based on comparing a mixture spectrum with a set of standard spectra. Standard calibration spectra are measured on quantitatively known mixtures of compounds. Each predicted component must be active in the given spectral region.

SIMPLE LINEAR MODEL

The simplest situation for quantitative spectroscopic analysis is when a studied system contains only one spectroscopically active component. The simple linear model can also be used to analyze multicomponent mixtures if spectral areas of individual components are sufficiently distinct and if spectroscopic signals of components do not overlap.

The single-component linear model assumes that the spectroscopic signal intensity of a studied system is a linear function of component concentration:

$$a = k \cdot c \tag{4.1}$$

In the first step of the analysis the value of the sensitivity coefficient k is determined from spectra of various samples (standards) with known component concentrations. This step is known as the calibration step. In prediction, the spectroscopic response of an unknown amount of the component is measured and the component's concentration is calculated from the model. Although the theoretical validity of the model (Equation 4.1) is assumed, in practice it is usually replaced by a more general model:

$$a = k_o + k \cdot c \tag{4.2}$$

where an absolute term k_o corresponds to a background.

The spectral response of M standard samples is measured in the calibration step, and M pairs of (a_i^c, c_i^c) are obtained, where a_i^c is the spectroscopic signal of a standard i with a concentration of c_i^c. The best estimation of the calibration function (Equation 4.2) parameters k_o and k, and their standard deviations s_{k_o} and s_k, are yielded by the least squares method[1] (see Appendix B). To apply the least squares method and estimate the calibration function coefficients, three assumptions must be satisfied:[2]

1. All experimental error is associated with the dependent variable, i.e., with the measured spectroscopic signal. Practically, it means that concentrations of standards must be defined as precisely as possible.
2. The repeated measurements of the spectroscopic response of each sample must be normally distributed.
3. The variance of the spectroscopic measurement must be independent of the sample concentration.

In analyzing an unknown sample, the average content of a compound $\bar{c}$ is obtained from N repeated spectroscopic measurements, using the analytical function

$$\bar{c} = \frac{\bar{a} - k_o}{k} \tag{4.3}$$

where $\bar{a}$ is the average value of the N spectroscopic measurements of the given sample. The standard deviation of a predicted concentration $\bar{c}$ with $M - 2$ degrees of freedom is[3]

$$s_{\bar{c}} = \frac{s_a}{k}\left[\frac{1}{M} + \frac{1}{N} + \left(\frac{s_k}{k\, s_a}\right)^2 \left(\bar{a} - \bar{a}^c\right)^2\right]^{\frac{1}{2}} \tag{4.4}$$

where s_a is the standard deviation of the spectroscopic measurements and where $\bar{a}^c$ is the average value of spectroscopic signals of all standards:

$$\bar{a}^c = \frac{1}{M}\sum_{i=1}^{M} a_i^c \tag{4.5}$$

If a very low value for the absolute term k_o is obtained, a test of the statistical significance of this parameter is recommended (see Appendix B). If k_o is not statistically different from zero, Equation 4.2 is replaced by the simpler calibration function (Equation 4.1). The standard deviation of the concentration prediction with $M - 1$ degrees of freedom then has the form

$$s_{\bar{c}} = \frac{s_a}{k}\left[\frac{1}{M} + \frac{1}{N} + \left(\frac{s_k}{k\, s_a}\right)^2 \bar{a}^2\right]^{\frac{1}{2}} \tag{4.6}$$

The expressions for standard deviation (Equations 4.4 and 4.6) show that the prediction accuracy of the simple linear calibration function increases with the number of the parallel spectroscopic measurements of a sample N. The number of calibration standards M should be close to 10 and the number of parallel determinations N should be at least 2.[3]

The constant k, the measure of method sensitivity, is the principal influence on prediction accuracy. The standard deviation of the spectroscopic measurement $s_{\bar{a}}$ also influences the value of $s_{\bar{c}}$. The term $\bar{a} - \bar{a}^c$ in Equation 4.4 causes prediction accuracy based on the model in Equation 4.2 to decrease as the measured values deviate from the center of the calibration interval.

The matrix effect is frequently a problem in spectroscopic determination by calibration functions. The interaction of the predicted component with the minority components causes the sensitivity coefficient k of an unknown sample to have a value different from the coefficient determined by the calibration standards.

If the linear model (Equation 4.1) is valid, it is possible to reduce matrix effects by the standard addition method.[3] The sample studied is divided into parts. Known amounts of the component to be estimated by the method of the standard additions are added to $L - 1$ parts of analyte so that its concentration is increased by c_i^p. It is assumed that the signals produced from the standard additions are not influenced by experimental error. The spectroscopic signal of such treated samples is measured. According to Equation 4.1,

$$a_i = k\left(c + c_i^p\right) \qquad i = 1 \rightarrow L \tag{4.7}$$

is valid. The spectroscopic signal of the original sample, of unknown concentration c, is a:

$$\left(a - a_i\right) = k\, c_i^p \tag{4.8}$$

The sensitivity coefficient k, reflecting the chemical environment's influence, is calculated from Equation 4.8 by the least squares method. The concentration of the component is then calculated from N parallel spectroscopic measurements of a sample:

$$\bar{c} = \frac{\bar{a}}{k} \tag{4.9}$$

The standard deviation in the standard addition method $s_{\bar{c}}$,

$$s_{\bar{c}} = \frac{s_a \sqrt{2}}{k}\left(\frac{1}{N} + \frac{s_b}{k\, s_a \sqrt{2}}\,\bar{a}^2\right)^{\frac{1}{2}} \tag{4.10}$$

has $L - 2$ degrees of freedom.

MULTICOMPONENT LINEAR MODEL

The multicomponent linear model is a generalization of the single-component model. It is supposed that the spectroscopic responses of individual components are additionally summed into a spectrum of a multicomponent system. The intensity a_i at the ith spectral coordinate (ith wavenumber, wavelength, mass coordinate, and so on) in an N_C component system is the sum

$$a_i = \sum_{l=1}^{N_C} c_l \, k_{il} \tag{4.11}$$

where c_l is the concentration of the lth component and where k_{il} is the sensitivity coefficient of the lth component at the ith spectral coordinate.

If the multicomponent system spectrum is digitized at N_W points, it can be described by the N_W dimensional column vector $\vec{A}$, the elements of which are signal intensities of spectra of the individual spectral coordinates. Equation 4.11 is valid for each element of the vector $\vec{A}$, and in matrix form this system can by written as

$$\vec{A} = \bar{\bar{K}} \, \vec{C} \tag{4.12}$$

$\vec{C}$ is the N_C dimensional column vector comprising individual component concentrations of a system. $\bar{\bar{K}}$ is the matrix of the sensitivity coefficients or the "calibration matrix". Spectra of pure components are placed in its columns.

Applying the multicomponent linear model (Equation 4.12) in quantitative spectroscopic analysis requires, as does the single-component linear model, fulfilling the following assumptions.[4]

1. For each spectral coordinate the spectral response is linearly dependent on the component concentration in the whole studied concentration interval.
2. The spectral responses of the individual components are independent of the other components.
3. A spectral response of zero corresponds to a concentration of each component of zero.
4. The sensitivity coefficients are not changed between calibration and the prediction of an unknown component.

These assumptions are not always precisely met for real multicomponent systems.

Classical Least Squares Method

The classical least squares method (CLS),[5-9] called the K-matrix method, begins with calibration, i.e., yielding the $\bar{\bar{K}}$ matrix elements. The calibration is experimentally realized by measuring spectra of M standard mixtures. The spectra of these standards are arranged into the columns of a matrix $\bar{\bar{A}}_C$ with dimension $N_W \times M$. Equation 4.12 is changed for the calibration to the form

$$\bar{\bar{A}}_C = \bar{\bar{K}} \cdot \bar{\bar{C}}_C \tag{4.13}$$

The rows of matrix $\bar{\bar{C}}_C$ contain the concentrations of individual standard mixtures.

The CLS solution of Equation 4.13 for unknown $\bar{\bar{K}}$ has the form

$$\bar{\bar{K}} = \bar{\bar{A}}_C \, \bar{\bar{C}}_C^T \left(\bar{\bar{C}}_C \, \bar{\bar{C}}_C^T \right)^{-1} \tag{4.14}$$

Equation 4.14 is soluble if the inequality

$$M \geq N_C \tag{4.15}$$

is satisfied, i.e., the number of calibration standards must be at least equal to the number of components.

Based on the calibration matrix $\bar{\bar{K}}$ obtained, the concentration vector of the analyzed mixture is calculated from Equation 4.12:

$$\vec{C} = \left(\bar{\bar{K}}^T \, \bar{\bar{K}} \right)^{-1} \vec{A} \tag{4.16}$$

In order to avoid an underdetermined solution according to Equation 4.16 the following equality must be valid:

$$N_W \geq N_C \tag{4.17}$$

Systems in which M or N_W is greater than N_C are called overdetermined with respect to calibration standards or overdetermined according to the number of spectral points. Enough information is available to solve overdetermined systems, and the solution error is minimized in the least squares procedure.

Inversion of two matrices, the matrix ($\bar{\bar{C}}_C \, \bar{\bar{C}}_C^T$) of dimensions $N_C \times N_C$ in Equation 4.14 and the matrix ($\bar{\bar{K}}^T \, \bar{\bar{K}}$) in Equation 4.16 are necessary in the CLS method. This is often thought to be a disadvantage of the CLS method, compared to the inverse least squares method (see Inverse Least Squares Method, below). With the computational power available in modern spectroscopic laboratories, such inversion is not a problem. In fact, N_W does not influence the dimension of inverted matrices. The CLS method can use spectral data in a relatively extended set of spectral coordinates. Because such sets can describe spectral characteristics of individual components better than sets collected over a more limited region, the analysis accuracy often improves.

Haaland and Easterling[8,9] proposed a modified version of the CLS method, named by McClure et al.[10] the Q-matrix method. This method is based on the idea that a spectrum of multicomponent mixtures can be expressed as a linear combination of standard mixture spectra:

$$\vec{A} = \overline{\overline{A}}_C \, \vec{Q} \tag{4.18}$$

The components of the column vector $\vec{Q}$ are the coefficients of the sample decomposition into individual standard spectra. If

$$N_W > M \tag{4.19}$$

the vector $\vec{Q}$ elements can be obtained by the least squares method:

$$\vec{Q} = \left(\overline{\overline{A}}^T \, \overline{\overline{A}} \right)^{-1} \overline{\overline{A}}^T \, \vec{A} \tag{4.20}$$

The values of the component concentration of the mixture are calculated from the vector $\vec{Q}$ and the concentration matrix of standards $\overline{\overline{C}}_C$:

$$\vec{C} = \overline{\overline{C}}_C \, \vec{Q} \tag{4.21}$$

The solution to Equation 4.21 has a physical meaning only if the conditions in Equations 4.15 and 4.17 are satisfied.

For the special case, if the standards are pure components at unit concentrations, the matrix $\overline{\overline{A}}_C$ will be identical with the matrix $\overline{\overline{K}}$ and the vector $\vec{Q}$ with the vector $\vec{C}$. From the practical point of view, application of component mixtures as standards is more effective. Such a method of calibration considers the interaction between components; it therefore compensates for the possible nonlinear deviations from the model (Equation 4.11) to some degree.

If various parts of standard spectra and a studied sample have different signal/noise ratios, it is efficient to use the weighted least squares method (see Appendix B) to calculate Equation 4.20. In this procedure, greater significance is attributed to spectral data with a more advantageous signal/noise ratio, and these data influence the solution more than data in which this ratio is less advantageous. The selection of weighting coefficients depends on the type of spectroscopy and instrumental conditions. With weighting coefficients, the relationship in Equation 4.20 has the form

$$\vec{Q} = \left(\overline{\overline{A}}_C^T \, \overline{\overline{F}}^{-1} \, \overline{\overline{A}}_C \right)^{-1} \overline{\overline{A}}_C^T \, \overline{\overline{F}}^{-1} \, \vec{A} \tag{4.22}$$

If the noise levels are mutually independent of the individual coordinates, $\overline{\overline{F}}$ is a diagonal matrix of dimension $N_W \times N_W$ having weighting factors for individual spectral coordinates for the sample studied on the diagonal. For infrared spectra, Haaland and Easterling[8] proposed deriving the weighting coefficients from the squares of the transmittance spectrum of a sample:

$$f_{ii} = T_i^{-2} \quad \textbf{(4.23)}$$

where T_i is the transmittance of the sample at the spectral coordinate (wavenumber) i.

The existence of non-zero baselines and variable backgrounds in standard spectra as well as in the spectra samples is a frequent problem in multicomponent spectroscopic analysis. In many cases it is possible to eliminate the background of the individual spectra (see Chapter 1, Elimination of Spectral Background). Another way to include the background directly in the spectral decomposition model makes use of Equation 4.18.

If the background is linear over the whole spectral coordinate range, the expression of the spectral decomposition (Equation 4.18) can be extended to

$$a_i = C + D_i \nu_i + \sum_{k=1}^{M} a_{ki}^c \, q_k \quad \textbf{(4.24)}$$

where a_i is the spectral intensity of the sample for the ith coordinate, ν_i is the value of the ith spectral coordinate, and where C and D are constants of the linear background function. In matrix form, Equation 4.24 is changed to

$$\vec{A} = \overline{\overline{U}}\, \vec{\theta} \quad \textbf{(4.25)}$$

The matrix $\overline{\overline{U}}$ with dimension $N_W \times (M + 2)$ contains values of 1 in the first column and values of the spectral coordinates in the second column. The other columns are identical with the columns of the matrix $\overline{\overline{A}}$:

$$\overline{\overline{U}} = \begin{pmatrix} 1 & \nu_1 & \\ 1 & \nu_2 & \\ \cdot & \cdot & \overline{\overline{A}}_C \\ \cdot & \cdot & \\ 1 & \nu_{N_W} & \end{pmatrix} \quad \textbf{(4.26)}$$

The first two elements of a vector $\vec{\theta}$ are the constants C and D, the other elements are identical with the elements of the vector $\vec{Q}$:

$$\vec{\theta} = \begin{pmatrix} C \\ D \\ \vec{Q} \end{pmatrix} \quad \textbf{(4.27)}$$

The elements of vector $\vec{\theta}$ are calculated in a manner similar to that used in obtaining the elements of vector $\vec{Q}$ in Equations 4.20 and 4.22, respectively.

The need to know and involve all spectroscopically active elements in the calibration are disadvantages of the CLS method for multicomponent spectroscopic analysis. Practically, it means that CLS methods (the K-matrix method and the Q-matrix method) are sensitive to the presence of secondary interfering components, even in low amounts (e.g., impurities). As Haaland and Easterling[8,9] showed, this sensitivity to interferences is diminished in cases where each spectral band of the sample is analyzed individually and the calculations of the individual bands are summed statistically. Another advantage of this procedure is the possibility of modeling the linear background in the regions of the individual spectral bands:

$$a_i = C^p + D_i^p \, \nu_i^p + \sum_{k=1}^{M} a_{ki}^{cp} \, q_k^p \tag{4.28}$$

where index p describes the contribution to the spectral band p. The components of vector $\vec{Q}$ are then obtained as the weighted average of the individual vector $\vec{Q}^p$ components. The weighted factors for the individual bands reflect the quality of decomposition (Equation 4.28) and are inversely proportional to the standard deviation of the element q_k^p for each corresponding component. The parts of a spectrum most suitable for the model (Equation 4.28), from the point of view of the assumption about the components present and the linear dependence of the intensity of the spectral response on the component concentration, have the largest influence on the results.

Haaland and Easterling applied the modified Q-matrix method to analyze the mixture of three xylene isomers in the ratio 1:1:1 using infrared spectroscopy.[8] Table 4.1 shows that the method can also analyze mixtures with unknown, perturbing components in relatively significant concentrations.

Another modification of the CLS method, one in which the assumption of a linear dependence of spectral response intensity on the component concentrations is less important, makes analyzing samples possible.[11]

Using FTIR spectroscopy to determine the polychlor biphenyls (PCBs) in transformer oils is another example of an application of the CLS calibration method.[12] Three samples of commercial PCB isomers in transformer oil were prepared for verifying this procedure (Table 4.2). The isomer solutions

Table 4.1 Relative Error of Component Concentrations from IR Analysis of o-, m-, and p-Xylene Mixtures (1:1:1) by the Q-Matrix Method According to Equation 4.28

Standard spectra used in calibration	Relative error (%)	
	o-Xylene	**m-Xylene**
o-, m-Xylene	0.9	0.7
o-Xylene	3.3	
m-Xylene		4.0

Table 4.2 Component Concentration of PCB Mixtures, Aroclor 1242 and Aroclor 1254, in Transformer Oil

	Concentration (wt %)		
Sample	**Transformer oil**	**Aroclor 1242**	**Aroclor 1254**
1	98.9999	0.5246	0.4755
2	98.9985	0.7964	0.2051
3	98.9980	0.9053	0.0967

in transformer oil and the pure transformer oil were used as the calibration standards (Figure 4.1). Spectra of both isomers, isolated by CLS calibration, are shown in Figure 4.2. The accuracy of analysis (Table 4.3) documents the effectiveness of the CLS calibration method when the spectral bands of system components represent practically only minor perturbations on the spectrum of the principal component, transformer oil in this case.

Inverse Least Squares Method

In the inverse least squares method (ILS),[13-15] also called the P-matrix method, the model (Equation 4.12) is transformed so that the concentrations of components are defined as functions of the spectral response intensity:

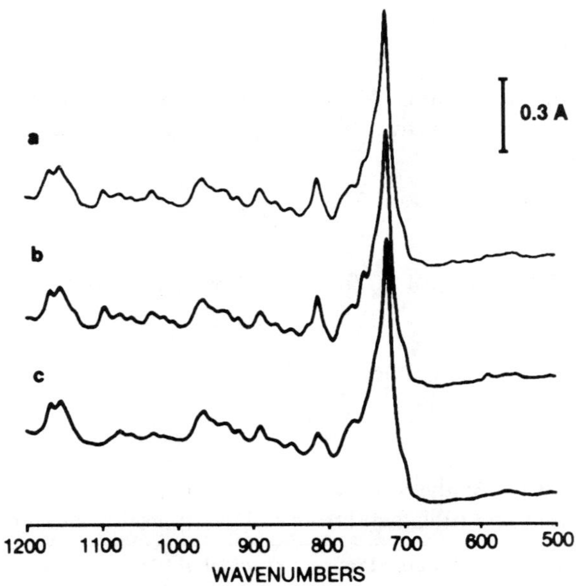

Figure 4.1 Infrared spectra of the calibration standards in 200 μm pathlength cells: **(a)** 1 wt % Aroclor 1254 in transformer oil, **(b)** 1 wt % Aroclor 1242 in transformer oil, **(c)** pure transformer oil. (From *Am. Lab.* 17(7), 1985. With permission.)

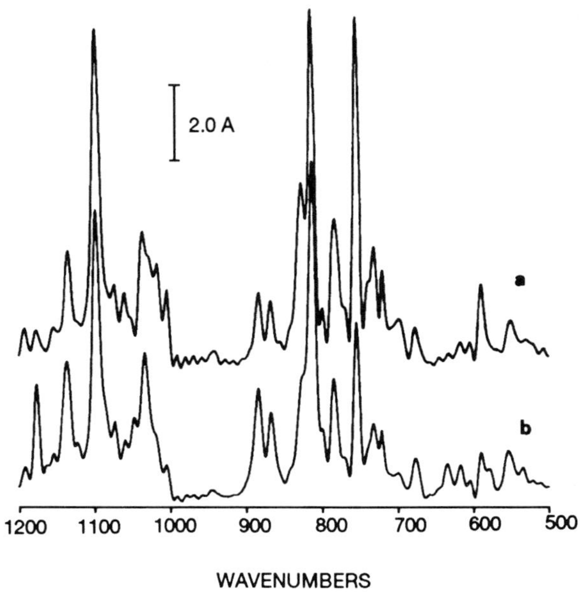

Figure 4.2 CLS estimated pure (100%) Aroclor PCB spectra in 200 μm pathlength cells: **(a)** Aroclor 1242, **(b)** Aroclor 1254. (From *Am. Lab.,* 17(7), 1985. With permission.)

$$\vec{C} = \bar{\bar{P}}\,\vec{A} \tag{4.29}$$

A matrix $\bar{\bar{P}}$ of dimensions $N_C \times N_W$ contains calibration coefficients of individual components in rows. The ILS method describes the calibration by the relationship

$$\bar{\bar{C}}_C = \bar{\bar{P}}\,\bar{\bar{A}}_C \tag{4.30}$$

The solution of Equation 4.30 for the calibration matrix $\bar{\bar{P}}$ is obtained by least squares:

$$\bar{\bar{P}} = \bar{\bar{C}}\,\bar{\bar{A}}^T\left(\bar{\bar{A}}\,\bar{\bar{A}}^T\right)^{-1} \tag{4.31}$$

Table 4.3 Analysis of PCB Mixtures in Transformer Oil — Absolute and (Relative) Errors of Determined Concentrations

Sample	Aroclor 1242	Aroclor 1254	Total PCB
1	–0.0017	–0.0019	–0.0036
	(–0.32)	(–0.40)	(–0.36)
2	0.0086	–0.0103	–0.0017
	(1.1)	(–5.0)	(–0.17)
3	0.0021	–0.0035	–0.0014
	(0.23)	(–3.6)	(–0.14)

The solution of Equation 4.31 is dependent upon satisfying the inequality

$$M > N_W \tag{4.32}$$

Based on the calculated matrix $\overline{\overline{P}}$, the component concentrations of the analyzed sample are determined as the components of vector $\vec{C}$ (Equation 4.29).

The need to invert only one matrix for calibration and prediction is an advantage of the ILS method over the CLS methods. The capacity of the ILS method to predict concentrations of the individual mixture components independent of those of the other components is the main advantage of this method. The calculation of the component k concentration as the kth element of the vector $\vec{C}$ can be expressed, in accordance with Equation 4.29 in the form

$$c_k = \sum_{i=1}^{N_W} p_{ki}\, a_i \tag{4.33}$$

This expression implies that only the kth row of the calibration matrix $\overline{\overline{P}}$ is necessary to determine the component k concentration. From Equation 4.31 it follows that

$$\vec{P}_k^T = \vec{C}_{Ck}^T\, \overline{\overline{A}}_C^T \left(\overline{\overline{A}}_C\, \overline{\overline{A}}_C^T \right)^{-1} \tag{4.34}$$

where $\vec{P}_k^T$ is the kth row of the matrix $\overline{\overline{P}}$ and $\vec{C}_{Ck}^T$ is the kth row of the matrix $\overline{\overline{C}}_C^T$. Therefore, only knowledge of the concentration of the standard mixtures is sufficient for calibration in the ILS method.

Satisfying the inequality (Equation 4.32) as a condition of solution in the ILS method requires a higher number of calibration standards than the number of points used in the sample spectrum. The practical realization of this requirement limits the range of the spectral data set used in the analysis. In this situation, especially in the case of complicated spectra, the spectral coordinates (wavenumbers, wavelengths, etc.) from the complete sample spectrum involved in analysis must be chosen carefully. Nonoptimal selection of the spectral coordinates, which does not involve all important spectral characteristics, decreases the accuracy of analysis.

Honigs et al.[16,17] proposed an algorithm for selecting an optimal, N_W-dimensional subset from the total set of sample spectral coordinates, which is optimal for the ILS method. This iterative process begins with the calculation of the matrix $\overline{\overline{P}}$ (Equation 4.31) for the set of the first N_W spectral coordinates. Based on the matrix $\overline{\overline{P}}$, the component concentrations in the standards are calculated, and the correlation coefficient of such calculated and real concentrations is evaluated. The spectral coordinate, to which the element p_{ij} with the maximal value corresponds, is eliminated from the set and is replaced by another spectral coordinate from the complete set. The correlation coefficient of the calculated and of the real standard concentrations are obtained in the same way. The iterative cycle is finished after two loops through the complete set. The set of N_W spectral coordinates, to which the greatest correlation coefficient corresponds, is used for ILS analysis.

The restrictions of the ILS method become advantages for applications requiring large sets of calibration standards to characterize the system studied, and for instances in which the relevant spectral information is contained in a relatively small fraction of the spectral data.[18]

Generalized Standard Addition Method

The influence of the matrix effect has been discussed above (see Simple Linear Model, above), i.e., the influence of the predicted component interaction with other components of a system on spectroscopic quantitative analysis of single-component systems. Obviously, similar effects can complicate multicomponent analysis as well. Also, in this situation the most suitable procedure for eliminating the matrix effect is the standard addition method. Saxberg and Kowalski[19] developed a variant for multicomponent systems and called this procedure the generalized standard addition method (GSAM).

The GSAM is experimentally realized by subsequent additions to an analyzed sample. Each addition contains one or several components, but each component of the analyzing sample must be added at least once individually. The spectrum of the sample $\vec{A}_m$ is measured after each addition. According to Equation 4.12 the following is valid:

$$\vec{A}_m = \bar{\bar{K}}\,\vec{C}_m \tag{4.35}$$

where

$$\vec{C}_m = \vec{C}_o + \Delta\vec{C}_m \tag{4.36}$$

is the concentration vector of the sample after the mth addition, $\vec{C}_o$ is the concentration vector of the original sample, and $\Delta\vec{C}_m$ is the vector of the changes of the component concentration of the sample after the mth addition. Using Equation 4.35 for all the additions can be associated with the following matrix equation:

$$\bar{\bar{A}} = \bar{\bar{K}}\,\bar{\bar{C}} \tag{4.37}$$

The columns of matrices $\bar{\bar{A}}$ and $\bar{\bar{C}}$ correspond to the individual additions.

The solution of Equation 4.37 for the calibration matrix $\bar{\bar{K}}$ is obtained after the decomposition of the matrices $\bar{\bar{A}}$ and $\bar{\bar{C}}$:

$$\bar{\bar{A}} = \bar{\bar{A}}_o + \Delta\bar{\bar{A}} \tag{4.38}$$

$$\bar{\bar{C}} = \bar{\bar{C}}_o + \Delta\bar{\bar{C}} \tag{4.39}$$

where $\bar{\bar{A}}_o$ and $\bar{\bar{C}}_o$ are matrices containing the spectral vector of the original sample $\vec{A}_o$, and the vector of the original concentration $\vec{C}_o$ in all columns. After the separation (Equations 4.38 and 4.39), Equation (4.37) becomes

$$\Delta\bar{\bar{A}} = \bar{\bar{K}}\,\Delta\bar{\bar{C}} \tag{4.40}$$

The solution of Equation 4.40 for the matrix $\bar{\bar{K}}$ is analogous to the solution of Equation 4.14:

$$\bar{\bar{K}} = \Delta\bar{\bar{A}}\,\Delta\bar{\bar{C}}^T\left(\Delta\bar{\bar{C}}\,\Delta\bar{\bar{C}}^T\right)^{-1} \tag{4.41}$$

The concentrations of components in the original sample are then determined as the elements of the vector $\vec{C}_o$:

$$\vec{C}_o = \left(\bar{\bar{K}}^T\,\bar{\bar{K}}\right)^{-1}\vec{A}_o \tag{4.42}$$

Because the original component concentrations of a sample are obviously not known, the elements of the matrix $\Delta\bar{\bar{C}}$ can be calculated only in the cases where the changes of the sample volume, after the additions, are neglible. For the opposite situation, it is necessary to introduce a simple correction transforming the component concentrations to absolute amounts, using the GSAM algorithm. Because the concentration is the quotient of the amount (mass) and the volume, it is possible to rewrite Equation 4.35:

$$\vec{A}_m = \left(\frac{1}{V_m}\right)\bar{\bar{K}}\,\vec{N}_m \tag{4.43}$$

where V_m is the volume of the system after the mth addition and $\vec{N}$ is the vector of the absolute amounts (i.e., mass) of the components in the sample after the mth addition. If the vector of the volume-corrected spectral response of the sample after the mth addition is defined as

$$\vec{R}_m = V_m\,\vec{A}_m \tag{4.44}$$

Equation 4.43 then becomes

$$\vec{R}_m = \bar{\bar{K}}\,\vec{N}_m \tag{4.45}$$

Then Equation 4.45 for $\bar{\bar{K}}$ is changed in the same way as Equation 4.41:

$$\bar{\bar{K}} = \Delta\bar{\bar{R}}\,\Delta\bar{\bar{N}}^T\left(\Delta\bar{\bar{N}}_m\,\Delta\bar{\bar{N}}^T\right)^{-1} \tag{4.46}$$

Principal Component Regression

The principal component regression method (PCR)[20-22] represents the combination of principal component analysis (PCA) (see Chapter 3, Abstract Factor Analysis and Theory of Errors in Factor Analysis) with the ILS method (see Inverse Least Squares Method, above).

The application of PCA to a matrix of standard mixture spectra $\bar{\bar{A}}_C$ gives the abstract factor decomposition of this matrix in the primary factor space:

$$\bar{\bar{A}}_C = \bar{\bar{Q}}_C^{\ddagger}\, \bar{\bar{U}}_C^{\ddagger}$$

Placing (4.47) into the calibration equation for the ILS method (Equation 4.30), the following is obtained:

$$\bar{\bar{C}}_C = \bar{\bar{P}}\, \bar{\bar{Q}}_C^{\ddagger}\, \bar{\bar{U}}_C^{\ddagger} \tag{4.48}$$

If $\bar{\bar{G}}$ is defined as the new calibration matrix,

$$\bar{\bar{G}} = \bar{\bar{P}}\, \bar{\bar{Q}}_C^{\ddagger} \tag{4.49}$$

Equation 4.48 is changed to

$$\bar{\bar{C}}_C = \bar{\bar{G}}\, \bar{\bar{U}}_C^{\ddagger} \tag{4.50}$$

Equation 4.50 describes the regression of the concentration of the calibration standards and the projections of the matrix $\bar{\bar{A}}_C$ to the primary factor space. According to the orthogonality of rows of the matrix $\bar{\bar{U}}_C^{\ddagger}$ (Equation 3.8), the solution for the matrix $\bar{\bar{G}}$ is obtained in the form

$$\bar{\bar{G}} = \bar{\bar{C}}_C\, \bar{\bar{U}}_C^{\ddagger T}\, \bar{\bar{\lambda}}^{\ddagger -1} \tag{4.51}$$

where $\bar{\bar{\lambda}}^{\ddagger}$ is the diagonal matrix of primary eigenvalues.

From Equations 4.29 and 4.49 the relationship for calculating the vector of the unknown component concentration follows:

$$\vec{C} = \bar{\bar{G}}\, \vec{U} \tag{4.52}$$

where

$$\vec{U} = \bar{\bar{Q}}_C^{\ddagger}\, \vec{A} \tag{4.53}$$

is the projection of the sample spectrum to the primary factor space.

The PCR method conserves the main advantage of the ILS method, i.e., the capability of quantifying the individual mixture components independently of other components. Because the number of primary abstract factors used in the regression is small — often identical to the number of components in the mixture — the number of mixtures required is not extensive, and the inequality (Equation 4.32) is easily satisfied. Therefore, PCR is able to treat the spectral data of relatively extended spectral coordinate sets, similar to those used in the CLS and PLS methods, which contributes to the higher accuracy of results. Finally, because PCR acts in the primary factor space of the dimension N_F, the influence of part of the experimental errors will be eliminated, just as in other factor analysis methods.

The application of the PCR method to solve a simple calibration problem will be illustrated in the model set of spectra in Figure 4.3. The reliability of the calibration model will be evaluated by the cross validation method.[23,24] This procedure repeats the calibration M times, each time on a subset of $M - 1$ standard samples, leaving one standard j, $j = 1 \rightarrow M$, out. The results of each calibration are used to "predict" the concentration of the standard left out. The calibration model then characterizes the root mean square error of cross validation (RMSECV):

$$\mathrm{RMSECV}_l = \left[\frac{1}{M} \sum_{j=1}^{M} \left(\hat{c}_{lj} - c_{lj}^{c} \right)^2 \right]^{\frac{1}{2}} \tag{4.54}$$

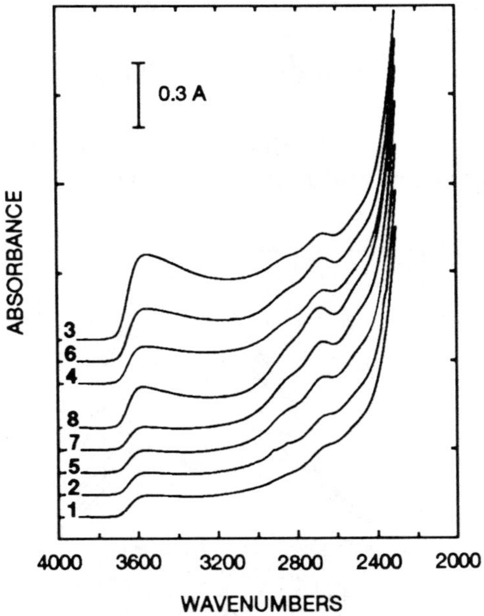

Figure 4.3 Spectra of the model calibration set, concentrations 0.26, 0.52, 0.73, 0.87, 1.09, 1.55, and 2.12 (in arbitrary units) correspond to the calibration standards 1 through 7.

Table 4.4 Principal Component Analysis of the Figure 4.3 Spectra Set

Size of factor space	RE	α_k[a] (%)	CV standard deviation of TFA projections
1	0.0266	<0.001	0.0115
2	0.0013	<0.001	0.0007
3	0.0007	9	0.0005
4	0.0005	19	0.0004
5	0.0003	28	0.0003
6	0.0003	52	0.0003

Note: $H_0{:}REV_k = \overline{REV}_k^{\,0}$ against $H_a{:}REV_k > \overline{REV}_k^{\,0}$
[a] α_k is the level of significance for rejecting the hypothesis.

where $\hat{c}_{lj}$ is the predicted value and c_{lj}^c is the true value of component l concentration in the jth eliminated standard.

Analysis of principal components (see Chapter 3, The Determination of the Number of Active Components in the Spectral Set) shows that, besides the major predicted component, a second minor factor is present in this set of spectra (Table 4.4). It is not possible to consider the one-component linear model of calibration (Equaion 4.2) reliable (RMSECV = 0.1, i.e., 10% of the average concentration of calibration standards). Its inconvenience is also proved by comparing the true and predicted concentrations (Figure 4.4).

The PCR calibration model (Equation 4.50) in a two-dimensional factor space for the predicted component has the following form:

$$c_j^c = 0.92\left(u_c^{\ddagger}\right)_{1,j} + 0.25\left(u_c^{\ddagger}\right)_{2,j} \tag{4.55}$$

Coefficient values show that the first PCA factor is more important for the concentration calibration and following prediction. The second factor can

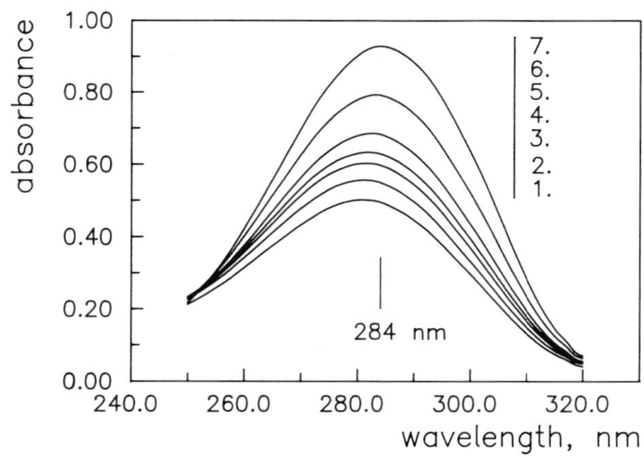

Figure 4.4 Predicted concentration vs. true concentration for the univariate linear calibration model for the spectral set of Figure 4.3 (solid line is the function $\hat{c} = c$).

indicate the significant disturbing effects, background and so on. For this calibration model, RMSECV is lower than 0.03% of the average concentration of calibration objects. Also, Figure 4.5 confirms the convenience of the calibration model (Equation 4.55).

Besides the determination of the multicomponent mixture composition, the PCR method is also used for quantification and modeling of various quantities, properties, and relationships[23] (e.g., Cowe and McNicol applied PCR to determine the moisture content and content of flour proteins based on the reflectance spectra in the NIR region[25] and Fredericks et al. correlated different properties of coal with FTIR spectra by means of PCR[26,27]).

Partial Least Squares Method

Abstract factor decomposition in the PCR calibration model (Equation 4.47) defines a set of significant factors based on the properties of a spectral data matrix of calibration mixtures $\bar{\bar{A}}_c$ exclusively. The information in the concentration matrix of standards $\bar{\bar{C}}_c$ is not used until the regression step (Equation 4.51). The PCA factors (a row of matrix $\bar{\bar{U}}_c^{\ddagger}$), important in describing a spectral data matrix, are not necessarily equally optimal for prediction.

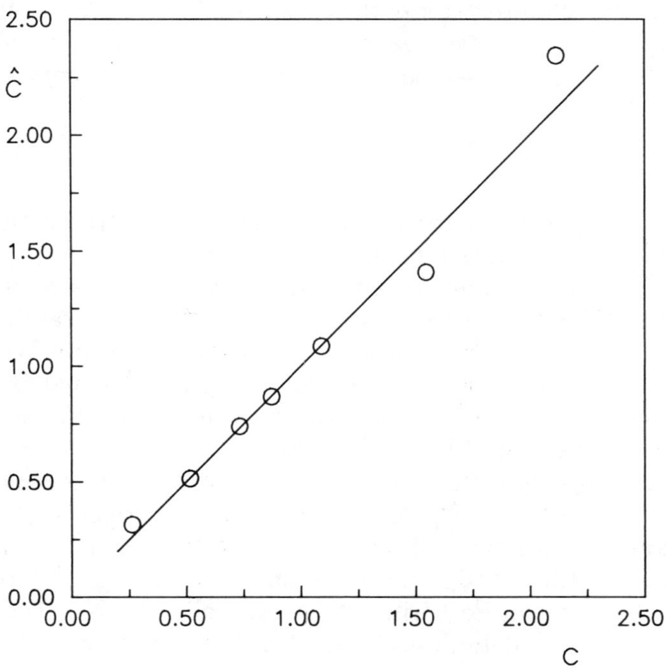

Figure 4.5 Predicted concentration vs. true concentration for the PCR calibration model in the spectral set of Figure 4.3 (solid line is the function $\hat{c} = c$).

The method of partial least squares (PLS),[23,28] also known as partial least squares regression, also involves the matrix of calibration standards $\bar{\bar{C}}_c$ in the factor decomposition of a calibration spectra matrix. The influence of the factors significant in PCA, which are irrelevant in the calibration modeling, is attenuated by considering the properties and structure of the matrix $\bar{\bar{C}}_c$. Therefore, the resulting PLS calibration model is often simpler than the PCR model.

The PLS calibration model is analogous to the PCR model:

$$\bar{\bar{X}} = \bar{\bar{T}} \cdot \bar{\bar{Q}}^{\ddagger T} \tag{4.56}$$

$$\bar{\bar{Y}} = \bar{\bar{T}} \cdot \bar{\bar{G}}^{T} \tag{4.57}$$

Matrices $\bar{\bar{X}}$ and $\bar{\bar{Y}}$ are mean-centered, transposed matrices $\bar{\bar{A}}_c$ and $\bar{\bar{C}}_c$, respectively:

$$\bar{\bar{X}} = \bar{\bar{A}}_c^T - \vec{1} \cdot \vec{X}_m^T \tag{4.58}$$

$$\bar{\bar{Y}} = \bar{\bar{C}}_c^T - \vec{1} \cdot \vec{Y}_m^T \tag{4.59}$$

where $\vec{X}_m$ is the column vector of the average calibration spectrum, $\vec{Y}_m$ is the column vector of the average component concentration in calibration standards, and $\vec{1}$ is an N_S-dimensional vector whose elements are unity. The matrix $\bar{\bar{T}}$ is

$$\bar{\bar{T}} = \bar{\bar{U}}^{\ddagger T} \tag{4.60}$$

The mean-centering of the matrices of spectral and concentration calibration data simplifies the calibration model and the algorithm.

The basic version of the PLS method, called PLS1,[28] is a method for modeling one component in a multicomponent system. Several PLS1 algorithms were developed;[23-33] the original algorithm of Svante Wold[23,28] is introduced in Tables 4.5 and 4.6. Because it gives the matrix $\bar{\bar{T}}$ with orthogonal columns, similarly to the orthogonal rows in $\bar{\bar{U}}^{\ddagger}$ Equation 3.8, it is called the orthogonalized PLS1 algorithm.

The effectivness of the PLS method depends on the correct choice for the dimension of a factor space N_F. As the PLS model describes only that part of the spectral data variance which is relevant for prediction, N_F determined by PCA tests (see Chapter 3, Determination of the Number of Active Components in the Spectral Set) will not be generally reliable. The optimal number of factors N_F is determined, using PLS methods, by cross validation. RMSECV (Equation 4.54) is calculated for each dimension of a factor space considered. N_F is then identical to that value k corresponding to the lowest RMSECV or the first k for which RMSECV is not statistically different from the lowest RMSECV.[23,24]

Table 4.5 Orthogonalized PLS1 Algorithm for One-Component Calibration

C1	$\overline{\overline{X}}_0 = X$; $\vec{y}_0$ is the corresponding column of $\overline{\overline{Y}}$ and N_F is the chosen dimension of factor space for each factor $k = 1 \rightarrow N_F$ — perform step C2.1 to C2.5	
C2.1	Model	$\overline{\overline{X}}_{k-1} = \vec{y}_{k-1}\, \vec{w}_k^T + \overline{\overline{E}}_x$
	LS solution	$\vec{w}_k = \overline{\overline{X}}{}^T_{k-1} \cdot \vec{y}_{k-1}$
		Normalize $\vec{w}_k$ to unity length
C2.2	Model	$\overline{\overline{X}}_{k-1} = \vec{t}\, \vec{w}_k^T + \overline{\overline{E}}_x$
	LS solution	$\vec{t} = \overline{\overline{X}} \cdot \vec{w}$
C2.3	Model	$\overline{\overline{X}}_{k-1} = \vec{t}_k\, \vec{p}_k^T + \overline{\overline{E}}_x$
	LS solution	$\vec{p} = \overline{\overline{X}}{}^T \cdot \vec{t}\Big/\left(\vec{t}^{\,T} \cdot \vec{t}\right)$
C2.4	Model	$\vec{y}_{k-1} = \vec{t}_k\, g_k + \overline{\overline{e}}_y$
	LS solution	$\vec{g}_k = \vec{y}_{k-1}^T \cdot \vec{t}_k\Big/\left(\vec{t}_k^T \cdot \vec{t}_k\right)$
C2.5	Calculate spectral $\left(\overline{\overline{E}}_x\right)$ and concentration $\left(\vec{e}_y\right)$ residuals:	
		$\overline{\overline{E}}_x = \overline{\overline{X}}_{k-1} - \vec{t}_k\, \vec{p}_k^T$
		$\vec{e}_y = \vec{y}_{k-1} - g_k\, \vec{t}_k$
	if $k < N_F$	$\overline{\overline{X}}_x = \overline{\overline{E}}_x$; $\vec{y}_k = \vec{e}_y$

From Martens, H. and Naes, T., *Multivariate Calibration,* John Wiley & Sons, Chichester, 1989.

Table 4.7 shows the cross validation for the PLS1 calibration of the spectral set in Figure 4.3. It is evident that $N_F = 1$ is the optimal dimension for this factor space. Figure 4.6 also proves the reliability of this calibration. The PLS1 calibration gives a simpler model than does PCR but has comparable accuracy.

The FTIR spectra of eight samples of glass are shown in Figure 4.7.[24,34] The main spectral effects in these samples are due to B_2O_3, OH groups, and various components of SiO_2. This set of spectra was analyzed by the CLS, PCR, and PLS1 methods. The B_2O_3 content and relative content of OH

Table 4.6 Orthogonalized PLS1 Algorithm for One-Component Prediction

For each prediction object l perform steps P1 to P3

P1 Center input spectral vector $\vec{A}_l$:

$$\vec{X}_{l,0} = \vec{A}_l - \vec{X}_m$$

for each factor $k = 1 \rightarrow N_F$ perform steps P2.1 to P2.2

P2.1 Find $t_{l,k}$ according to the formula in C2.2:

$$t_{l,k} = \vec{X}^T_{l,k-1} \cdot \vec{w}_k$$

P2.2 Compute residuals:

$$\vec{X}_{l,k} = \vec{X}_{l,k-1} - t_{l,k} \cdot \vec{p}^T_k$$

P3 Predict $\hat{y}_i$:[a]

$$\hat{y}_i = y_m + \sum_{k=1}^{N_F} t_{l,k} \cdot q_k$$

[a] y_m is the mean concentration of standard concentrations, i.e., corresponding element of $\vec{y}_m$

From Martens, H. and Naes, T., *Multivariate Calibration,* John Wiley & Sons, Chichester, 1989.

groups were determined. Each of these methods was used for each component individually, i.e., other components were not considered directly in the calibration. The values of RMSECV for individual methods are in Table 4.8. It is obvious that for this set of spectra, in which not all components are known, the results of the PCR and PLS methods are significantly more reliable than the results of the CLS method.

PLS1 can also be used to analyze multicomponent mixtures one component at a time. In some cases, especially if individual components of mixtures strongly intercorrelate, global calibration of the multicomponent system (PLS2) is more suitable. This algorithm is an iterative modification of the PLS1 algorithm.[23,30,33,35]

Table 4.7 RMSECV for PLS1 Calibration of the Spectra Shown in Figure 4.3

Number of factors, k	RMSECV
1	0.007
2	0.007
3	0.007

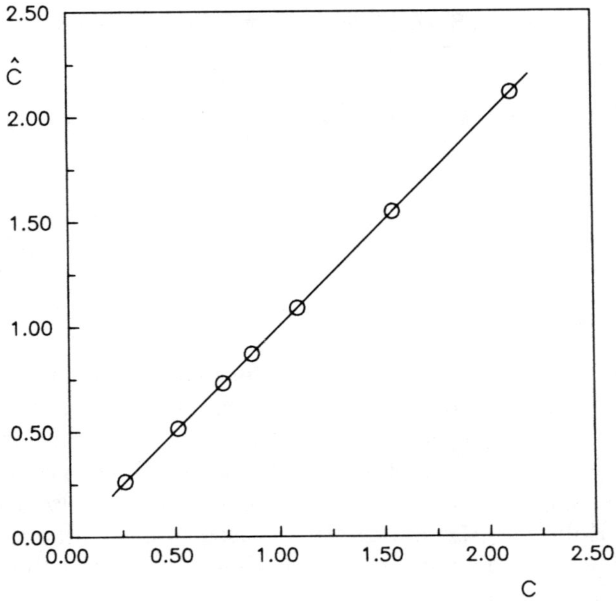

Figure 4.6 Predicted concentration vs. true concentration for PLS1 calibration $N_F = 1$ for the spectra shown in Figure 4.3 (solid line is the function $\hat{c} = c$).

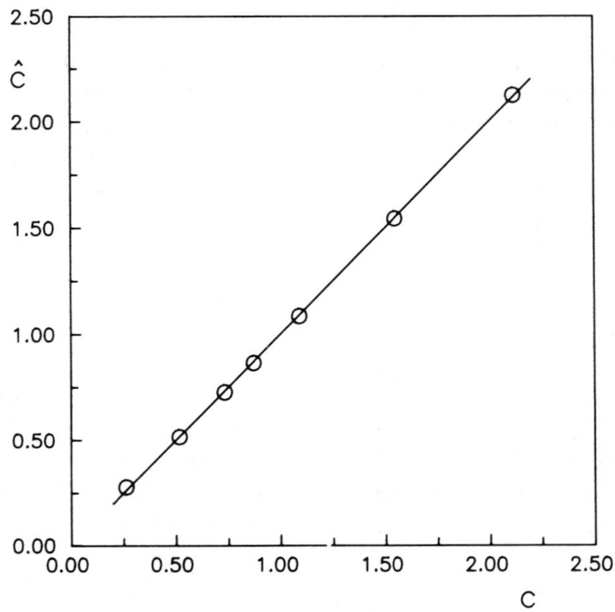

Figure 4.7 FTIR spectra of 5-μm *S* glass samples. (From Haaland, D. M. and Thomas, E. V., *Anal. Chem.* 60, 1202, 1988. With permission.)

Table 4.8 FTIR Analysis of Samples of Glass

	RMSECV	
Method	**B_2O_3 (wt %)**	**OH (rel. conc.)**
CLS	0.65	0.07
PCR	0.085	0.010
PLS	0.076	0.009

KALMAN FILTERING

Poulisse[36] proposed applying a Kalman filter to determine individual sample concentrations in a multicomponent system. The method is based on the knowledge of individual spectra of all components possibly present in the system and on the spectrum of a studied mixture. Besides concentrations of individual components, their standard deviations are also the results of calculation. Kalman filtering can also be used to determine the number of components present in a mixture, as was shown by Didden and Poulisse[37] in their examples of numerically simulated spectra. Statistically insignificant concentration values are obtained for components not present in a mixture.

A vector $\vec{A}$ with elements a_i equal to a spectral intensity at the ith spectral coordinate (wavelength, etc.) of a digitized mixture spectrum $i = 1, 2, \ldots N_W$ results from measuring the spectrum of a mixture.

$$\vec{A} \equiv \left(a_1, a_2, \ldots a_m\right) \tag{4.61}$$

The presence of the maximal number of N_C components with known spectra, arranged in a matrix $\bar{\bar{A}}$ of dimensions $N_C \times N_W$ is assumed in a mixture so that the spectrum of a component k is placed in the kth row. If columns of the matrix $\bar{\bar{A}}$ are labeled as vectors $\vec{h}_k$, $k = 1, 2, \ldots N_C$ and a column vector of predicted concentrations $\vec{C}$ is introduced, then for a mixture spectrum the following can be written as

$$a_k = \vec{h}_k \cdot \vec{C} + v_k \tag{4.62}$$

or in matrix form

$$\vec{A} = \bar{\bar{A}}^T \cdot \vec{C} + \vec{v} \tag{4.63}$$

where the components of the column vector $\vec{v}$ are the values of spectral noise v_k at individual spectral coordinates. It is supposed that the vector $\vec{v}$ elements are mutually independent and have normal probability distribution with a zero mean value and constant covariance r_k:

$$\hat{E}\left(v_k\right) = 0, \qquad k = 1, 2, \ldots N_W \tag{4.64}$$

$$\hat{E}\left(v_k v_j\right) = \delta_{k,j} r_k, \quad k.\ j = 1, 2, \ldots N_W \tag{4.65}$$

where $\hat{E}$ is the mean value operator and $\delta_{k,j}$ is the Kronecker delta function.

The Kalman filtering algorithm is based on sequential treatment of the individual points of a mixture spectrum, in which the $(k-1)$th estimate of a concentration vector is improved recursively by the treatment at the kth point:

$$\vec{C}_k = \vec{C}_{k-1} + \vec{k}_k\left(\vec{A}_k - \vec{h}_k^T \cdot \vec{C}_{k-1}\right) \quad \textbf{(4.66)}$$

$$\vec{k}_k = \overline{\overline{P}}_{k-1} \cdot \vec{h}_k\left(\vec{h}_k^T \cdot \vec{P}_{k-1} \cdot \vec{h}_k + r_k\right)^{-1} \quad \textbf{(4.67)}$$

$$\overline{\overline{P}}_k = \overline{\overline{P}}_{k-1} - \vec{h}_k \cdot \vec{h}_k^T \cdot \overline{\overline{P}}_{k-1} \quad \textbf{(4.68)}$$

The vector $\vec{C}_k$ is the "new" estimate of concentrations based on spectral data $a_1, a_2, \ldots a_k$. The correction that is added to the "old" estimate of $\vec{C}_{k-1}$ (Equation 4.66) during the calculation consists of two parts. The scalar term in parentheses represents the difference ν_k between the measured value a_k and the value calculated in accordance with the "old" estimate of concentrations:

$$\nu_k = a_k - \sum_{i=1}^{N_C} a_{ik} \cdot \left(\vec{C}_{k-1}\right)_i = a_k - \vec{h}_k^T \cdot \vec{C}_{k-1} \quad \textbf{(4.69)}$$

The Kalman gain, an N_C-membered column vector $\vec{k}_k$, which is evaluated based on the results of the previous $(k-1)$th step according to Equation 4.67 is multiplied by the scalar value ν_k. A square matrix $\overline{\overline{P}}$ of dimensions $N_C \times N_C$ represents a covariance matrix characterizing a statistical distribution of estimation $\vec{C}_k$ errors according to accurate values $\vec{C}_k^*$:

$$\overline{\overline{P}}_k = \hat{E}\left[\left(\vec{C}_k - \vec{C}_k^*\right)\left(\vec{C}_k - \vec{C}_k^*\right)^T\right] \quad \textbf{(4.70)}$$

Thus, the diagonal elements $(\overline{\overline{P}}_k)_{jj}$ reflect uncertainty in the estimated jth component $(\vec{C}_k)_j$ concentration. The correlation coefficients ρ_{ij} between the estimated concentrations of components i and j can be determined from the offdiagonal elements of the matrix $\overline{\overline{P}}$:

$$\rho_{ij} = \frac{\left(\overline{\overline{P}}_k\right)_{ij}}{\left[\left(\overline{\overline{P}}_k\right)_{ii}\left(\overline{\overline{P}}_k\right)_{jj}\right]^{\frac{1}{2}}} \quad \textbf{(4.71)}$$

Those values ρ_{ij} statistically different from zero indicate situations where it is not possible to determine individual concentrations unambiguously, but it is possible to determine their linear combination.

Equations (4.67) and (4.68) show, that the matrix $\overline{\overline{P}}$ is determined only by spectral characteristics of individual pure components, and that it is not dependent on the measured values a_k. In other words, the accuracy of the resulting concentration estimates and the existence of linear relations between them depend exclusively on the measure of the difference of their individual spectra.

The calculation procedure implied in Equations 4.66 and 4.68 is relatively fast, and contrary to the least squares method, it does not require matrix inversion (the value in parentheses of Equation 4.67 is scalar).

At the beginning, it is necessary to determine the starting values of concentration estimates $\vec{C}_o$ and a matrix $\bar{\bar{P}}_0$. If any preliminary information about a system is not available, the starting concentrations can be chosen to have zero values:

$$\vec{C}_o = \vec{0} \tag{4.72}$$

The accuracy of an estimate $\vec{C}_o$ does not have any influence on the resulting values of concentrations if the spectra of pure components are sufficiently different; it only determines the rate by which these values are stabilized during the recursive calculation.

Starting values of the matrix $\bar{\bar{P}}$ are estimated using

$$\bar{\bar{P}}_0 = \sigma_0^2 \, \bar{\bar{I}} \tag{4.73}$$

where $\bar{\bar{I}}$ is the unit diagonal matrix of dimensions $N_C \times N_C$ and σ_0^2 is the estimate of the mean variance of concentration values. Generally, the value σ_0^2 is chosen to be relatively large, but the algorithm of calculation is sufficiently robust over a fairly large range in this estimate.

The values r_k for the given experimental arrangement can be relatively easily estimated, and based on the matrix $\bar{\bar{P}}$, they can be calculated in advance. Therefore, the suitability of the chosen system of spectral quantitative analysis, which can be optimized in accordance with the selectivity of spectral range, density and accuracy of spectra digitization and so on, can be judged *a priori*.

Figure 4.8A shows the spectrum of a six-component mixture. The spectra of six components (solid lines) and two other compounds (dotted lines) are shown in Figure 4.8B. Kalman filtering of this mixture spectrum was first attempted with the application of a model made with spectra of only six components (Figure 4.9A). The second trial used a model made from the spectra of six components and two other compounds not present in the mixture (Figure 4.9B). The results of filtering for both examples are shown in Table 4.9. The Kalman filtering correctly differentiates components present in a mixture from other components included in the model but not present in the mixture.

Solving the given concentration problem is only a special case of Kalman filtering. The description and principles of general linear discrete Kalman filtering follow; the method and symbols are taken from the review article by Brown.[38]

Operation in the time domain is the most common use for Kalman filtering. This is convenient for data produced by analytical instrumentation considered as a specific time series. The time domain for the discrete filter is represented by the discrete time series $t_0 < t_1 < t_2 < \ldots < t_k$.

Time domain approaches have a special terminology — the problem being studied is called a "system". Mathematical descriptions of this system in a time-domain state-space offer a number of mathematical and notational advantages over the frequency-domain approach.

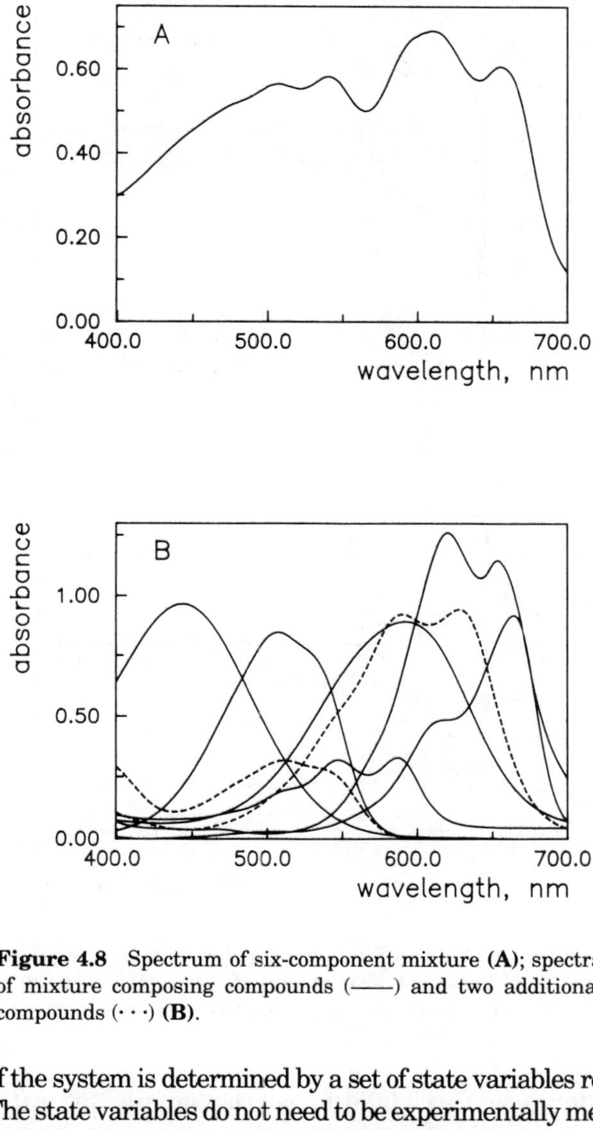

Figure 4.8 Spectrum of six-component mixture **(A)**; spectra of mixture composing compounds (——) and two additional compounds (· · ·) **(B)**.

A state of the system is determined by a set of state variables represented by a vector $\vec{X}$. The state variables do not need to be experimentally measurable, but they must describe accurately the temporal evolution of the process being filtered. The choice of the state variables can be arbitrary according to linear transformation. Any set of variables that completely describes the dynamics of the system is convenient. The number of variables determines the order of the system.

The estimation of a variable value x at time t_k with the knowledge of its value in the first j intervals is labeled as $x(k/j)$. During filtering the variable value is estimated based on the entire previous history of the system, as well as the result of a measurement at time t_k. This corresponds to the notation $x(k/k)$.

If the value of the variable x at the point k is estimated according to the known values of $k - 1$ points, it is called a one-step-ahead prediction, which has the label $x(k/k - 1)$.

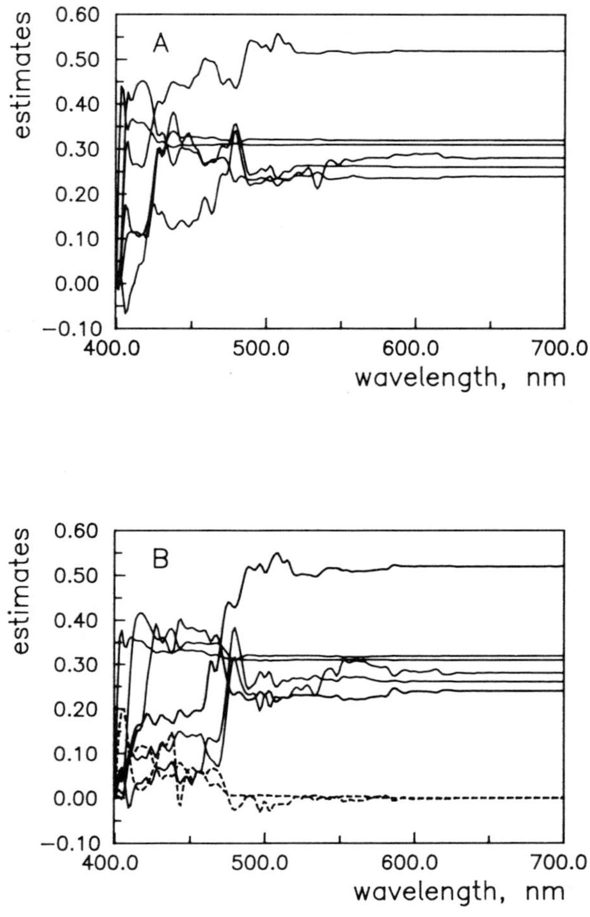

Figure 4.9 Kalman filtering estimates for a six-component mixture of Figure 4.8: only spectra of six compounds used in filtering **(A)**; spectra of six compounds and two other compounds used in filtering **(B)**.

Linear filters use a weighted linear combination of past and present data to filter the result of a current measurement. If these weights are chosen to minimize a cost function given by the average squared difference between the estimates and true values, these filters are called minimum variance estimators. Kalman filtering is one such filter. Recursive calculation of the representative weights is another characteristic of Kalman filtering.

Application of Kalman filtering is based on a mathematical representation of the system known as a system model. Discrete Kalman filtering requires a linear, discrete system model. Continuous, nonlinear systems can be transformed by digitization (see Chapter 1, Scanning and Acquisition of Spectral Data) or by linearization (extended Kalman filtering) to discrete linear systems.

Table 4.9 Kalman Filtering of the Spectrum in Figure 4.8A

	Composition[a]		
Compound	**True**	**A[b]**	**B[c]**
1	0.280	0.280	0.280
2	0.320	0.320	0.320
3	0.310	0.310	0.310
4	0.260	0.260	0.260
5	0.240	0.240	0.240
6	0.520	0.520	0.520
7	—	—	−0.001
8	—	—	0
Standard deviation		0.0004	0.0004

[a] Compositions in arbitrary units.
[b] Only spectra of six compounds used in filtering.
[c] Spectra of six compounds and two other compounds used in filtering.

The system studied is defined by two basic equations. The first determines the dynamics of the system in terms of the state variables:

$$\vec{X}(k) = \bar{\bar{F}}(k, k-1)\,\vec{X}(k-1) + \bar{\bar{G}}(k-1)\,\vec{w}(k-1) \tag{4.74}$$

where $\vec{X}(i)$ is the value of a state vector at time t_i and $\bar{\bar{F}}$ is a matrix called a system dynamics matrix, which determines the time dependence of the system in the absence of any external noise input. The matrix $\bar{\bar{F}}$ is generally the function of times t_k, t_{k+1}. The vector $\vec{w}$ contains random noise, sometimes called modeling noise. The influence of modeling noise on the temporal evolution of a state vector is determined by the matrix $\bar{\bar{G}}$. Models with nonzero elements $\bar{\bar{G}}(k-1)\ \vec{w}(k-1)$ are labeled stochastic. Deterministic models, in which this term is zero, are more often encountered in analytical chemistry.

The vector $\vec{Z}(k)$ represents measured values at times t_k. The second defining equation of Kalman filtering is the measurement model that defines the linear relationship between measured quantities and state variables:

$$\vec{Z}(k) = \bar{\bar{H}}(k)\,\vec{X}(k) + \vec{v}(k) \tag{4.75}$$

where the general time dependent matrix $\bar{\bar{H}}$ is known as an observation matrix and the vector $\vec{v}$ describes the measurement noise present in $\vec{Z}$.

For Kalman filtering to work optimally, it is necessary that the noise sequences $\vec{w}(k)$ and $\vec{v}(k)$ be independent white-noise sequences, each with zero mean. Covariances of modeling and measurement noise are represented by the matrices $\bar{\bar{Q}}$ and $\bar{\bar{R}}$, defined by the relationships

$$E\left[\vec{w}(k)\vec{w}^{T}(j)\right] = \vec{Q}\,\delta_{jk} \tag{4.76A}$$

$$E\left[\vec{v}(k)\vec{v}^{T}(j)\right] = \vec{R}\,\delta_{jk} \tag{4.76B}$$

$$E\left[\vec{w}(k)\vec{v}^{T}(j)\right] = 0 \tag{4.77}$$

where $E[\ ]$ is an expectation value and δ is Kronecker's δ function.

Estimated quantities obtained from the filter that fulfill these conditions are statistically unbiased. If noises $\vec{w}$ and $\vec{v}$ have Gaussian distribution, minimum variance estimates obtained by Kalman filtering are the best possible estimates.

Measurement errors of the types commonly experienced in analytical chemistry often come close to satisfying these noise requirements. If the conditions in Equations 4.76A, 4.76B, and 4.77 are satisfied, time, as an independent variable, can be replaced by any independent variable, for example, wavelength of frequency for spectroscopic applications.

The algorithm used in Kalman filtering is based on recursive calculation of the estimate $\hat{\vec{X}}(k/k)$ of the state vector and the error connected with this estimate. The error is determined by the covariance matrix:

$$\begin{aligned}\bar{\bar{P}}(k) &= E\left[\left(\vec{X}(k) - \hat{\vec{X}}(k)\right)\left(\vec{X}(k) - \hat{\vec{X}}(k)^{T}\right)\right] \\ &= E\left[\vec{e}(k)\vec{e}^{T}(k)\right]\end{aligned} \tag{4.78}$$

where $\vec{e}(k)$ is the estimated error of point k.

The minimization of diagonal elements of the matrix $\bar{\bar{P}}$, which corresponds to minimization of the mean squared estimation error, is the aim of filtering. From this minimization condition, the Kalman filter results; the Kalman filter is a recursive algorithm consisting of the following steps:

- State estimate extrapolation:

$$\hat{\vec{X}}(k/k-1) = \bar{\bar{F}}(k, k-1)\,\hat{\vec{X}}(k-1/k-1) \tag{4.79}$$

- Error estimate extrapolation:

$$\bar{\bar{P}}(k/k-1) = \bar{\bar{F}}(k, k-1)\,\bar{\bar{P}}(k-1/k-1)\,\bar{\bar{F}}^{T}(k, k-1) \tag{4.80}$$

- Kalman gain matrix calculation (i.e., calculation of optimal filter weights, corresponding to the minimum variance fit of the model to data):

$$\bar{\bar{K}}(k) = \bar{\bar{P}}(k/k-1)\bar{\bar{H}}^{T}(k)\left[\bar{\bar{H}}(k)\,\bar{\bar{P}}(k/k-1)\bar{\bar{H}}^{T}(k) + \bar{\bar{R}}(k)\right]^{-1} \tag{4.81}$$

- State estimate update:

$$\hat{\vec{X}}(k/k) = \hat{\vec{X}}(k/k-1) + \bar{\bar{K}}(k)\left[\vec{Z}(k) - \bar{\bar{H}}(k)\,\hat{\vec{X}}(k/k-1)\right] \tag{4.82}$$

- Error covariance update:

$$\bar{\bar{P}}(k/k) = \left[\bar{\bar{I}} - \bar{\bar{K}}(k)\bar{\bar{H}}(k)\right] \bar{\bar{P}}(k/k-1) \tag{4.83}$$

where $\bar{\bar{I}}$ is a unit matrix.

Starting the calculation, it is necessary to define initial conditions, i.e., the initial estimate of a state vector $\vec{X}_0$:

$$E\left[\vec{X}(0)\right] = \vec{X}_0 \tag{4.84A}$$

and the initial estimate $\bar{\bar{P}}_0$ of an error covariance matrix:

$$E\left[\left(\vec{X}(0) - \vec{X}_0\right)\left(\vec{X}(0) - \vec{X}_0\right)^T\right] = \bar{\bar{P}}_0 \tag{4.84B}$$

These values are used to calculate the first estimates:

$$\hat{\vec{X}}(1/0) = F(1,0)\vec{X}_0 \tag{4.85}$$

$$\bar{\bar{P}}(1/0) = F(1,0)\bar{\bar{P}}_0\, F^T(1,0) \tag{4.86}$$

The solution of the concentration problem represented by Equations 4.66 through 4.68 is a special example of Kalman filtering when system dynamics are defined by the relationship

$$\vec{X}(k) = \vec{X}(k-1)$$

and a measurement model matrix $\bar{\bar{H}}$ consists of absorption coefficients. For the initial conditions $\vec{X}_0 = 0$ and $\bar{\bar{P}}_0 = \infty$, this solution is equivalent to the solution of a predetermined set of linear equations by the method of weighted least squares.

General formulation of Kalman filtering also solves dynamic problems when the concentration changes of individual compounds are time-dependent.

The most important limitations of the Kalman filtering method are based on computer round-off problems, modeling errors, and system observability problems. The first kind of error emerges during recursive calculation of the $\bar{\bar{P}}$ matrix in cases where measurements are relatively noise-free. Some authors recommend the application of the modified equation for the error covariance update in this case:

$$\bar{\bar{P}}(k/k) = \left[\bar{\bar{I}} - \bar{\bar{K}}(k)\bar{\bar{H}}(k)\right] \bar{\bar{P}}(k/k-1) \left[\bar{\bar{I}} - \bar{\bar{K}}(k)\bar{\bar{H}}(k)\right]^T + \bar{\bar{K}}(k)\bar{\bar{R}}(k)\bar{\bar{K}}^T(k) \tag{4.87}$$

and some other generally more pretentious calculation methods (e.g., propagation of the square root of the covariance matrix).

The second type of limitation is based on the application of unsuitable or insufficient general models. Other than the trivial solution where a more correct model can be sought, the adaptive Kalman filter, used to modify the model during calculation, can be applied as well.

The last source of errors in Kalman filtering of applications in analytical chemistry occurs when some state variables, or their linear combinations, cannot be determined from measurements made of the system.

Many applications of Kalman filtering in analytical chemistry involve problems of noise removal, peak resolution, detection of and compensation for instrumental drift, and model identification and improvement. Various reviews discuss this topic.[38-40]

REFERENCES

1. Sharaf, M. A., Illman, D. L., and Kowalski, B. R., *Chemometrics,* John Wiley & Sons, New York, 1986.
2. Agterdenbos, J., *Anal. Chim. Acta,* 108, 315, 1979.
3. Doerffell, K. and Eckschlager, K., *Optimale Strategien in der Analytik,* VEB Deutcher Verlag für Grundstoffindustrie, Leipzig, 1988.
4. Osten, D. W. and Kowalski, B. R., Multicomponent calibration and quantitation methods, in *Computarized Quantitative Infrared Analysis,* McClure, G. L., Ed., ASTM Spec. Techn. Publ., Philadelphia, 1987, p. 6.
5. Blackburn, J. A., *Anal. Chem.,* 37, 1000, 1965.
6. Antoon, M. K., Koenig, J. H., and Koenig, L. J., *Appl. Spectrosc.,* 31, 518, 1977.
7. Haaland, D. M. and Easterling, R. G., *Appl. Spectrosc.,* 34, 539, 1980.
8. Haaland, D. M. and Easterling, R. G., *Appl. Spectrosc.,* 36, 665, 1982.
9. Haaland, D. M. and Easterling, R. G., *Appl. Spectrosc.,* 39, 73, 1985.
10. McClure, G. L., Roush, P. B., Williams, J. F., and Lehman, C. A., Application of computerized quantitative infrared spectoroscopy of the principal lipids found in blood serum, in *Computarized Quantitative Infrared Analysis,* McClure, G. L., Ed., ASTM Spec. Techn. Publ., Philadelphia, 1987, p. 1316.
11. Haaland, D. M., Methods to include Beers Law nonlinearities in quantitative spectral analysis, in *Computarized Quantitative Infrared Analysis,* McClure, G. L., Ed., ASTM Spec. Techn. Publ., Philadelphia, 1987, p. 78.
12. Haaland, D. M. and Barbour, R. L., *Am. Lab.,* 17, 14, 1985.
13. Brown, C. W., Lynch, P. F., Obremski, R. J., and Lavery, D. S., *Anal. Chem.,* 54, 1472, 1982.
14. Maris, M. A., Brown, C. W., and Lavery, D. S., *Anal. Chem.,* 55, 1694, 1983.
15. Kisner, H. J., Brown, C. W., and Kavarnos, G. J., *Anal. Chem.,* 55, 1703, 1983.
16. Honigs, D. E., Freelin, J. M., Hieftje, G. M., and Hirschfeld, T. B., *Appl. Spectrosc.,* 37, 491, 1983.
17. Honigs, D. E., Hieftje, G. M., and Hirschfeld, T. B., *Appl. Spectrosc.,* 38, 844, 1984.

18. Geladi, P., MacDougall, D., and Martens, H., *Appl. Spectrosc.,* 39, 491, 1985.

19. Saxberg, B. E. and Kowalski, B. R., *Anal. Chem.,* 51, 1031, 1979.

20. Beebe, K. R. and Kowalski, B. R., *Anal. Chem.,* 59, 1007A, 1987.

21. Brown, C. W., Obremski, R. J., and Anderson, P., *Appl. Spectrosc.,* 40, 734, 1986.

22. Cahn, F. and Compton, S., *Appl. Spectrosc.,* 42, 865, 1988.

23. Martens, H. and Naes, T., *Multivariate Calibration,* John Wiley & Sons, Chichester, 1989.

24. Haaland, D. M. and Thomas, E. V., *Anal. Chem.,* 60, 1202, 1988.

25. Cowe, I. A. and McNicol, J. W., *Appl. Spectrosc.,* 39, 257, 1985.

26. Fredericks, P. M., Lee, J. B., Osborn, P. R., and Swinkels, D. A. J., *Appl. Spectrosc.,* 39, 303, 1985.

27. Fredericks, P. M., Lee, J. B., Osborn, P. R., Swinkels, D. A. J., *Appl. Spectrosc.,* 39, 311, 1985.

28. Wold, S., Martens, H., and Wold, H., The multivariate calibration problems in chemistry solved by the PLS method, in *Proceedings of the Conference on Matrix Pencils,* Ruhe, A. and Kagström, P., Eds., Springer-Verlag, Heidelberg, 1983.

29. Haaland, D. M. and Thomas, E. V., *Anal. Chem.,* 60, 1193, 1988.

30. Geladi, P. and Kowalski, B. R., *Anal. Chim. Acta,* 185, 1, 1986.

31. Manne, R., *Chemom. Intell. Lab. Syst.,* 2, 187, 1987.

32. Lindberg, W., Persson, J.-A., and Wold, S., *Anal. Chem.,* 55, 643, 1983.

33. Lorber, A., Wangen, L. E., and Kowalski, B. R., *J. Chemometric,* 1, 19, 1987.

34. Haaland, D. M., *Anal. Chem.,* 60, 1208, 1988.

35. Beebe, K. R. and Kowalski, B. R., *Anal. Chem.,* 59, 1007A, 1987.

36. Poulisse, H. N. J., *Anal. Chim. Acta,* 112, 361, 1979.

37. Didden, C. B. M. and Poulisse, H. N. J., *Anal. Lett.,* 13(A11), 921, 1980.

38. Brown, S. D., *J. Chemometrics,* 5, 147, 1991.

39. Brown, S. D., *Anal. Chim. Acta,* 181, 1, 1986.

40. Rutan, J., *J. Chemometrics,* 1, 1, 1987.

Chapter 5

Reflectance Spectroscopy

Reflectance spectroscopy measures the dependence of electromagnetic radiation flux, reflected by the phase interface, on the angle of incident radiation and wavelength. A reflection spectrum forms a record of the dependence of the R ratio of the reflected and incident radiation flux on the wavelength or the radiation frequency for certain geometric configurations. The absorption spectra, refractive index dispersion, absorbing compound concentrations of mixtures, dye characteristics, thickness of thin films, and so on can be obtained from reflection spectra using numerical methods in reflectance spectroscopy.

According to the quality of an interface that reflects electromagnetic radiation, reflection spectroscopy can be classified as two basic types. The known specular reflection is observed, i.e., the radiation is reflected exactly in the defined direction determined by the reflection rule at the ideal smooth (glossy) interface. On the other hand, the radiation is reflected uniformly in all directions of the hemisphere on an ideal matte surface. The first is called specular reflectance spectroscopy, the second is diffuse reflectance spectroscopy.

The different mechanisms producing specular and diffuse spectra lead to differing ways of numerical evaluation and affect the range and accuracy of information obtained about systems studied.

Numerical methods of acquiring reflection spectra based on the measurements of the reflection ratio R are described in this chapter. Methods of ellipsometry and interferometry, although widely used, are not discussed among the following methods due to space limitations. On the contrary, some physical principles are introduced to point out specific differences between reflection spectra and common molecular absorption spectra. Special attention is paid to the physical background, which gives the mathematical details for the reflection ratio dependence on radiation polarization and on the characteristics of the compound and geometry.

PROPAGATION OF ELECTROMAGNETIC RADIATION IN A MEDIUM

The electromagnetic radiation flux via nonabsorbing dielectric is characterized by equations that describe the dependence of the intensity vector of the electric field $\vec{E}$ and magnetic intensity $\vec{H}$ on the position vector $\vec{r}$ and the time t:[1,2]

$$\vec{E} = \vec{A} \cdot \exp\left[i\left(\omega t - \vec{k} \cdot \vec{r}\right)\right] = \vec{A} \cdot \exp\left[i\omega\left(\frac{t - \vec{\ell} \cdot \vec{r}}{v}\right)\right] \quad \textbf{(5.1)}$$

$$\vec{H} = \vec{\ell} \times \vec{A}' \exp\left[i\left(\omega t - \vec{k} \cdot \vec{r}\right)\right] = \vec{\ell} \times \vec{A}' \exp\left[i\omega\left(\frac{t - \vec{\ell} \cdot \vec{r}}{v}\right)\right] \quad \textbf{(5.2)}$$

where i is the imaginary unit, $\vec{\ell}$ is the unit vector in the direction of the propagation of radiation, $\vec{k}$ is the wave vector, $\vec{A}$ and $\vec{A}'$ are constant and collinear vectors, perpendicular to the vector $\vec{\ell}$, ω is the y angular frequency connected with the frequency ν, the wavelength λ and the speed of the radiation propagation v, through the relationships

$$\omega = 2\pi\nu \quad \textbf{(5.3)}$$

$$v = \lambda\nu \quad \textbf{(5.4)}$$

The wave vector is collinear with the vector $\vec{\ell}$, determining the direction of the radiation propagation. Its dimension is described by the wavelength λ:

$$k = \frac{2\pi}{\lambda} = \frac{2\pi\nu}{v} = \frac{\omega}{v} \quad \textbf{(5.5)}$$

$$\vec{k} = \frac{\omega}{v}\vec{\ell} \quad \textbf{(5.6)}$$

The ratio of the speed of electromagnetic radiation propagation in a vacuum c ($c = 2.99792458 \times 10^8$ m/s) and in a certain material v is labeled as the (absolute) refractive index of the given environment:

$$n = \frac{c}{v} \quad \textbf{(5.7)}$$

A material with a greater index of refraction is considered to be the optically denser medium, if two of them are compared. Simultaneously, the refraction index (Equation 5.7) explains the ratio of the wavelength λ_0, which corresponds to the electromagnetic radiation of a given frequency in a vacuum, to the wavelength in a certain medium. The relationship between the reflection index and the dielectric constant ε is

$$n = \varepsilon^2 \quad \textbf{(5.8)}$$

for frequency areas of molecular spectroscopy in which the orientation polarization can be neglected.

The ratio of vectors $\vec{A}$ and $\vec{A}'$ magnitudes is determined by the magnetic permeability μ and the dielectric constant ε of the environment:

$$A' = \sqrt{\frac{\varepsilon}{\mu}}\, A \tag{5.9}$$

The radiant flux of the electromagnetic radiation (intensity) is proportional to the square of the magnitude of vector $\vec{A}$.

Generally vector $\vec{A}$ is complex and vector $\vec{E}$ describes an ellipse in a plane perpendicular to the radiation direction $\vec{\ell}$. The movement forming the vector $\vec{E}$ projection to that plane is called polarization. Movement along this line is known as linear polarization, and movement on a path around the ellipse is called circular or elliptic polarization. In the right-handed coordinate system with unit vectors $\vec{\imath}$, $\vec{\jmath}$, and $\vec{\ell}$ (with vectors $\vec{\imath}$ and $\vec{\jmath}$ in the principal axes of the ellipse described by a vector $\vec{E}$), the polarization is unambiguously determined by a vector $\vec{p}$, for which[2]

$$\vec{p} \cdot \vec{p}^{\,*} = 1 \tag{5.10}$$

$$\vec{A} = \left|\vec{A}\right| \vec{p} \tag{5.11}$$

$$\vec{p}(\delta) = \vec{\imath}\cos\left(\frac{\delta}{2}\right) + i\,\vec{\jmath}\sin\left(\frac{\delta}{2}\right) \tag{5.12}$$

Linear polarization in direction $\vec{\imath}$, respective to $\vec{\jmath}$, is obtained by setting δ equal to 0, and the right-handed, and left-handed polarization by setting δ equal to π/2 and 3π/2, respectively (Figure 5.1).

Each polarization can be considered a superposition of two elliptically polarized waves, with axes arbitrarily chosen to be congruent with the ellipse's principal axes, determined by the vector $\vec{A}$:

$$\vec{A} = \left(\vec{A} \cdot \vec{p}^{\,*}\right)\vec{p} + \left(\vec{A} \cdot \vec{m}^{\,*}\right)\vec{m} \tag{5.13}$$

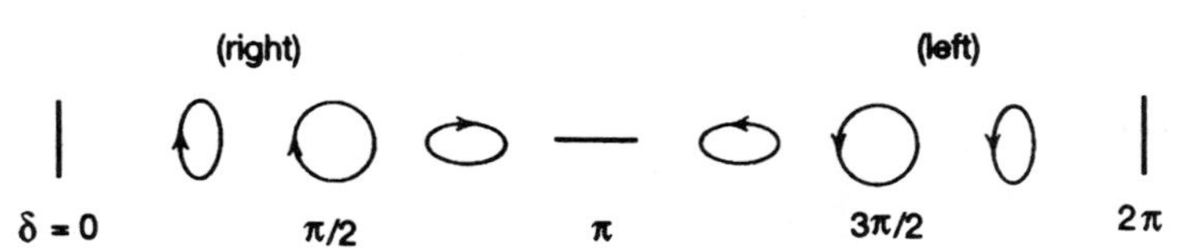

Figure 5.1 The dependence of the type of electromagnetic radiation polarization on the angle δ: δ = 0, π, 2π — linearly polarized radiation; δ = π/2, 3π/2 — circularly polarized radiation; other cases of elliptic polarization.

where, for a vector $\vec{m}$, determining the second elliptically polarized component the following is valid:

$$\vec{m}(\delta) = \vec{p}(\delta + \pi) = i\,\vec{i}\,\sin\left(\frac{\delta}{2}\right) - \vec{j}\,\cos\left(\frac{\delta}{2}\right) \tag{5.14}$$

and where vectors $\vec{p}$ and $\vec{m}$ are mutually perpendicular:

$$\vec{p}(\delta)\cdot\vec{m}^*(\delta) = \vec{p}^*(\delta)\cdot\vec{m}(\delta) = 0 \tag{5.15}$$

For example, if $\vec{A}$ represents the linear polarization in a plane forming an angle θ with a plane ($\vec{i}$, $\vec{\ell}$) and δ = 0, vectors $\vec{p}$ and $\vec{m}$ correspond to the linear polarization along $\vec{i}$ respective to $\vec{j}$, and components of vector $\vec{A}$ have dimensions of $A\cos\theta$ and $-A\sin\theta$ in the direction of these axes, respectively.

SPECULAR REFLECTANCE AT NONABSORBING MATERIAL INTERFACE

The reflection of electromagnetic radiation from a plane interface between two environments characterized by the refraction indexes n_1 and n_2 is shown in Figure 5.2. The reflecting interface is identical to a plane yz that uses the Cartesian coordinate system, and the incident radiation beam is represented by a vector in a plane xz so that an axis z creates a line normal to the incidence of radiation at the interface. Following incidence, some of the radiation is reflected and some is refracted, continuing to another medium.

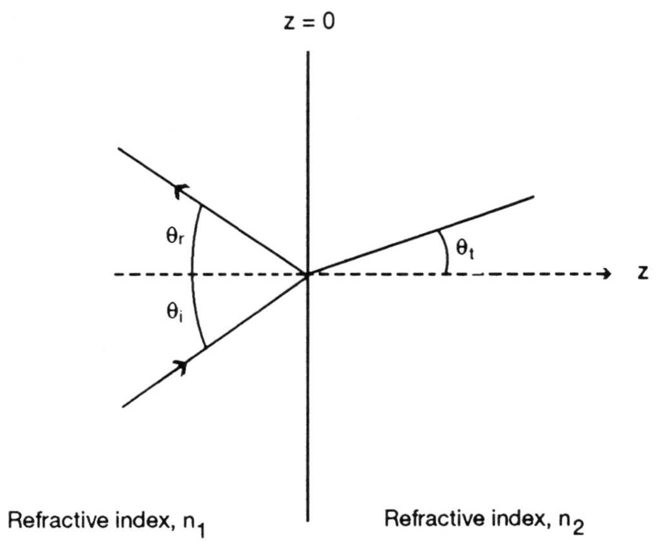

Figure 5.2 The scheme of the reflection and refraction of the electromagnetic radiation on the plane phase interface.

Indices i, r, and t denote incident, reflected, and transmitted radiation. Acute angles formed by single rays with a line normal to the boundary interface are labeled the incidence angle θ_i, the reflection angle θ_r, and the refraction angle θ_t. Relative intensities of reflected and refracted beams are dependent on the refraction indices n_1 and n_2, respectively, on the incidence angle θ_i, and the degree of polarization of the incident radiation. Using the procedure introduced in the previous section, each beam can be divided into two components. A component with polarization parallel to an incident plane determined by the beam direction and a line normal to an interface in the place of incidence is labeled with the index p (in literature the designation $\|$ is used also) and a component perpendicular to an incidence plane with the index s (known sometimes as $\perp$):

$$\vec{A}_i = A_{ip}\left(\vec{i} \cdot \cos\theta_i - \vec{k} \cdot \sin\theta_i\right) + A_{is}\vec{j} \tag{5.16}$$

$$\vec{A}_r = A_{rp}\left(\vec{i} \cdot \cos\theta_r + \vec{k} \cdot \sin\theta_r\right) + A_{rs}\vec{j} \tag{5.17}$$

$$\vec{A}_t = A_{tp}\left(\vec{i} \cdot \cos\theta_t - \vec{k} \cdot \sin\theta_t\right) + A_{ts}\vec{j} \tag{5.18}$$

where $\vec{i}$, $\vec{j}$, and $\vec{k}$ are unit vectors in the direction of the coordinate axes x, y, and z (Figure 5.2).

The continuity conditions for tangential vector components of electric and magnetic intensities and normal vector components of electric and magnetic inductions must be fulfilled at the boundary of the place of incidence of the radiation. Under these conditions, radiation does not change its frequency at reflection and refraction. Also, the equality of the incidence and reflection angles are described by the reflection law and by Snell's rule of refraction:

$$n_1 \sin\theta_i = n_2 \sin\theta_t \tag{5.19}$$

The relationships between the amplitudes of perpendicular and parallel components of the incident and reflected (respective to refracted) rays, described by the Fresnel equations:[1-3]

$$A_{rp} = -A_{ip}\frac{\tan(\theta_i - \theta_t)}{\tan(\theta_i + \theta_t)} \tag{5.20}$$

$$A_{rs} = -A_{is}\frac{\sin(\theta_i - \theta_t)}{\sin(\theta_i + \theta_t)} \tag{5.21}$$

$$A_{tp} = A_{ip}\frac{2\sin\theta_t \cos\theta_i}{\sin(\theta_i + \theta_t) \cdot \cos(\theta_i - \theta_t)} \tag{5.22}$$

$$A_{ts} = A_{is} \frac{2 \sin\theta_t \cos\theta_i}{\sin(\theta_i + \theta_t)} \quad \textbf{(5.23)}$$

are the last consequence.

The radiation flux is proportional to the square of the amplitude of an electrical intensity vector. From Equations 5.20 and 5.21, the coefficients of reflectance R_s and R_p, defined by the ratio of a radiation flux of reflected and incident radiation to the given polarization, give

$$R_p = \left(\frac{A_{rp}}{A_{ip}}\right)^2 = \left\{ \frac{\left(\frac{n_2}{n_1}\right)^2 \cos\theta_i - \left[\left(\frac{n_2}{n_1}\right)^2 - \sin^2\theta_i\right]^{\frac{1}{2}}}{\left(\frac{n_2}{n_1}\right)^2 \cos\theta_i + \left[\left(\frac{n_2}{n_1}\right)^2 - \sin^2\theta_i\right]^{\frac{1}{2}}} \right\}^2 \quad \textbf{(5.24)}$$

$$R_s = \left(\frac{A_{rs}}{A_{is}}\right)^2 = \left\{ \frac{\cos\theta_i - \left[\left(\frac{n_2}{n_1}\right)^2 - \sin^2\theta_i\right]^{\frac{1}{2}}}{\cos\theta_i + \left[\left(\frac{n_2}{n_1}\right)^2 - \sin^2\theta_i\right]^{\frac{1}{2}}} \right\}^2 \quad \textbf{(5.25)}$$

or for unpolarized radiation:

$$R = \frac{1}{2}\left(R_p + R_s\right) \quad \textbf{(5.26)}$$

Figure 5.3 shows the course of dependencies R, R_s, and R_p for the ratio $n_2/n_1 = 1.5$, approximately corresponding to the air–glass interface. The significant point of the R_p curve is a local minimum, $R_p = 0$, obtained by satisfying the condition

$$\theta_i + \theta_t = \frac{\pi}{2} \quad \textbf{(5.27)}$$

when the denominator of the right side of Equation 5.20 has as its limit infinity, $+\infty$. The incident angle, for which the condition in Equation 5.27 is satisfied, is called the polarization or Brewster angle, and the following equation results from Snell's rule:

$$\theta_i = \theta_B = \arctan\left(\frac{n_2}{n_1}\right) \quad \textbf{(5.28)}$$

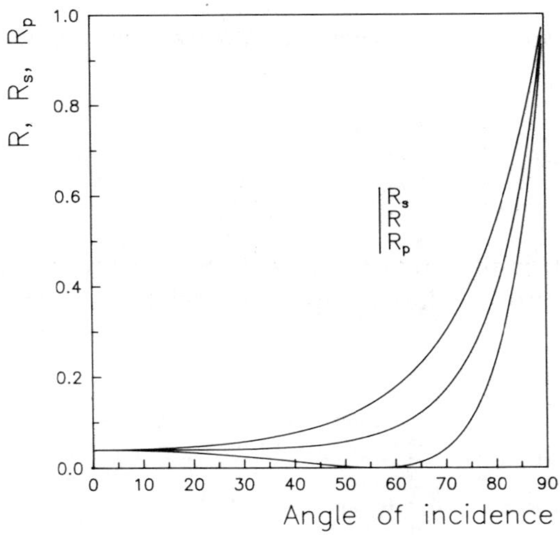

Figure 5.3 The dependence of the reflectance coefficients (R_p, R_s, R) on the incident angle of the electromagnetic radiation at the material interface with relative refractive index n_2/n_1 = 1.5.

The numerical evaluation of spectral measurement results is based on determining the value n_2/n_1, which is the only unknown variable if the incident angle θ_i known and the only one experimentally determined of the values R_s, R_p, and R. The most practical case is the reflection at the interface between the air and the compound studied. Then, $n_1 \cong 1$ and the value of the refraction index of the compound studied n_2 is a direct result of numerical evaluation.

If any R values for several values of the incidence angle θ_i are measured, the best nonaligned estimate of n_2 can be calculated by minimizing the sum of the squares of the deviations (see Appendix C):

$$U(n_2) = \sum_{\theta_i} \left[R(\theta_i) - R^{\exp}(\theta_i) \right]^2 = \min \qquad \textbf{(5.29)}$$

where $R^{\exp}(\theta_i)$ is the experimental value and $R(\theta_i)$ is the value calculated according to equations 5.24 to 5.26. In the case where R_s and R_p are measured simultaneously, the sum of the squares of the deviations is similarly minimized:

$$U(n_2) = \sum_{\theta_i} \left\{ \left[R_p(\theta_i) - R_p^{\exp}(\theta_i) \right]^2 + \left[R_s(\theta_i) - R_s^{\exp}(\theta_i) \right]^2 \right\} = \min \qquad \textbf{(5.30)}$$

For the reflection at the interface with an optically thinner medium ($n_2 < n_1$), total reflection occurs for incidence angles greater than the critical value:

$$\theta_c = \arcsin\left(\frac{n_2}{n_1} \right) \qquad \textbf{(5.31)}$$

The imaginary value for the cosine of a refraction angle:

$$\cos\theta_t = \left[1 - \left(\frac{n_1}{n_2}\right)^2 \sin\theta_i^2\right]^{\frac{1}{2}} = i\left[\left(\frac{n_1}{n_2}\right)^2 \sin^2\theta_i - 1\right]^{\frac{1}{2}} \quad \textbf{(5.32)}$$

for angles $\theta_i > \theta_c$ results from Snell's rule. Substituting into Equations 5.20 and 5.21, the complex relationships are obtained for total reflection:

$$A_{rs} = A_{is} \frac{\left(\frac{1}{n}\right)\cos\theta_i - i\left[\left(\frac{1}{n^2}\right)\sin^2\theta_i - 1\right]^{\frac{1}{2}}}{\left(\frac{1}{n}\right)\cos\theta_i + i\left[\left(\frac{1}{n^2}\right)\sin^2\theta_i - 1\right]^{\frac{1}{2}}} \quad \textbf{(5.33)}$$

$$A_{rp} = A_{ip} \frac{\cos\theta_i - \left(\frac{i}{n}\right)\left[\left(\frac{1}{n^2}\right)\sin^2\theta_i - 1\right]^{\frac{1}{2}}}{\cos\theta_i + \left(\frac{i}{n}\right)\left[\left(\frac{1}{n^2}\right)\sin^2\theta_i - 1\right]^{\frac{1}{2}}} \quad \textbf{(5.34)}$$

where n is the relative refraction index:

$$n = \frac{n_2}{n_1} \quad \textbf{(5.35)}$$

Numerators and denominators of fractions in equations 5.33 and 5.34 can be mutually complex conjugated, which leads to unit values of reflectances:

$$R_s = \left(\frac{A_{rs}}{A_{is}}\right)\left(\frac{A_{rs}}{A_{is}}\right)^* = 1 \quad \textbf{(5.36)}$$

$$R_p = \left(\frac{A_{rp}}{A_{ip}}\right)\left(\frac{A_{rp}}{A_{ip}}\right)^* = 1 \quad \textbf{(5.37)}$$

If Equations 5.33 and 5.34 are rewritten in the form

$$A_{rs} = A_{is}\exp\left(-i\delta_s\right) \quad \textbf{(5.38)}$$

$$A_{rp} = A_{rp}\exp\left(-i\delta_p\right) \quad \textbf{(5.39)}$$

the phase delays δ_s and δ_p of both polarized components:

$$tg\left(\frac{\delta_s}{2}\right) = \frac{\left[\left(\frac{1}{n^2}\right)\sin^2\theta_i - 1\right]^{\frac{1}{2}}}{\left(\frac{1}{n}\right)\cos\theta_i} \quad \textbf{(5.40)}$$

$$tg\left(\frac{\delta_p}{2}\right) = \frac{\left(\frac{1}{n}\right)\left[\left(\frac{1}{n^2}\right)\sin^2\theta_i - 1\right]^{\frac{1}{2}}}{\cos\theta_i} \quad \textbf{(5.41)}$$

result.

From the discussion above, it is apparent that radiation with elliptic polarization, given by the phase difference $(\delta_p - \delta_s)$ between the shifts of the perpendicular and parallel components, results from the complete reflection of linear polarized radiation, so that

$$tg\left(\frac{\delta_p - \delta_s}{2}\right) = \frac{\cos\theta_i\left[\left(\frac{1}{n^2}\right)\sin^2\theta_i - 1\right]^{\frac{1}{2}}}{\left(\frac{1}{n}\right)\sin^2\theta_i} \quad \textbf{(5.42)}$$

If the phase shift introduced (ellipticity) is determined experimentally for various values of θ_i, a predetermined system of equations with only one unknown variable, n, is obtained from Equation 5.42.

The relationships in Equations 5.24 and 5.25 are significantly simplified for the special case of perpendicular incidence; when $\theta_i = 0$, the following is obtained:

$$R = R_p = R_s = \left(\frac{n_2 - n_1}{n_2 + n_1}\right)^2 = \left(\frac{n-1}{n+1}\right)^2 \quad \textbf{(5.43)}$$

SPECULAR REFLECTANCE AT ABSORBING MATERIAL INTERFACE

If an axis x is set identical to the direction of the electromagnetic radiation propagation, the decrease of radiant flux Φ along the geometric trajectory x in the absorbing environment is expressed by the Lambert-Beer rule in the form

$$\Phi = \Phi_0 \exp\left[-\left(\frac{4\pi\kappa}{\lambda}\right)x\right] = \Phi_0 \exp[-\alpha x] \quad \textbf{(5.44)}$$

where κ is the absorption index, ν is the wavelength, and Φ is the radiant flux at the point $x = 0$. The Napierian linear absorption coefficient α is defined by the following relationship:

$$\alpha = \frac{4\pi}{\lambda}\kappa = \frac{4\pi n}{\lambda_0}\kappa = \frac{4\pi}{\lambda_0}\kappa_0 \tag{5.45}$$

where λ_0 is the wavelength of the electromagnetic radiation in a vacuum and where κ_0 is known as the extinction coefficient. If the complex refraction index:

$$N = n(1 - i\kappa) = n - i\kappa_0 \tag{5.46}$$

is defined, a formal analogous mathematical description, as in the case of a transparent environment, is obtained for propagation of radiation in an absorbing environment:[1-5]

$$\vec{E} = \vec{A}\exp\left[i\omega\left(t - \frac{N}{c}\vec{n}\cdot\vec{r}\right)\right] \tag{5.47}$$

For such a refraction index all relationships derived for the nonabsorbing environment are valid. For simplicity, the following study will be limited to the interface between vacuum and the absorbing compound, or to the interface between air and the absorbing compound. Let the absorbing compound be characterized by the real refraction index n and by the absorption index κ. From Snell's rule the following complex values of trigonometric functions of the refraction index θ_t result:

$$\sin\theta_t = \frac{\sin\theta_i}{n(1 - i\kappa)} \tag{5.48}$$

$$\cos\theta_t = \frac{1}{n(1 - i\kappa)}\left[n^2(1 - i\kappa)^2 - \sin^2\theta_i\right]^{\frac{1}{2}} \tag{5.49}$$

Applying the substitution

$$N\cos\theta_t = \left(N^2 - \sin^2\theta_i\right)^{\frac{1}{2}} = a - ib \tag{5.50}$$

from the Fresnel equations (Equations 5.5 and 5.6), the following equations are obtained for the reflection coefficients of the perpendicular (R_s) and parallel (R_p) polarized radiation:

$$R_s = \left(\frac{A_{rs}}{A_{is}}\right)\left(\frac{A_{rs}}{A_{is}}\right)^* = \frac{a^2 + b^2 - 2a\cos\theta_i + \cos^2\theta_i}{a^2 + b^2 + 2a\cos\theta_i + \cos^2\theta_i} \tag{5.51}$$

$$R_p = \left(\frac{A_{rp}}{A_{ip}}\right)\left(\frac{A_{rp}}{A_{ip}}\right)^* = R_s \frac{a^2 + b^2 - 2a\sin\theta_i\tan\theta_i + \sin^2\theta_i\tan^2\theta_i}{a^2 + b^2 + 2a\sin\theta_i\tan\theta_i + \sin^2\theta_i\tan^2\theta_i} \quad \textbf{(5.52)}$$

or, for the nonpolarized radiation:

$$R = \frac{1}{2}\left(R_s + R_p\right) = R_s \frac{a^2 + b^2 + \sin^2\theta_i\tan^2\theta_i}{a^2 + b^2 + 2a\sin\theta_i\tan\theta_i + \sin^2\theta_i\tan^2\theta_i} \quad \textbf{(5.53)}$$

For the values a and b the following expressions are valid:

$$a^2 + b^2 = \frac{1}{2}\left[n^2\left(1-\kappa^2\right) - \sin^2\theta_i\right] + \frac{1}{2}\left\{\left[n^2\left(1-\kappa^2\right) - \sin^2\theta_i\right]^2 + 4n^4\kappa^2\right\}^{\frac{1}{2}} + \frac{2n^4\kappa^2}{\left[n^2\left(1-\kappa^2\right) - \sin^2\theta_i\right] + \left\{\left[n^2\left(1-\kappa^2\right) - \sin^2\theta_i\right]^2 + 4n^2\kappa^2\right\}^{\frac{1}{2}}} \quad \textbf{(5.54)}$$

$$a = \left(\frac{1}{2}\left[n^2\left(1-\kappa^2\right) - \sin^2\theta_i\right] + \frac{1}{2}\left\{\left[n^2\left(1-\kappa^2\right) - \sin^2\theta_i\right]^2 + 4n^4\kappa^2\right\}^{\frac{1}{2}}\right)^{\frac{1}{2}} \quad \textbf{(5.55)}$$

Therefore, the values a and b can be directly evaluated if the values θ_i, n, and κ are known.

The behavior of the functions $R_s(\theta_i)$, $R_p(\theta_i)$, and $R(\theta_i)$ for $n = 1.5$ and $\kappa = 0.2$ are shown in Figure 5.4. R_p does not intercept the origin and curves are shifted to higher values of reflectance as the value of κ increases, which is the significant consequence of absorption. The values n and κ can be, in principle, determined from two experimentally obtained values for the reflectance (e.g., from the values R_x for two incidence angles, or from the values R_s and R_p for one incidence angle). Similar to that done for nonabsorbing media, the predetermined system of equations can be solved by least squares when more than two experimental values are known. The accuracy of values of κ by this approach is not always satisfactory. As shown by Fahrenfort,[6] the accuracy of determination is not sufficient even for $\kappa = 0.2$ corresponding to a statistically significant, high value of the absorption coefficient α. It therefore follows that it is not possible to determine the optical constants of the majority of organic compounds by the procedure discussed here.

For the perpendicular incidence of radiation on the interface, Equations 5.51 to 5.55 simplify to only one equation:

$$R = R_s = R_p = \frac{(1-n)^2 + n^2\kappa^2}{(1+n)^2 + n^2\kappa^2} \quad \textbf{(5.56)}$$

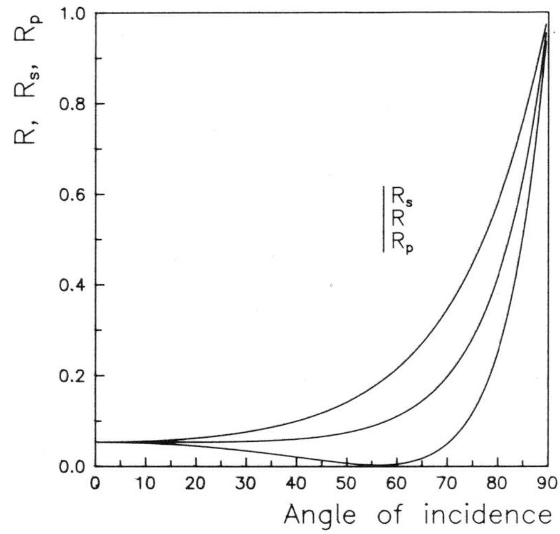

Figure 5.4 The dependence of the reflectance coefficients (R_p, R_s, R) at the incident angle of the electromagnetic radiation on the vacuum plane interface with the absorbing medium characterized by the refractivity index $n = 1.5$ and absorption index $\kappa = 0.2$.

For many strongly absorbing compounds, e.g., metals in the visible region of the spectrum, the following inequality is satisfied:

$$n^2 + \kappa^2 \gg 1 \tag{5.57}$$

The equalities (Equations 5.51 to 5.55) can be simplified to

$$R_p = \frac{n^2\left(1+\kappa^2\right)\cos^2\theta_i - 2n\cos\theta_i + 1}{n^2\left(1+\kappa^2\right)\cos^2\theta_i + 2n\cos\theta_i + 1} \tag{5.58}$$

$$R_s = \frac{n^2\left(1+\kappa^2\right) - 2n\cos\theta_i + \cos^2\theta_i}{n^2\left(1+\kappa^2\right) + 2n\cos\theta_i + \cos^2\theta_i} \tag{5.59}$$

The values n and κ can be determined from these relationships in a manner similar to what was described previously, by application of least squares.

SPECTROSCOPY INVOLVING ATTENUATION OF TOTAL REFLECTION

As was shown in the previous section, specular reflection on the interface of the nonabsorbing media with optically denser absorbing material can be utilized to determine the optical constants n and κ of the optically denser media only for relatively large values of κ.

The opaque situation arises if the reflection at the interface of the optically denser nonabsorbing medium and the optically thinner absorbing medium is investigated, i.e., when

$$n_1 > n_2 \quad \textbf{(5.60)}$$

$$\kappa_1 = 0 \quad \textbf{(5.61)}$$

$$\kappa_2 = \kappa > 0 \quad \textbf{(5.62)}$$

In this case, the absorption of radiation at the interface with the absorbing material decreases the reflectance values R_s and R_p in the region of incidence angles greater than the critical angle θ_c. In this region total reflection would appear in the nonabsorbing, optically thinner medium, and values R_s and R_p would be equal to 1. This fact gives rise to the name attenuated total reflectance (ATR) spectroscopy.

In the case introduced, the basic equations determining the values R_s and R_p are obtained from Equations 5.20 and 5.21 after the introduction of the complex refraction index:

$$N = n_2(1 - i\kappa) = n_2 - i\kappa_0 \quad \textbf{(5.63)}$$

similarly, as for the previous case (see Specular Reflectance at Absorbing Material Interface, above):

$$R_p = \left| \frac{n_1 \cos\theta_i - N \cos\theta_t}{n_1 \cos\theta_i + N \cos\theta_t} \right|^2 \quad \textbf{(5.64)}$$

$$R_s = \left| \frac{N \cos\theta_i - n_1 \cos\theta_t}{N \cos\theta_i + n_1 \cos\theta_t} \right|^2 \quad \textbf{(5.65)}$$

and the complex value $\cos\theta_t$ is determined by Snell's rule:

$$\cos\theta_t = \frac{1}{N}\left[N^2 - n_1^2 \sin^2\theta_i\right]^{\frac{1}{2}} \quad \textbf{(5.66)}$$

If the relative refraction index:

$$n = \frac{n_2}{n_1} \quad \textbf{(5.67)}$$

is introduced, the solution of Equations 5.64 and 5.65 are obtained in the form of Equations 5.50 to 5.55.

Young and Jones[7] summarized the basic principles of numerical evaluation for the optical constants n and κ from ATR spectra. Fahrenfort and Visser[8,9] proposed the explicit method based on measurement of the value R_s for two different angles of incidence θ_i. The method, in principle, consists of putting Equation 5.65 in the form of the quadratic equation

$$-\frac{1}{4} AM^2 + 4BM - 4CuM - 16Du^2 = 0 \tag{5.68}$$

in which

$$A = \frac{(1 - R_s)}{R_s} \tag{5.69}$$

$$B = (1 + R_s)^2 \cos^4 \theta_i \tag{5.70}$$

$$C = (1 - R_s)^2 \cos^2 \theta_i \tag{5.71}$$

$$D = R_s \cos^4 \theta_i \tag{5.72}$$

$$u = 1 - u^2 + u^2 \kappa^2 \tag{5.73}$$

$$M = u^2 + w^2 \tag{5.74}$$

$$w = 2\, u^2\, \kappa \tag{5.75}$$

and n is defined by the relationship given in Equation 5.67. For the values R_{s1} and R_{s2}, corresponding to the incidence angles θ_{i1} and θ_{i2}, a system of two quadratic equations will be obtained, and from it, by eliminating the u^2 equation:

$$u = -\frac{1}{16} PM + Q \tag{5.76}$$

where

$$P = \frac{A_1\, D_2 - A_2\, D_1}{C_1\, D_2 - C_2\, D_1} \tag{5.77}$$

$$Q = \frac{B_1\, D_2 - B_2\, D_1}{C_1\, D_2 - C_2\, D_1} \tag{5.78}$$

and where the indices 1 and 2 determine which one from the value R_{s1}, R_{s2}, respective to θ_{i1}, θ_{i2}, was used to calculate the functions A, B, C, and D. After substituting Equation 5.76 into Equation 5.68, the quadratic equation results for M with only one physically permitted solution:

$$M = \frac{-X - \left(Y^2 - 4XZ\right)^{\frac{1}{2}}}{2X} \tag{5.79}$$

where

$$X = \frac{1}{4}\left(A_1 - C_1 P + \frac{1}{4} D_1 P^2\right) \tag{5.80}$$

$$Y = 4\left(-B_1 + C_1 Q - \frac{1}{2} D_1 PQ\right) \tag{5.81}$$

$$Z = 16 D_1 Q \tag{5.82}$$

Substituting M into Equation 5.76, the value u is obtained, and by substituting into Equation 5.74 the value w is obtained as a positive root:

$$w = \left(M - u^2\right)^{\frac{1}{2}} \tag{5.83}$$

Finally the values n and κ are calculated from Equations 5.73 and 5.75:

$$\kappa = \frac{-(1-u) + \left[(1-u)^2 + w^2\right]^{\frac{1}{2}}}{w} \tag{5.84}$$

$$n = \left(\frac{w}{2\kappa}\right)^{\frac{1}{2}} \tag{5.85}$$

Sometimes the numerical calculation of the root M of the quadratic equation (Equation 5.68) cannot be sufficiently accurate. Hansen[10] showed that the calculation will be more accurate if the common formula (Equation 5.79) used to solve for the root of the quadratic equation is replaced by the formally equivalent relationship:

$$M = \frac{-2Z}{Y - \left(Y^2 - 4XZ\right)^{\frac{1}{2}}} \tag{5.86}$$

in which the influence of the numerical error of subtraction below the root decreases. The numerical accuracy can subsequently increased if the expressions for A_1, A_2, B_1, B_2, C_1, C_2, D_1, and D_2 are placed directly into Equation 5.76. The program for calculating optical constants by this algorithm is published in Cameron et al.[11] in the FORTRAN language.

The Fahrenfort-Visser calculation method is elegant and simple, but because of the availability of modern computing facilities and the efficiency of nonlinear least squares algorithms, the iterative method of optical constant calculation is more suitable.

Besides the problem of selecting suitable angles θ_1 and θ_2 to secure optimal accuracy in calculating n and κ, the losses caused by using only transversely polarized components of radiation and by measuring the minimal radiant flux significantly decreases the efficiency of the spectrometer. Compensating for the radiant flux attenuation by using a broader slot decreases the spectral resolution and at the same time increases dispersion of the incident angles, and thus influences the measured spectrum by a greater experimental error.

Minimizing the sum of squared differences of values R_s, R_p, or R, measured for the series of values of the incidence angle θ_i and corresponding values calculated from Equations 5.64 and 5.65, is the goal of iterative evaluation. The sum of squares is minimized with respect to the values of n_2 and κ. The starting values for the iterative calculations are estimated or calculated by the Fahrenfort-Visser algorithm. The software for the iterative procedure based on the measurement of the unpolarized radiation reflectance is published in Cameron et al.[11] The calculated values R^{calc} are determined from the relationship

$$R^{calc} = \frac{f(\nu)}{f(\nu)+1} R_s + \frac{1}{f(\nu)+1} R_p \tag{5.87}$$

in which $f(\nu)$ is the experimentally determined coefficient for the polarization recognition ability of the device, defined as the ratio for the transversely and parallelly polarized radiation with the frequency ν.

Published software is used to evaluate data from the spectrometer, in which an ATR adapter creates the hemicylinder from material with a high refraction index, connected to the goniometer (Figure 5.5). The sample is

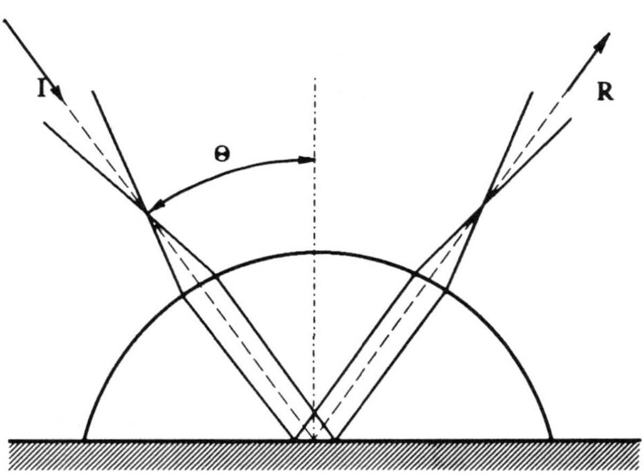

Figure 5.5 The scheme of the experimental arangement during the ATR spectrum measurement.

placed below the rectangular base of the hemicylinder, and the radiation detector is placed behind the adapter on the movable spoke of the goniometer. Because the reflectance strongly depends on the incidence angle, the angle scale of the goniometer is precisely calibrated. Another complication appears when the beam incident on the interface of the measured sample with the hemicylider is not a point but forms a trace of finite width that corresponds to a range of the incidence angles that must be taken into consideration during calculation. If the dependence R on θ_i is recorded experimentally at the hemicylinder–air interface (i.e., without sample), the dependence corresponding to the theoretical dependence determined by Equations 5.64 and 5.65 will not be obtained in the vicinity of the critical angle θ_c, but the course of R corresponding to the convolution of the theoretical dependence with the function determining the incidence angle distribution will be obtained.

Practically, a symmetric, triangular distribution function with base b is assumed. In the case of asymmetry of the theoretical dependence, the convolution is also shifted, correcting the average value of the incidence angle subtracted from the goniometer. If the convolution function based on the five-point Bode approximation is calculated[12]

$$\int_{x_0}^{x_4} f(x)\,dx = \frac{2h}{45}\left[7f(x_0) + 32f(x_1) + 12f(x_2) + 32f(x_3) + 7f(x_4)\right] \qquad \textbf{(5.88)}$$

in which

$$h = x_{i+1} - x_i = \text{constant for } i = 0, 1, 2, 3 \qquad \textbf{(5.89)}$$

the convoluted function

$$R^c(\theta_i) = \frac{4}{11}\left[R\left(\theta_i - \frac{b}{4}\right) + R\left(\theta_i + \frac{b}{4}\right)\right] + \frac{3}{11}R(\theta_i) \qquad \textbf{(5.90)}$$

will be obtained, where $R(\theta_i)$ are the theoretical values determined from Equations 5.64 and 5.65. The angle correction of the goniometer with the value b is obtained using nonlinear minimization of the sum of the squares of the deviations:

$$F = \sum_{\theta_i}\left[R^c(\theta_i) - R^{\exp}(\theta_i)\right]^2 \qquad \textbf{(5.91)}$$

where $R^{\exp}(\theta_i)$ is the experimental value.

KRAMERS-KRONIG TRANSFORMATION

From the Fresnel equations (see Specular Reflectance at Nonabsorbing Material Interface and Specular Reflectance at Absorbing Material Interface, above), the relationships between the amplitudes of the incident and reflected radiation generally determine the complex coefficients r_s and r_p, which can be expressed by the real reflectances R_s and R_p and corresponding phase shifts ϕ_s and ϕ_p:

$$r_s = \frac{A_{rs}}{A_{is}} = R_s^{\frac{1}{2}} \exp\left(i\,\theta_s\right) \qquad \textbf{(5.92)}$$

$$r_p = \frac{A_{rp}}{A_{ip}} = R_p^{\frac{1}{2}} \exp\left(i\,\theta_p\right) \qquad \textbf{(5.93)}$$

Kramers[13] and Kronig[14] derived general dispersion relationships expressing the mutual dependence of real $f_1(\nu)$ and imaginary $f_2(\nu)$ components of a function $f(\nu)$. The relationship determines the linear, time independent response of a physical system to the input stimulus at frequency ν

$$f_2(\nu) = \left(\frac{2\nu}{\pi}\right) \int_0^{\infty} \frac{f_1(\nu)}{\nu'^2 - \nu^2}\, d\nu' = \left(\frac{2\nu}{\pi}\right) I \qquad \textbf{(5.94)}$$

where I is the infinite integral. A complex dielectric constant:

$$\varepsilon(\nu) = \varepsilon_1(\nu) - i\varepsilon_2(\nu) \qquad \textbf{(5.95)}$$

a complex refractive index:

$$N(\nu) = n(\nu)\,\left(1 - i\kappa(\nu)\right) = n(\nu) - i\kappa_0(\nu) \qquad \textbf{(5.96)}$$

or a reflective coefficient logarithm:

$$\ln r_s = \frac{1}{2} \ln R_s + i\phi_s \qquad \textbf{(5.97)}$$

$$\ln r_p = \frac{1}{2} \ln R_p + i\phi_p \qquad \textbf{(5.98)}$$

can be used as examples of physical variables.

The change in dependence of a phase shift on frequency $\phi_x(\nu)$ for a corresponding polarized component can be determined from the change in reflectance $R_x(\nu)$

$$\phi_x(\nu) = \left(\frac{\nu}{\pi}\right) \int_0^{\infty} \frac{\ln R_x(\nu')}{\nu'^2 - \nu^2}\, d\nu' \qquad x = s, p \qquad \textbf{(5.99)}$$

The values $n(\nu)$ and $\kappa_0(\nu)$ can then be calculated from the Fresnel equations and from the known values $\phi_x(\nu)$ and $R_x(\nu)$. For example, for $x = s$ from the relationship

$$r_s = \frac{a - ib - \cos\theta_i}{a - ib + \cos\theta_i} = \sqrt{R_s}\,\exp\left(i\phi_s\right) \tag{5.100}$$

the values a and b will be determined. The values n and κ are evaluated from the relationship

$$\left[n^2\left(1 - i\kappa\right)^2 - \sin^2\theta_i\right]^{\frac{1}{2}} = a - ib \tag{5.101}$$

Application of Kramers-Kronig analysis for reflection spectra is also known as Robinson-Price analysis, according to the authors of the first papers in this field.[15,16]

Up to now several numerical procedures for calculating an integral I from digitized spectral data,[17-21] including the fast Fourier transformation,[22-27] have been proposed. Ohta and Ishida[28] numerically compared methods using the Mac Laurin formula, trapezoidic integration, the Simpson formula, and successive double Fourier transformation. It was found that the Mac Laurin integration, which does not require any special approximation for the pole calculation at integration, gives the most precise results, with relatively short calculating time.

The function contained in the integral in Equation 5.99 has a pole at $\nu = \nu'$. Based on the equality

$$\int_0^\infty \frac{\ln R_x(\nu)}{\nu'^2 - \nu^2}\,d\nu' = 0 \tag{5.102}$$

Equation 5.99 can be changed to

$$\phi_x(\nu) = \left(\nu/\pi\right)\int_0^\infty \frac{\ln R_x\left(\nu'\right) - \ln R_x\left(\nu\right)}{\nu'^2 - \nu^2}\,d\nu' \tag{5.103}$$

This function contained in the integral is an infinite expression of type 0/0 for $\nu' = \nu$, and a value for the corresponding limit is determined by the L'Hospital rule:

$$\lim_{\nu\to\nu'} \frac{\ln R_x\left(\nu'\right) - \ln R_x\left(\nu\right)}{\nu'^2 - \nu^2} = \lim_{\nu\to\nu'} \frac{d\ln R_x\left(\nu\right)}{2\nu\,d\nu} \tag{5.104}$$

The derivative value in the numerator can be determined by applying any method of numerical differentiation (see Chapter 1, Numerical Differentiation of Spectra) to the spectrum $R_x(\nu)$.

Another problem that must be solved by Kramers-Kronig analysis results from the fact that the reflection spectrum is measured over a restricted interval of frequency ν:

$$0 < \nu_l \leq \nu \leq \nu_u < \infty \tag{5.105}$$

The integral form of Equation 5.99 can be formally rewritten in the following terms:

$$I = \int_0^{\infty} = \int_0^{\nu_l} = \int_{\nu_l}^{\nu_u} = \int_{\nu_u}^{\infty} = I_{0l} + I_{1u} + I_{u\infty} \tag{5.106}$$

Suitable, physically allowable extrapolation of the function $R_x(\nu)$ into the areas $\nu < \nu_l$ respective to $\nu > \nu_u$ is necessary to calculate integrals I_{0l} and $I_{u\infty}$. The constant value of $R_x(\nu) = R_x(\nu_l)$ is usually assumed in the area $\nu < \nu_l$. The asymptotic relationships for an electron gas, according to which

$$\begin{aligned} \varepsilon_1 - 1 &\approx \nu^{-2} \\ \varepsilon_2 &\approx \nu^{-3} \\ R_x &\approx \nu^{-4} \end{aligned} \tag{5.107}$$

can be used for the high frequency region $\nu > \nu_u$. Based on these assumptions, Klucker and Nielsen[29] developed a numerical procedure and software in the FORTRAN programming language.

The experimental determination of a function $f_1(\nu)$ for p of generally nonequidistant frequencies:

$$\nu_l = \nu_1 < \nu_2 < \ldots < \nu_i < \ldots < \nu_{p-1} < \nu_p = \nu_u \tag{5.108}$$

is the starting point of the evaluation. $\varepsilon_1(\nu)$ or 1/2 ln $R_x(\nu)$ can be used for the function $f_1(\nu)$. Then the calculation of the integrals I_{0l} and I_{1u} is the same for both cases, but the different asymptotic behaviors of ε_1 and R_x must be taken into consideration to calculate the integral $I_{u\infty}$.

The function $f_1(\nu)$ can be extended to the infinite power series in intervals $\nu_i < \nu \leq \nu_{1+1}$:

$$f_1(\nu) = \sum_{j=0}^{\infty} a_{ji}\, \nu^j \qquad \text{pre} \quad i = 1, 2, \ldots p \tag{5.109}$$

For the contribution $I_{i,i+1}$ of the given interval to the integral I_{1u}, to calculate a value F_2 at a point ν_k, after replacing

$$I_{i,i+1} = s_{1i} \ln \frac{\nu_{i+1} + \nu_k}{\nu_i + \nu_k} + \sum_{j=2}^{\infty} \frac{s_{ji}}{j-1} \left(\nu_{i+1}^{j-1} - \nu_i^{j-1} \right)$$

$$+ \left(s_{0i} + \nu_k s_{1i} \right) \frac{1}{2\nu_k} \left(\ln \left| \frac{\nu_{i+1} - \nu_k}{\nu_i - \nu_k} \right| - \ln \frac{\nu_{i+1} - \nu_k}{\nu_i - \nu_k} \right) \tag{5.110}$$

is obtained, and for $\nu_k \neq \nu_i$, ν_{i+1}, and for s_{ji} the following is valid:

$$s_{ji} = \sum_{l=0}^{\infty} a_{2l+j,i} \, \nu_k^{2l} \tag{5.111}$$

At sufficiently dense spectral digitization, the following linear approximation is adequate:

$$a_{ji} = 0 \qquad \text{for} \quad j > 1 \tag{5.112}$$

therefore

$$s_{0i} = a_{0i}$$

$$s_{1i} = a_{1i}$$

$$s_{ji} = 0 \qquad \text{for} \quad j > 1 \tag{5.113}$$

Applying these assumptions, Equation 5.110 is significantly simplified to

$$I_{i,i+1} = \frac{1}{2\nu_k} \left[\left(a_{0i} + \nu_k a_{1i} \right) \ln \left| \frac{\nu_{i+1} - \nu_k}{\nu_i - \nu_k} \right| \right]$$

$$- \left(a_{0i} - \nu_k a_{1i} \right) \ln \frac{\nu_{i+1} + \nu_k}{\nu_i + \nu_k} \tag{5.114}$$

The extrapolation for $\nu \to 0$ necessary for evaluating the integral I_{0l} is performed simply by adding the point for $i = 0$ with $\nu_i = \nu_0 = 0$ and $f(\nu_0) = f(\nu_1)$. Then

$$I_{0l} = I_{0,1} = \frac{f(\nu_1)}{2\nu_k} \ln \frac{\nu_k - \nu_1}{\nu_k + \nu_1} \tag{5.115}$$

If $f_1(\nu) = 1/2 \ln R_x(\nu)$, the integral $I_{u\infty}$ is calculated assuming a valid extrapolation:

$$R_x(\nu) = R_x(\nu_p) \left(\frac{\nu_p}{\nu} \right)^{\alpha} \qquad \text{pre} \quad \nu \geq \nu_p \tag{5.116}$$

where $\nu_p = \nu_u$, and according to Equation 5.107 $\alpha = 4$. Replacing Equation 5.116 and 5.106 the following results:

$$I_{u\infty} = \frac{1}{2} \int_{\nu_u}^{\infty} \frac{\ln R_x(\nu_p) + \alpha \ln\left(\frac{\nu_p}{\nu}\right)}{\nu^2 - \nu_k^2} d\nu \quad \textbf{(5.117)}$$

and after the integration:

$$I_{u\infty} = \frac{1}{4\nu_k}\left(\ln R_x(\nu_p) \ln \frac{1-q}{1+q} + \alpha F(q)\right) \quad \textbf{(5.118)}$$

where $q = \frac{\nu_k}{\nu_p}$ and where

$$F(q) = \begin{pmatrix} 2\sum_{j=1}^{\infty} \frac{q^{2j-1}}{(2j-1)^2} \\ \\ \frac{1}{4}\pi^2 - \ln q \ln \frac{1-q}{1+q} - 2\sum_{j=1}^{\infty} \frac{\left(\frac{1-q}{1+q}\right)^{2j-1}}{(2j-1)^2} \end{pmatrix} \quad \textbf{(5.119)}$$

given q from the interval $\sqrt{2} - 1 < q < 1$, the first of the alternate series introduced converges better than does the other. The general expression (Equation 5.118) enables values of $\alpha \neq 4$ to be used for evaluation.

If $f_1(\nu) = \varepsilon_1(\nu)$, for $\nu > \nu_u = \nu_p$, the following dependence is assumed:

$$f_1(\nu) = 1 + \left[f_1(\nu_p) - 1\right]\left(\frac{\nu_p}{\nu}\right)^2 \quad \text{pre } \nu > \nu_u \quad \textbf{(5.120)}$$

and then for $I_{u\infty}$:

$$I_{u\infty} = \int_{\nu_u}^{\infty} \frac{1 + \left[f_1(\nu_p) - 1\right]\left(\frac{\nu_p}{\nu}\right)^2}{\nu^2 - \nu_k^2} d\nu \quad \textbf{(5.121)}$$

after integration:

$$I_{u\infty} = \frac{1}{2\nu_k}\left\{1+\left(\frac{\nu_p}{\nu_k}\right)^2\left[f_1(\nu_p)-1\right]\right\}$$

$$\cdot \ln\left[\frac{(\nu_p+\nu_k)}{(\nu_p-\nu_k)}\right]-\left(\frac{\nu_p}{\nu_k^2}\right)\left[f_1(\nu_p)-1\right] \tag{5.122}$$

To summarize, the relationships obtained for the linear approximation in intervals $\nu_i \le \nu < \nu_{i+1}$ can be written in the form

$$f_2(\nu_k) = \frac{T + dF(q)}{2\pi} \tag{5.123}$$

where

$$T = \sum_{i=1}^{p}\left(a_{1i} - a_{1,i-1}\right)\left(\nu_i + \nu_k\right)\ln\left(\nu_i + \nu_k\right)$$

$$+\sum_{\substack{i=1\\ i\neq k}}^{p}\left(a_{1i} - a_{1,i-1}\right)\left(\nu_k - \nu_i\right)\ln\left(\nu_k - \nu_i\right) \tag{5.124}$$

$$a_{li} = \frac{f_1(\nu_{i+1}) - f_1(\nu_i)}{\nu_{i+1} - \nu_i} \tag{5.125}$$

$$a_{10} = a_{1u} = 0 \tag{5.126}$$

$$q = \frac{\nu_k}{\nu_p} \tag{5.127}$$

When $F_2(\nu) = \phi_x(\nu)$:

$$f_1(\nu) = \ln R_x(\nu) \tag{5.128}$$

$d = \alpha$ and $F(q)$ is determined by Equation 5.119

When $f_2(\nu) = \varepsilon_2(\nu)$:

$$f_1(\nu) = 2\,\varepsilon_1(\nu) \tag{5.129}$$

$$d = 2\left[\varepsilon_1(\nu_p) - 1\right] \tag{5.130}$$

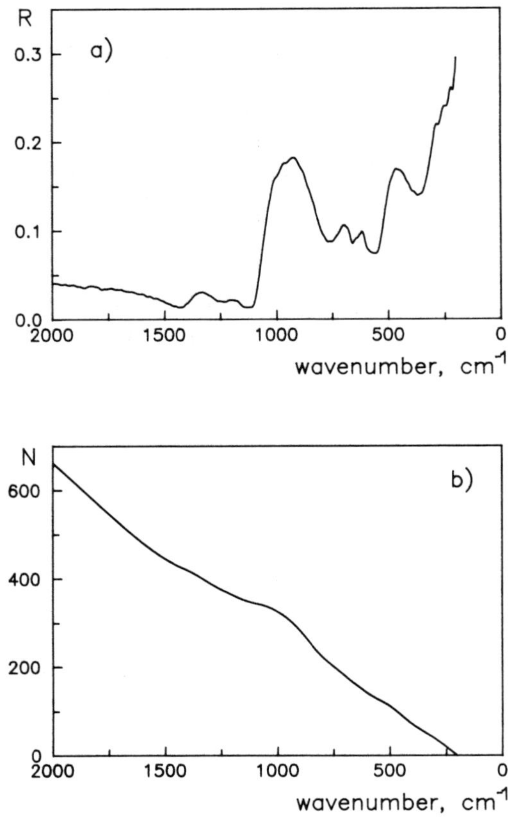

Figure 5.6 Illustration of the application of Kramers-Kronig analysis: **(a)** the reflection spectrum of the perpendicularly polarized component of the electromagnetic radiation; **(b)** the calculated spectrum of the absorbing index.

$$F(q) = \left\{ \begin{array}{ll} \dfrac{2}{q} + \left(\dfrac{1-1}{q^2} \right) \ln \dfrac{1+q}{1-q} & \text{for} \quad q > 0,2 \\ 4q \displaystyle\sum_{j=1}^{\infty} \dfrac{q^{2j-2}}{4j^2 - 1} & \text{for} \quad q \leq 0,2 \end{array} \right\} \tag{5.131}$$

The application of this method to analyze the perpendicularly polarized component of the reflection spectrum of aluminum silicate glass is illustrated in Figure 5.6.

DIFFUSE REFLECTANCE SPECTRA

Reflected radiation consists of two components when light is reflected from a matte (rough) surface. A regular (specular) component is dependent on the reflection index n and on the absorption coefficient κ, according to the Fresnel equations (see Specular Reflectance at Nonabsorbing Material Interface, above). The diffuse reflected component is created from radiation penetrating the sample. A part of this radiation comes back, and absorption from a sample and multireflection at the boundaries of individual components, from which the compound is composed, appears. Attenuation of this diffuse part of the reflected radiation by absorption in the compound is governed by the Bourguer-Lambert rule[2,3]

$$\Phi = \Phi_0 \exp\left(-\kappa \, \bar{d}\right) \quad \textbf{(5.132)}$$

where $\bar{d}$ is the average penetration, i.e., the average layer thickness of radiation penetration, respective to the average geometric trajectory of radiation in a compound. Here κ denotes the Napier absorption coefficient (previously denoted as α in Equations 5.44 and 5.45).

Whereas specular reflection from dielectrics is completely described by the Fresnel equations (Equations 5.20 to 5.23), up to now none of the theories developed for diffuse reflection is generally valid.

A matte surface irradiated by a constant flux is similarly bright at observation from all angles. Based on this, the Lambert cosine law (LKL) was derived, according to which for radiant flux B reflected by a nonabsorbing matte surface:

$$B = \left(\frac{\Phi_0}{\pi}\right) \cos\alpha \, \cos\Theta \quad \textbf{(5.133)}$$

where Φ_0 is the radiant flux falling perpendicularly to the surface, α is the incidence angle, and Θ is the angle of observation.

According to Equation 5.133 the reflected radiation is symmetric to the line normal to the surface at any incidence angle. When reflected radiation is polarized, there is always polarization of incident radiation as well. The LKL is valid only for low values of angles α and Θ, and deviations are significant for absorbing compounds, which was proved experimentally,[2,30,31] the reason being multiple reflection on the microcrystal surfaces. Therefore, reflected radiation always contains a partially polarized component too. Pokrowski[32] expressed the portion B_{11} of specular components by averaging the Fresnel relationships for a plane with statistically uniformly oriented specularity:

$$B_{11} = \frac{a}{2}\left[\frac{\sin^2(\alpha' - \beta)}{\sin^2(\alpha' + \beta)} + \frac{\tan^2(\alpha' - \beta)}{\tan^2(\alpha' + \beta)}\right]$$

$$= \frac{a}{2} f(\alpha', \beta) \quad \textbf{(5.134)}$$

where a is the constant. For the angles α' and β:

$$\alpha' = \frac{\alpha + \Theta}{2} \tag{5.135}$$

$$\beta = \arcsin\left[\left(\frac{1}{n}\right)\sin\alpha'\right] \tag{5.136}$$

is valid. Combining LKL for the diffusion component with Equation 5.134 for the reflected radiant flux, the following relationship is obtained:

$$B = b\cos\Theta + \left(\frac{a}{2}\right) f(\alpha', \beta) \tag{5.137}$$

which also explains the partial polarization of reflected radiation. Practically, only partial coincidence of the Pokrowian model with the experiment was observed. Deviations occur, caused especially by the preference orientation of microcrystals.

In absorbing compounds, a portion of a specular component also depends on the absorption coefficient value. For a large absorption coefficient value the reflection coefficient is close to 1 (see Specular Reflectance at Absorbing Material Interface, above). Reflected, practically pure diffuse radiation can be obtained by grinding weakly absorbing compounds very finely while strongly absorbing compounds always give reflected radiation with a significant specular component.

The most commonly used approach for describing diffusion reflection spectra is based on the solution of the system of first order differential equations for the dispersion and absorption of the radiation. The first such solutions introduced were published in 1905 by Schuster[33] and in 1931 by Kubelka and Munk.[34] Subsequently, various authors derived other solutions,[2] similar to that of Kubelka and Munk (KM) in that they contain two parameters. As was shown by Ingle[35] and Kubelka,[36] the general KM solution involves all of the other solutions as special cases. Solutions involving a greater number of parameters were developed by, for example, Silberstein[37] (three parameters), Ryde[38,39] (four parameters), and Duntley[40] (8 parameters). Further attention will be focused on the universal KM solution, which has the most extended practical applications.

Let us consider a layer of randomly and uniformly distributed absorbing components with negligible dimensions compared to the layer thickness. Because surface layer dimensions are very high, boundary effects are not observed on its edges. Assume that monochromatic electromagnetic radiation is falling on the layer. Let d be the layer thickness and x a coordinate direction perpendicular to the layer with an origin on the layer surface. Let us label the radiant flux in the direction of the x axis as I and in the opposite direction as J. The radiant flux incident on the layer is labeled I_0. The beam radiation was investigated as a function of the angle ϕ (measured from the line normal to the surface) through an infinitesimal layer with a thickness dx. The average trajectory of the beam in the layer is

$$d\xi_I = dx \int_0^{\frac{\pi}{2}} \frac{\partial I}{I\,\partial\phi} \frac{d\phi}{\cos\phi} = u\,dx \tag{5.138}$$

where $\partial I/\partial\phi$ is the angular distribution of radiation in the direction $+x$. For radiation in the direction $-x$ the following expression is given:

$$d\xi_J = dx' \int_0^{\frac{\pi}{2}} \frac{\partial J}{J\,\partial\phi} \frac{d\phi}{\cos\phi} = v\,dx' \tag{5.139}$$

where $dx' = -dx$. Let ε be the fraction of radiation absorbed at the unit trajectory in the layer and σ be the fraction diffused. If only radiation in the direction $+x$ is considered, the decreased intensity in the layer dx is

$$(\varepsilon + \sigma)\, I\, d\xi_I = (\varepsilon + \sigma)\, u\, I\, dx' \tag{5.140}$$

where $\varepsilon\, u\, I\, dx$ is absorbed radiation and $\sigma\, u\, I\, dx$ is diffused radiation. For the J component the following is given:

$$(\varepsilon + \sigma)\, J\, d\xi_J = (\varepsilon + \sigma)\, v\, J\, dx \tag{5.141}$$

The radiation diffusing from one direction contributes to radiant flux in the opposite direction:

$$dI = -(\varepsilon + \sigma)\, u\, I\, dx + \sigma\, v\, J\, dx \tag{5.142}$$

$$dJ = (\varepsilon + \sigma)\, v\, J\, dx - \sigma\, u\, I\, dx \tag{5.143}$$

For an ideal dispersing material:

$$u = v = 2 \tag{5.144}$$

is valid. Then, Equations 5.142 and 5.143 become

$$dI = -(k + s)\, I\, dx + s\, J\, dx \tag{5.145}$$

$$dJ = (k + s)\, J\, dx - s\, I\, dx \tag{5.146}$$

where $k = 2\varepsilon$ and $s = 2\sigma$. Integrating Equations 5.145 and 5.146 for radiant flux, the following is obtained:

$$I = A(1-\beta)\,e^{\kappa x} + B(1+\beta)\,e^{-\kappa x} \tag{5.147}$$

$$J = A(1+\beta)\,e^{\kappa x} + B(1-\beta)\,e^{-\kappa x} \tag{5.148}$$

where

$$\kappa = \left[k(k+2s)\right]^{\frac{1}{2}} \tag{5.149}$$

$$\beta = \left[\frac{k}{(k+2s)}\right]^{\frac{1}{2}} \tag{5.150}$$

The values A and B result from the marginal conditions: $I = 0$ for $x = x_0$ and $I = I_d$, and $J = 0$ for $x = d$, i.e.,

$$A = -\frac{(1-\beta)\,e^{-\kappa d}}{(1+\beta)^2\,e^{\kappa d} - (1-\beta)^2\,e^{-\kappa d}}\,I_0 \tag{5.151}$$

$$B = \frac{(1+\beta)\,e^{\kappa d}}{(1+\beta)^2\,e^{\kappa d} - (1-\beta)^2\,e^{-\kappa d}}\,I_0 \tag{5.152}$$

Relationships for practically significant variables can be determined from:

$$\begin{aligned} T = \frac{I_d}{I_0} &= \frac{4\beta}{(1+\beta)^2\,e^{\kappa d} - (1-\beta)^2\,e^{-\kappa d}} \\ &= \frac{2\beta}{(1+\beta)^2\,\sinh(\kappa d) + 2\beta\cosh(\kappa d)} \end{aligned} \tag{5.153}$$

If the limit for $s = 0$, i.e., for a nondispersing medium, the equalities $\kappa = k$, $\beta = 1$, and $T = e^{-kd}$ (which is consistent with the Lambert-Beer rule) hold.

The layer thickness can be considered infinite with sufficiently thick samples (1 to 3 mm for powders), and with that limit the following equations can be derived from Equations 5.153 and 5.154:

$$\begin{aligned} R = \frac{J_0}{I_0} &= \frac{\left(1-\beta^2\right)\left(e^{\kappa d} - e^{-\kappa d}\right)}{(1+\beta)^2\,e^{\kappa d} - (1-\beta)^2\,e^{-\kappa d}} \\ &= \frac{\left(1-\beta^2\right)\sinh(\kappa d)}{\left(1+\beta^2\right)\sinh(\kappa d) + 2\beta\cosh(\kappa d)} \end{aligned} \tag{5.154}$$

$$T_\infty = 0 \tag{5.155}$$

$$R_\infty = \frac{1-\beta}{1+\beta} = \frac{1-\left(\dfrac{k}{k+2s}\right)^{\frac{1}{2}}}{1+\left(\dfrac{k}{k+2s}\right)^{\frac{1}{2}}} \qquad \textbf{(5.156)}$$

respectively:

$$f(R_\infty) = \frac{(1-R_\infty)^2}{2R_\infty} = \frac{k}{s} \qquad \textbf{(5.157)}$$

where f is called the Kubelka-Munk function or remission function, and lower index ∞ denotes infinite layer thickness.

Schreyer[41] showed that the KM equation fails for strongly absorbing materials in which radiant flux from the sample is dependent on direction.

The application examples of diffuse reflectance spectroscopy for measuring colors and dyeing receipt calculation are introduced in the next two sections.

MEASUREMENT OF COLORS

Spectroscopic methods in the visible region enable unambiguous and objective quantification of color.[42-45] After incidence, a color sensation at the human retina creates an electromagnetic field in the visible region, which corresponds to the interval of wavelengths λ = 380 to 780 nm. Color is determined by relative energy distribution of wavelengths, i.e., a visible spectrum of observed radiation. The interval of 380 to 600 nm corresponds to blue, 500 to 600 nm to green, and 600 to 780 nm to red.

A radiation spectrum is determined by the spectral characteristics of an illuminant and by a reflection spectrum from a surface during the observation of a colored object in reflected light. A spectral characteristic of an illuminant is combined with the transmittance spectrum of an observed object (Figure 5.7) in the transmitted light. A spectrum (reflection and transmittance, respectively) of a colored compound in the visible spectral region is an unambiguously defined physical property entering into the creation of color perception. To quantify a color perception it is necessary to standardize other factors that are simultaneously operating during its creation, e.g., light illuminant and the spectral sensitivity of the human eye. In 1931, the International Commission of Illumination (CIE) labeled three standard illuminants, *A*, *B*, and *C*. An illuminant *A* has the spectral distribution of a Planck emitter at a temperature of 2854°K and can be realized by a lamp with a tungsten fiber filament. An illuminant *B* represents average daylight with the major component being direct solar radiation and a correlated color temperature (corresponding temperature of the blackbody) of 4879°K. It can be produced from illuminant *A* using two liquid

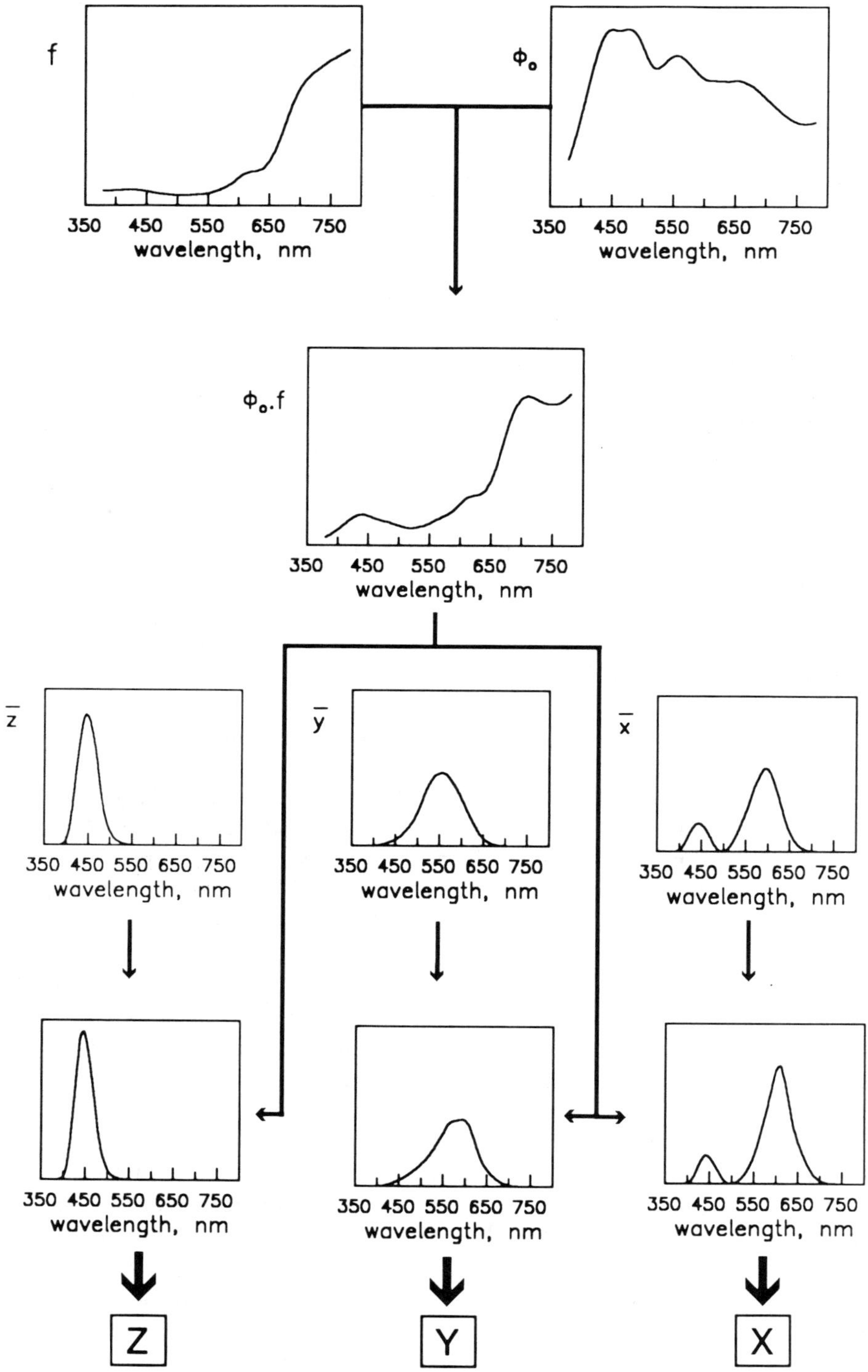

Figure 5.7 Observation of colors: $f(\lambda)$ is $\beta(\lambda)$ for observation in reflected light and $\tau(\lambda)$ for observation in direct light.

filters. An illuminant C, until recently the most used, has a temperature of 6770°K and represents average daylight without direct solar radiation. It is possible to realize this by means of an illuminant A and two color liquid filters.

In 1966, the CIE recommended illuminant $D65$, representing typical daylight with a correlated color temperature of 6500°K. It was anticipated that at some future date illuminants B and C would be dropped from the list of recommended standard illuminants.

The first condition for objective assessment and measurement of color was created by standardizing and applying the standard illuminant.

The need to standardize human eye properties and processes during the formation of a color perception arose from the subjective nature of color perception. In 1931, a standard colorimetric observer was defined by the CIE. The function of a standard observer eye is characterized by the color matching functions $\bar{x}(\lambda)$, $\bar{y}(\lambda)$, and $\bar{z}(\lambda)$, which are derived from experimental results yielded by visual colorimetry from a 2° angle of view. These functions represent the relative amounts of the three spectral primaries (X, Y, Z) needed by the standard observer to match a color of a given wavelength λ in an equal energy spectrum. The 2° angle is too small for the practical evaluation of colors. That was the reason why CIE accepted the complementary standard observer in 1964. The color matching functions $\bar{x}_{10}(\lambda)$, $\bar{y}_{10}(\lambda)$, and $\bar{z}_{10}(\lambda)$ were determined from measurements from a 10° viewing angle (Figure 5.8).

Definition of a standard colorimetric observer enables the objective description of color using the values X, Y, and Z, the so-called tristimulus values defined by the relationships

$$X = k \int_{\lambda} \beta(\lambda)\, \Phi_0(\lambda)\, \bar{x}(\lambda)\, d\lambda \tag{5.158}$$

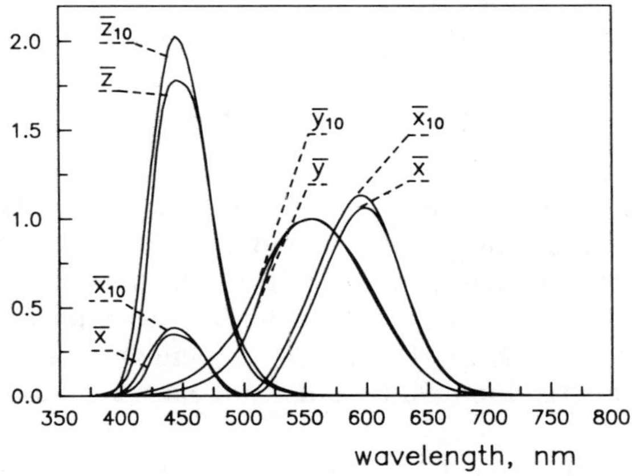

Figure 5.8 Color matching functions of the standard and complementary standard observer.

$$Y = k \int_{\lambda} \beta(\lambda)\, \Phi_0(\lambda)\, \tilde{y}(\lambda)\, d\lambda \quad \textbf{(5.159)}$$

$$Z = k \int_{\lambda} \beta(\lambda)\, \Phi_0(\lambda)\, \tilde{z}(\lambda)\, d\lambda \quad \textbf{(5.160)}$$

where $\Phi_0(\lambda)$ is the spectral characteristic of the chosen standard illuminant and $\beta(\lambda)$ is the diffuse reflectancy, also labeled $R(\lambda)$ (see Figure 5.7). For observations in transparent light, the transmittance $\tau(\lambda)$ is used instead of $\beta(\lambda)$. The standardizing factor k is conventionally defined so that the tristimulus value Y of an ideally white, diffuse plane is equal to a value of 100

$$k = \frac{100}{\int_{\lambda} \Phi_0(\lambda)\, \tilde{y}(\lambda)\, d\lambda} \quad \textbf{(5.161)}$$

Integration in Equations 5.158 to 5.161 is done over the wavelengths of the spectral visible region. Applying equidistantly digitized spectral data, the integration is replaced by a summation, e.g.,

$$X = k \sum_{i} \beta(\lambda_i)\, \Phi_0(\lambda_i)\, \tilde{x}(\lambda_i) \quad \textbf{(5.162)}$$

$$k = \frac{100}{\sum_{i} \Phi_0(\lambda_i)\, \tilde{y}(\lambda_i)} \quad \textbf{(5.163)}$$

If the entire wavelength region of the visible spectrum is not involved in the digitized spectrum, the values of β are replaced at a spectral boundary by a last measured value:

$$\beta(\lambda) = \beta(\lambda_{\mathrm{min}}) \qquad \text{for} \quad 380 \le \lambda \le \lambda_{\mathrm{min}} \quad \textbf{(5.164)}$$

$$\beta(\lambda) = \beta(\lambda_{\mathrm{max}}) \qquad \text{for} \quad 770 \ge \lambda \ge \lambda_{\mathrm{max}} \quad \textbf{(5.165)}$$

where values λ_{min} and λ_{max} define the measurement region of a spectrum. The values of color matching functions and the values of spectral characteristics of standard illuminants are taken from the literature, in which they are usually incremented in steps of 10 or 5 nm.[42] To simplify the calculation, the values of products $\Phi_0(\lambda)\ \bar{x}(\lambda)$, $\Phi_0(\lambda)\ \bar{y}(\lambda)$, and $\Phi_0(\lambda)\ \bar{z}(\lambda)$ are tabulated for various standard illuminants and for a normal and supplementary standard observer.

In the first approach, the meaning of the representative measure of three basic component colors making up an observed color can be attributed to tristimulus values. Component X determines the amount of red, component Y, green, and Z, blue. Simultaneously, a chromaticity coordinate Y represents lightness of the color. Therefore, just one point in the Cartesian coordinate system created by tristimulus values corresponds to each color, but not all points of the tristimulus values space represent real colors.

From the linearity of transformation (Equations 5.158 to 5.161), the transformation of reflectance (transmittance) spectra into the Cartesian coordinate system is defined according to the rules for mixing colors. Points on a line connecting two points representing real colors correspond to colors resulting from the mixing of these two real colors. On the other hand, a certain color can be obtained by mixing any pair of real colors placed on opposite half-lines of lines crossing the certain color.

Each color can be characterized by its lightness, shade, and excitation purity, corresponding to content of neutral grey. If single colors can be imagined as vectors connecting the coordinate origin of a tristimulus value system to a point (X, Y, and Z), then the coordinates of the intersection points of lines crossing these vectors with the unit plane, are the so-called chromaticity coordinates x, y, and z defined by the relationship

$$x = \frac{X}{(X + Y + Z)} \tag{5.166}$$

$$y = \frac{Y}{(X + Y + Z)} \tag{5.167}$$

$$z = \frac{Z}{(X + Y + Z)} \tag{5.168}$$

Chromaticity coordinates do not contain information about color lightness, and according to their linear dependence ($z = 1 - x - y$), they explain the plane representation of colors characterized by the color shade and excitation purity. A set of all real colors defined in a plane x, y is a horseshoe figure, the so-called color triangle or CIE chromaticity diagram (Figure 5.9). Pure spectral colors corresponding to monochromatic visible radiation are placed on its convex perimeter. Only one wavelength can be unambiguously associated with each point of the convex circumference. The purpuric colors, which do not have a monochromatic equivalent, lie at the bottom line restriction of the color triangle. A point of colorless light, C, corresponding to a chosen illuminant, lies inside the color triangle. The color shade represented by point F is given by the position of the intersection, O, of the half-line crossing point F from point C. (Figure 5.9). If this intersection point is placed on the convex section, the color shade is characterized by the so-called dominant wavelength, which is equal to the characteristic wavelength of the corresponding perimeter point. If the intersection point lies on the bottom line segment, the color shade is characterized by the so-called complementary wavelength, given by the intersection point of the half-line reflecting through

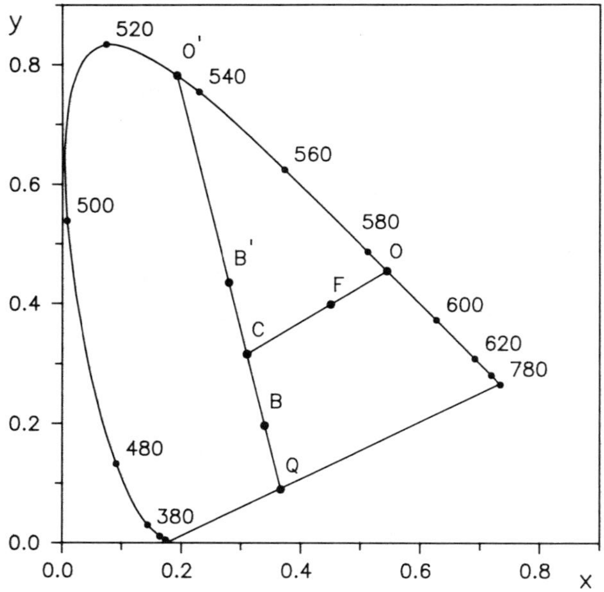

Figure 5.9 Color triangle (CIE chromaticity diagram).

the point C and the convex circumference section. The Helmholtz excitation purity p of the color defined by the point F is given by the relative distance from the colorless point

$$p = \overline{\mathrm{CF}} / \overline{\mathrm{CO}} \tag{5.169}$$

Therefore, spectral colors and colors of the purpuric line have unit excitation purity. The colorless (chromaticity) point has zero excitation purity.

The table of values (λ_i, x_{0i}, and y_{0i}) for $i = 1, 2, \ldots N$, determining the convex part of the color triangle, is used for numerical calculation. Points are placed so that $\lambda_{i+1} < \lambda_i$, i.e., the perimeter point circulates in a counterclockwise direction. These values are not dependent on the choice of a standard illuminant. After the selection of the standard illumination source, the coordinates x_c, y_c of the colorless point C are obtained. Based on the relationships

$$\sin \omega_i' = \frac{(y_{0i} - y_c)}{d_i} \tag{5.170}$$

$$\cos \omega_i' = \frac{(x_{0i} - x_c)}{d_i} \tag{5.171}$$

$$d_i = \left[(x_{0i} - x_c)^2 + (y_{0i} - y_c)^2 \right]^{\frac{1}{2}} \tag{5.172}$$

the distance d_i from the colorless point can be defined for each point i, and the color angle value ω_i' from the interval $0 \rightarrow 2\pi$ is measured in a counterclockwise direction from the line $y = y_c$ to the oriented half-line $\overline{CO}$. The values of the color angle can then be transformed in accordance with the following relationship:

$$\omega_i = \omega_i' - \omega_1' + 2\alpha\pi \tag{5.173}$$

where the integer α is determined so that $\omega_i \in (0, 2\pi)$ is valid. The color angles ω_F' and ω_F can be analogously determined for the color given by the point $F \neq C$ with the chromaticity coordinates x_F, y_F. If $\omega_F > \omega_N$ is valid, then it is the color with the complementary wavelength; in the opposite case it is the color with the dominant wavelength. The simplest way to determine the dominant wavelength is to find $i \in (1, N-1)$ for which

$$\omega_i \le \omega_F \le \omega_{i+1} \tag{5.174}$$

is valid, and the value of the dominant wavelength $\lambda_d(F)$ can be determined by linear interpolation:

$$\lambda_d(F) = \lambda_i + \frac{\lambda_{i+1} - \lambda_i}{\omega_{i+1} - \omega_i}\left(\omega_F - \omega_i\right) \tag{5.175}$$

The more effective algorithm is based on the preliminary analytical expression (e.g., the spline function) of the dependence $\lambda = \lambda(\omega)$ and replacing the value ω_F of this relationship.[46,47]

If the color given by point B (Figure 5.9) does not have the dominant wavelength, the coordinates of point B', obtained by reflecting point B through the colorless point C, are calculated first:

$$x_{B'} = 2x_c - x_B \tag{5.176}$$

$$y_{B'} = 2y_c - y_B \tag{5.177}$$

The complementary wavelength is determined as the negative value of the dominant wavelength of point B'

$$\lambda_k(B) = -\lambda_d(B') \tag{5.178}$$

The Helmholtz excitation purity of the color F (with the dominant wavelength) is obtained by the relationship

$$\frac{p_e(F) = \left[\left(x_F - x_c\right)^2 + \left(y_F - y_c\right)^2\right]^{\frac{1}{2}}}{d\left(\omega_F\right)} \tag{5.179}$$

where $d(\omega_F)$ is the distance $\overline{CO}$ determined by linear interpolation:

$$d(\omega_F) = d_i + \frac{d_{i+1} - d_i}{\omega_{i+1} - \omega_i}(\omega_F - \omega_i) \tag{5.180}$$

or by applying the created spline function $d = d(\omega)$.

The excitation purity of the color B (with the complementary wavelength) can be obtained in similar manner:

$$p_e(B) = \frac{\left[(x_B - x_c)^2 + (y_B - y_c)^2\right]^{\frac{1}{2}}}{d(\omega_B)} \tag{5.181}$$

and $d(\omega_B)$ labels the distance of point C from the intersection point of the half-line that starts from point C and crosses point B with the boundary line segment of purpuric colors (i.e., with the connecting line of the (x_{01}, y_{01}) and (x_{0N}, y_{0N}) points).

Based on the defined variables, a particular color can be described in three ways: (1) by giving the tristimulus values X, Y, and Z; (2) by giving the pair of chromaticity coordinates x and y and a tristimulus value Y representing lightness; and (3) by Helmholtz coordinates, i.e., by the dominant (λ_d), or by complementary (λ_k) wavelengths, excitation purity p_e, and a tristimulus value Y. A disadvantage of these coordinate systems (color spaces) is that they cannot be used for direct comparison of color differences. The different visual perception of the magnitude of color difference, according to the location of the pair to be judged in the color space, corresponds to the same distance between the points representing two different colors. The color recognition ability of a human eye can be illustrated in the color triangle by the MacAdam discrimination ellipses, which represent the standard deviations of subjective color settings, (the ellipses in Figure 5.10 are enlarged 10 times). Therefore, various transformations of the tristimulus value space were developed:

$$\xi = \xi(X, Y, Z)$$

$$\eta = \eta(X, Y, Z)$$

$$\zeta = \zeta(X, Y, Z) \tag{5.182}$$

transforming the tristimulus value space to an approximately uniform color space (ξ, η, ζ). The transformation (Equation 5.182) is based either on the Munsell color system or the MacAdam and Brown measurements of human eye sensitivity to color differences in various areas of the color space.[44,45]

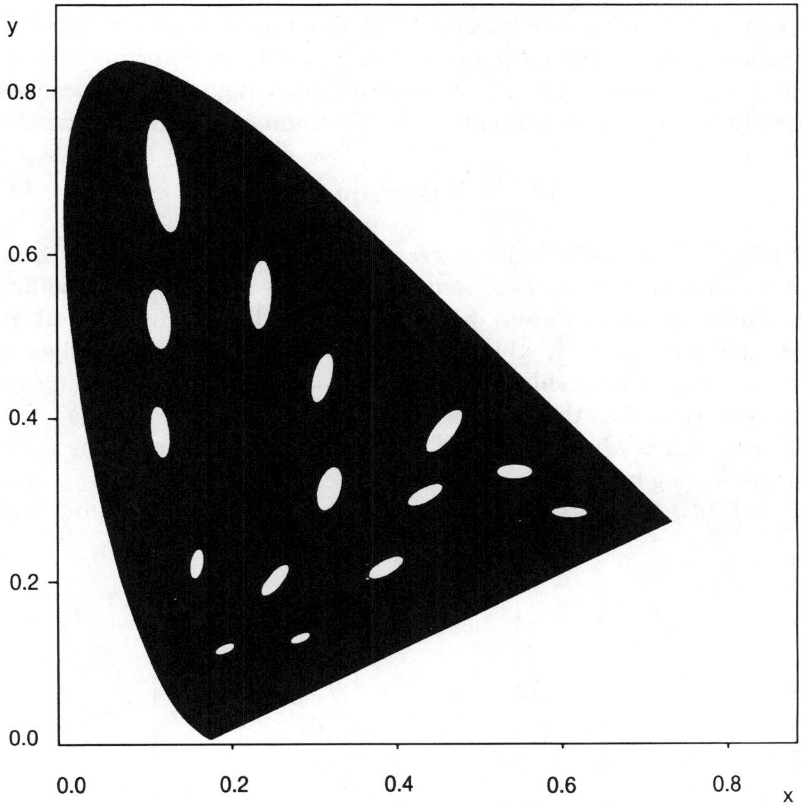

Figure 5.10 MacAdam discrimination ellipses (10 times enlarged).

THE CALCULATION OF DYEING RECIPE

The calculation of dyeing recipe[43] represents the determination of the dye concentration c_i, by which the substrate must be colored, so that it gives the same color perception as the defined color pattern for the chosen standard light illuminant and method of observation. The calculation is based on the relationship that expresses the corresponding spectral characteristic (reflectance curve $\beta[\lambda]$ respective to the transmittance curve $\tau[\lambda]$ in the visible region) as a function of the concentration vector $\vec{c}$. The Lambert-Beer rule is such a relationship for cases of real solutions and glasses that can be considered the real solutions of coloring oxides (see Chapter 4). For textile substrates, the Kubelka-Munk equation (KM) is used:

$$F\left[\beta(\lambda)\right] = \frac{\left[1-\beta(\lambda)\right]^2}{2\beta} = \frac{k_\lambda}{s_\lambda} \qquad \textbf{(5.183)}$$

where β is the diffuse reflectance and the function β(λ) is labeled the reflectance curve. If values of a reflectance coefficient determine the material of the color substrate, and all dyes and the single substrate contribute additively to the absorption coefficient, Equation 5.183 can be rewritten as

$$F\left[\beta(\lambda)\right] = F\left[\beta_S(\lambda)\right] + \sum_i A_{i,\lambda}\, c_i \tag{5.184}$$

where $\beta_S(\lambda)$ is the reflectance curve of the uncolored substrate, c_i is the concentration of the ith dye, and $A_{i,\lambda}$ are the non-negative coefficients determining the contribution of the ith dye to the KM function at wavelength λ. Simultaneously, the mutual interaction of individual dyes is ignored, i.e., the relationship (Equation 5.184) does not contain any mixed terms proportional to the product of concentrations $c_i\, c_j$.

The values of the coefficients $A_{i,\lambda}$ are determined from the series of substrate dyeing by the only dye with a different concentration (Figure 5.11 through 5.14). If a constant degree of exhaustion of the dyeing bath is

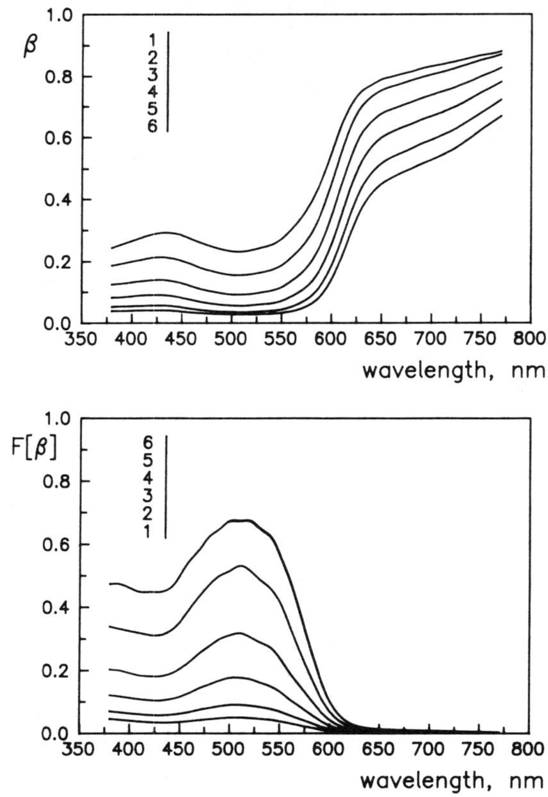

Figure 5.11 Concentration dependence of the luminiscence curve β(λ) and Kubelka-Munk function $F[\beta(\lambda)]$ for the dye Lanaset Red G on cotton substrate. Concentrations are expressed in mass %: 1 to 0.1, 2 to 0.2, 3 to 0.4, 4 to 0.75, 5 to 1.4, 6 to 2.0.

assumed, then values for c_i can be expressed by the amount of dye in the dyeing bath as it is related to the mass of the dyeing substrate. For more complex expressions of concentration dependence in the relationship (Equation 5.184), e.g., the quadratic function:

$$F\left[\beta(\lambda)\right] = F\left[\beta_S(\lambda)\right] + \sum_i \left(A_{i,\lambda}\ c_i + B_{i,\lambda}\ c_i^2\right) \quad \textbf{(5.185)}$$

can be used. For everyday applications in the textile industry, Equation 5.184 is usually sufficient, and the necessary correction for dyeing receipt can be calculated. If the jth concentration in the series of M_i concentrations of individual dyeing by the ith dye (concentrations of other dyes are zero) is labeled c_{ij}, then the predetermined ($M > 1$) system of linear equations will be obtained for coefficients $A_{i,\lambda}$

$$F\left[\beta_{ij}(\lambda)\right] = F\left[\beta_S(\lambda)\right] + A_{i,\lambda}\ c_{ij} \qquad \text{for} \quad j = 1, 2, \ldots M_i \quad \textbf{(5.186)}$$

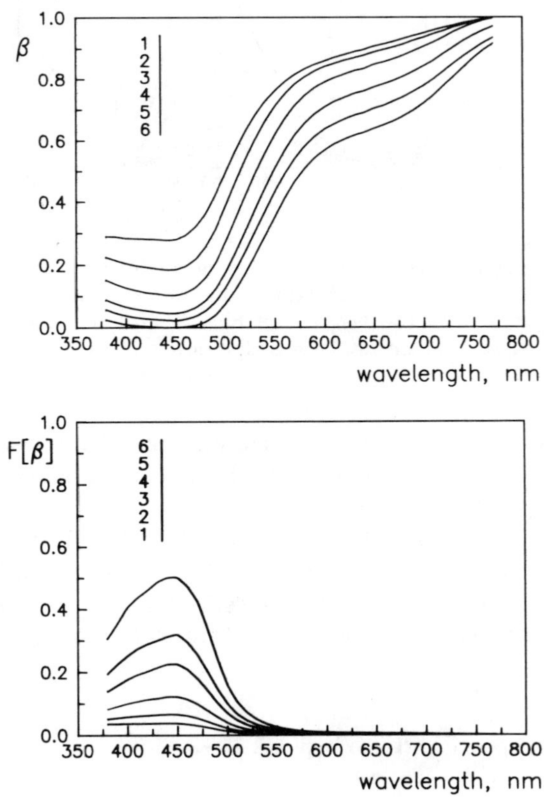

Figure 5.12 Concentration dependence of the luminiscence curve $\beta(\lambda)$ and Kubelka-Munk function $F[\beta(\lambda)]$ for the dye Lanaset Yellow 2R on cotton substrate. Concentrations are expressed in mass %: 1 to 0.1, 2 to 0.2, 3 to 0.4, 4 to 0.75, 5 to 1.4, 6 to 2.0.

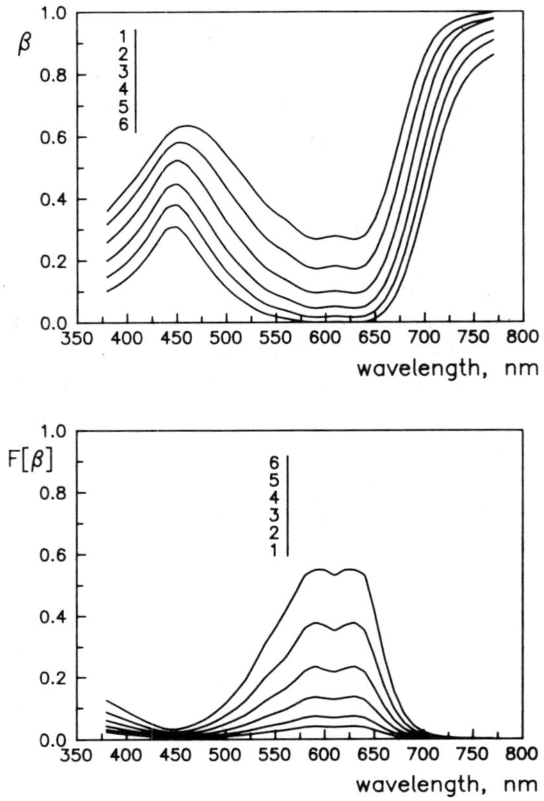

Figure 5.13 Concentration dependence of the luminiscence curve β(λ) and Kubelka-Munk function $F[\beta(\lambda)]$ for the dye Lanaset Blue 2R on cotton substrate. Concentrations are expressed in mass %: 1 to 0.1, 2 to 0.2, 3 to 0.4, 4 to 0.75, 5 to 1.4, 6 to 2.0.

where $\beta_{ij}(\lambda)$ is the corresponding reflectance curve. The best estimations of coefficients $A_{i,\lambda}$ can be calculated by the standard linear least squares method (see Appendix C).

The characterization of the dye pattern by the reflectance curve $\beta_p(\lambda)$, to which three tristimulus values X_p, Y_p, Z_p correspond for a chosen standard illuminant, is assumed. The reflectance curve $\beta(\lambda, \vec{c})$ of dyeing with the concentration vector $\vec{c}$ is obtained by solving the quadratic equation:

$$\beta(\lambda, \vec{c}) = 1 - \left\{F^2\left[\beta(\lambda, \vec{c})\right] + 2\,F\left[\beta(\lambda, \vec{c})\right]\right\}^{\frac{1}{2}} \tag{5.187}$$

where the value $F[\beta(\lambda, \vec{c})]$ is determined by Equation 5.184. Placing $\beta(\lambda, \vec{c})$ into Equations 5.158 to 5.161 gives the tristimulus values of dyeing $X(\vec{c})$, $Y(\vec{c})$, and $Z(\vec{c})$ as the function of concentration vector $\vec{c}$. Then the dyeing receipt can be calculated by minimizing the quadratic form:

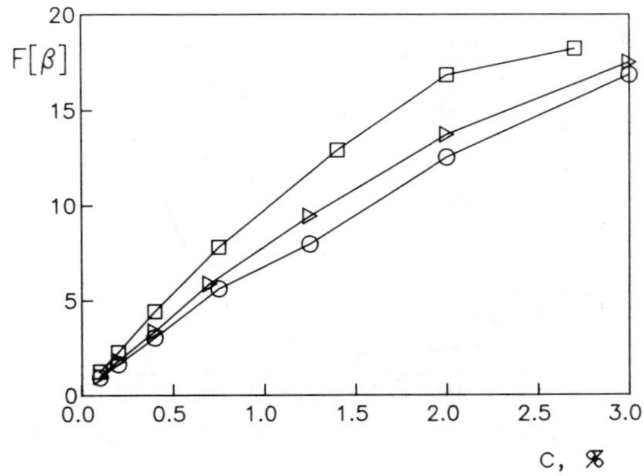

Figure 5.14 The dependence of the Kubelka-Munk function $F[\beta(\lambda)]$ on the dye concentration on cotton substrate at wavelength λ_{min} of the reflectance curve minimum: Lanaset Red G, λ_{min} = 500 nm **(1)**; Lanaset Yellow 2R, λ_{min} = 450 nm **(2)**; and Lanaset Blue 2R, λ_{min} = 590 nm **(3)**.

$$U(\vec{c}) = \left[X(\vec{c}) - X_p\right]^2 + \left[Y(\vec{c}) - Y_p\right]^2 + \left[Z(\vec{c}) - Z_p\right]^2 \quad \textbf{(5.188)}$$

Minimization of this can be advantageously accomplished by the simplex method (see Appendix C), which does not require any knowledge of the derivatives of the objective function with respect to individual concentrations.

The minimization problem defined can yield several solutions, because the same color perception (the same tristimulus values) can correspond to different reflectance curves. This effect is known as metametry. Equivalence of the given dyeing is associated with the selected standard illuminant. The metametry index, defined by the distance of the color points corresponding to two various light illuminants, quantitatively judges the color perception stability of the calculated receipt at the illumination change.

Figure 5.15 shows two metametry dyeings obtained by the combination of dyes defined in Figures 5.11 to 5.14 at the dyeing receipt evaluation.

The minimization task (Equation 5.188) requires a "reasonable" starting estimation $\vec{c}^{\,0}$ of the concentration vector, which can be obtained by solving the linear least squares problem for the KM functions of L wavelengths of λ_j:

$$F\left[\beta\left(\lambda_j, \vec{c}\right)\right] = F\left[\beta_p\left(\lambda_j\right)\right] \qquad \text{for} \quad j = 1, 2, \ldots L \quad \textbf{(5.189)}$$

Thus, the predetermined system of linear equations:

$$\sum_i A_{i,\lambda_j}\, c_i^0 = F\left[\beta_p\left(\lambda_j\right)\right] - F\left[\beta_S\left(\lambda_j\right)\right] \qquad \text{for} \quad j = 1, 2, \ldots L \quad \textbf{(5.190)}$$

is solved.

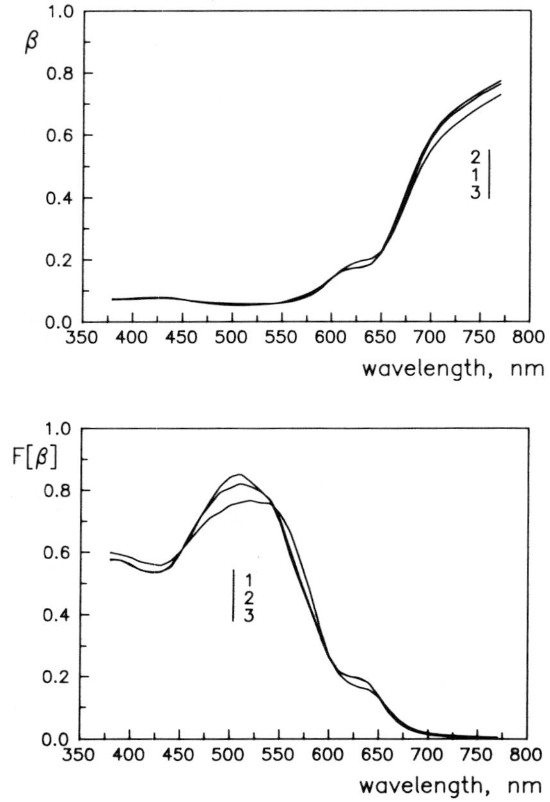

Figure 5.15 The results of the dyeing receipt calculation: relfectance curve of the pattern **(1)** and reflectance curves of metametry solutions of the minimization task (Equation 5.188) (**2** and **3**).

REFERENCES

1. Silvester, P., *Modern Electromagnetic Fields,* Prentice-Hall, New Jersey, 1968.
2. Wendtland, W. Wm. and Hecht, H. G., *Reflectance Spectroscopy,* Interscience Publishers, New York, 1966.
3. Kortüm, G., *Reflectance Spectroscopy,* Springer, New York, 1969.
4. Dunken, H. H. and Stephanowitz, R., *Z. Chem.,* 23, 353, 1983.
5. Dunken, H. H., *Ceramics,* 31, 151, 1987.
6. Fahrenfort, J., *Spectrochim. Acta,* 17, 698, 1961.
7. Young, R. P. and Jones, R. N., *Chem. Rev.,* 71, 219, 1971.
8. Fahrenfort, J. and Visser, W. M., *Spectrochim. Acta,* 18, 1103, 1962.
9. Fahrenfort, J. and Visser, W. M., *Spectrochim. Acta,* 21, 1433, 1965.

10. Hansen, W. N., *Spectrochim. Acta,* 21, 209, 1965.

11. Cameron, D. G., Escolar, D., Goplen, T. G., and Jones, R. N., *NRCC Bulletin No. 17,* 1, 1977.

12. Ambrowitz, M. and Stegun, I., *Handbook of Mathematical Functions, NBS Applied Mathematics Series No. 55,* 1964, 886.

13. Kramers, H. A., *Atti Congr. Intern. Fis. Como,* 2, 545, 1927.

14. Kronig, R., *J. Opt. Soc. Am.,* 12, 547, 1926.

15. Robinson, T. S., *Proc. Phys. Soc. (London),* 65B, 910, 1952.

16. Robinson, T. S. and Price, W. S., *Proc. Phys. Soc. (London),* 66B, 969, 1953.

17. Hawranek, J. P., Neelakantan, P., Young, R. P., and Jones, R. N., *Spectrochim. Acta,* 32A, 85, 1976.

18. Allara, D. L., Baca, A., and Pryde, C. A., *Macromolecules,* 11, 1215, 1987.

19. Goplen, T. G., Cameron, D. G., and Jones, R. N., *Appl. Spectrosc.,* 34, 652, 1980.

20. Graf, R. T., Koenig, J. L., and Ishida, H., *Appl. Spectrosc.,* 39, 405, 1985.

21. Knolle, W. R. and Allara, D. L. *Appl. Spectrosc.,* 40, 1046, 1986.

22. Peterson, C. W. and Knight, W. B., *J. Opt. Soc. Am.,* 63, 1238, 1973.

23. Hansen, W. N. and Abdou, W. A., *J. Opt. Soc. Am.,* 67, 1537, 1977.

24. Hansen, W. N. and Abdou, W. A., *J. Phys. Coll. (Paris),* 38(C5), 207, 1977.

25. King, F. W., *J. Opt. Soc. Am.,* 68, 994, 1987.

26. Bardwell, J. A. and Dingam. M. J., *Anal. Chim. Acta,* 172, 101, 1985.

27. Bertie, J. E. and Eysel, H. H., *Appl. Spectrosc.,* 39, 392, 1985.

28. Ohta, K. and Ishida, H., *Appl. Spectrosc.,* 42, 952, 1988.

29. Klucker, R. and Nielsen, U., *Comp. Phys. Commun.,* 6, 187, 1983.

30. Wright, N., *Ann. Physik,* 1, 17, 1900.

31. Agnew, J. T. and McQuistan, R. B., *J. Opt. Soc. Am.,* 43, 999, 1953.

32. Pokrowski, G. J., *Z. Physik,* 30, 66, 1924; 32, 523, 1925; 35, 34, 1926; 35, 390, 1926; 36, 472, 1926.

33. Schuster, A., *Astrophys. J.,* 21, 1, 1905.

34. Kubelka, P. and Munk, F., *Z. Techn. Physik,* 12, 593, 1936.

35. Ingle, G. W., *ASTM Bull., No. 32,* May, 1942.

36. Kubelka, P., *J. Opt. Soc. Am.,* 38, 448, 1948; 38, 1067, 1948.

37. Silberstein, L., *Phil. Mag.,* 4, 129, 1927.

38. Ryde, J. W., *Proc. R. Soc. London,* A90, 219, 1931.

39. Ryde, J. W., *J. Soc. Glass Technol.,* 16, 408, 1932.

40. Duntley, S. Q., *J. Opt. Soc. Am.,* 32, 61, 1942.

41. Schreyer, G., *Z. Physik. Chem. Frankfurt,* 12, 359, 1957; 12, 375, 1957; 18, 123, 1958.

42. Bouma, P. J., *Physical Aspects of Colour,* MacMillan, Eindhoven, 1971.

43. Judd, D. B. and Wyszecki, G., Color, in *Business, Science and Industry,* 2nd. ed., John Wiley & Sons, New York, 1963.

44. McDonald, R., *J. Oil Col. Chem. Assoc.,* 65, 43, 1982.

45. McDonald, R., *J. Oil Col. Chem. Assoc.,* 65, 93, 1982.

46. Leete, C. G. and Lytle, J. R., *Col. Eng.,* Jan./Feb., 27, 1966.

47. Steinus, A. S., *Col. Eng.,* May/June, 19, 1971.

Chapter 6

Methods of Numerical Analysis of Electron Spin Resonance Spectra

Electron spin resonance (ESR), sometimes called electron paramagnetic resonance (EPR), gives valuable and detailed information about the structure and properties of paramagnetic systems in gaseous, liquid, and solid phase, i.e., for systems that have nonzero electron magnetic moments.

ESR has the characteristics of a resonance spectroscopy, because electromagnetic radiation is absorbed between the levels which are split by an external magnetic field. The resonance condition can be obtained in two ways. The first uses a constant external magnetic field with induction B to change the electromagnetic radiation frequency until the resonance condition is fulfilled:

$$\hbar\omega = \Delta E \quad \mathbf{(6.1)}$$

where ω is the frequency of electromagnetic radiation, $\hbar$ is the Planck constant h divided by 2π ($\hbar = h/2\pi$), and ΔE is the difference of energetic levels split due to the influence of an external field. The second way uses a constant frequency of electromagnetic radiation and changes the induction of an external magnetic field to yield a resonance absorption. This method is used more often because of its experimental advantages.

Due to the heterogeneity and anisotropy of the magnetic properties of compounds, ESR spectra have complicated shapes and often are strongly anisotropic.

INTRODUCTION TO ELECTRON SPIN RESONANCE SPECTROSCOPY

Electron spin resonance spectroscopy works with energy levels whose distances are, maximally, a few reciprocal centimeters. For the interpretation of ESR spectra only the levels, which are degenerate or nearly degenerate without the presence of an external magnetic field, are interesting. The known spin Hamiltonian is used to determine the energies of such levels. Its specific form depends on the symmetry and character of internal fields. Generally, the spin Hamiltonian can be written in the following form:[1-7]

$$\hat{H}_S = \hat{H}_{CF} + \hat{H}_Z + \hat{H}_N + \hat{H}_{EN} + \hat{H}_{EE} + \hat{H}_Q + \hat{H}_J + \hat{H}_D \quad \textbf{(6.2)}$$

where $\hat{H}_{CF}$ is the Hamiltonian of a crystal field (order 10^5 cm^{-1}), $\hat{H}_Z$ is the Hamiltonian of Zeeman interaction (order 10^{-3} cm^{-1}), $\hat{H}_N$ is the Hamiltonian of nuclear Zeeman interaction (order 10^3 cm^{-1}), $\hat{H}_{EN}$ is the Hamiltonian of hyperfine interaction (order 10^{-1} to 10^{-3} cm^{-1}), $\hat{H}_{EE}$ is the Hamiltonian of spin–spin interaction (order 10^0 cm^{-1}), $\hat{H}_Q$ is the Hamiltonian of quadruple interaction (order 10^{-5} cm^{-1}), $\hat{H}_J$ is the Hamiltonian of exchange interaction (order 10^0 cm^{-1}), and $\hat{H}_D$ is the Hamiltonian of diamagnetic interaction (order 10^{-5} to 10^{-7} cm^{-1}).

Generally, the individual terms of this Hamiltonian have a tensor character, dependent on the orientation of a system according to an external magnetic field.[1-5]

Due to the tensor character of the spin Hamiltonian, the selection of a suitable coordinate system is very important for unambiguous interpretation. McGavin[6] discusses the constraints on EPR spin Hamiltonian parameters for all crystallographic point symmetries possible for a paramagnetic center.

The Hamiltonian of a crystal field $\hat{H}_{CF}$ includes the influence of the electrical potential of surroundings (the crystal field) of a paramagnetic article on its energy levels. Its specific form depends on the symmetry and intensity of a crystal field. Generally, it can be expressed in the following form:

$$\hat{H}_{CF} = \sum_{q} \sum_{i=x,y,z} B_{q,i}\, \hat{O}_{q,i} \quad \textbf{(6.3)}$$

where $B_{q,i}$ are the sequence coefficients and $\hat{O}_{q,i}$ are suitable combinations of spin operators, concrete forms dependent only on the crystal field symmetry.

The term of the Zeeman interaction, $\hat{H}_Z$, includes the energy of mutual interaction of the magnetic moments of unpaired electrons with an external magnetic field. It also includes the contribution of mutual interaction of orbital and spin moments of unpaired electrons. It can be written in a tensor expression as follows:

$$\hat{H}_Z = \beta\, \vec{B}^T\, \overline{\overline{g}}\, \hat{\vec{S}} \quad \textbf{(6.4)}$$

where β is the Bohr magneton, $\vec{B}$ is the vector of external magnetic field induction, $\vec{S}$ is the operator of electron spin momentum, and $\bar{\bar{g}}$ is the so-called g-tensor, which is a tensor of the second order. The contribution of the Zeeman interaction is, as a rule, the dominant contribution in ESR spectra.

The operator of spin–spin interaction $\hat{H}_{EE}$ takes into consideration the influence of mutual dipole–dipole interaction of unpaired electrons. This term of the spin Hamiltonian is nonzero only if the system contains two or more unpaired electrons. In tensor form it can be written as

$$\hat{H}_{EE} = \hat{\vec{S}}^T \, \bar{\bar{D}} \, \hat{\vec{S}} \tag{6.5}$$

where $\bar{\bar{D}}$ is the tensor of spin–spin (dipole–dipole) interaction of unpaired electrons. The tensor of spin–spin interaction is symmetrical with zero trace, and therefore this contribution of the spin Hamiltonian is averaged to value of zero in the ESR spectra of liquid phase.

The Hamiltonian of the nuclear Zeeman interaction $\hat{H}_N$ is the measure of interaction of a magnetic momentum of nuclei with an external magnetic field. It can be expressed as

$$\hat{H}_N = -g_I \beta_I \hat{\vec{I}}^T \vec{B} \tag{6.6}$$

where g_I is the nuclear g-factor (the constant value for a given isotope), β_I is known as the nuclear magneton, and $\vec{I}$ is the operator of a nuclear spin momentum. Due to the low value, this contribution can be neglected for most systems.

The Hamiltonian of hyperfine interaction, which includes the interaction of electron and nuclear magnetic momenta, is an important term of the spin Hamiltonian. It can be expressed in tensor formalism as follows:

$$\hat{H}_{EN} = \hat{\vec{S}}^T \bar{\bar{A}} \hat{\vec{I}} \tag{6.7}$$

where $\bar{\bar{A}}$ is the tensor of hyperfine interaction. This term is dominant in isotropic ESR spectra of free radicals in the liquid phase.

The Hamiltonian of quadruple interaction takes into account the interaction of electric quadruple momenta of nuclei, which have spin $I > 1/2$, with the gradients of electrical fields of an atom, a molecule, or a crystal. In a tensor expression it can be written in the following form:

$$\hat{H}_Q = \hat{\vec{I}}^T \bar{\bar{\chi}} \hat{\vec{I}} \tag{6.8}$$

where $\bar{\bar{\chi}}$ is the tensor of quadruple interaction. Its contribution is usually low, and as a rule it can be observed only in the spectra of single crystals.

If the system contains two or more mutually interacting paramagnetic particles, the term of their mutual interaction $\hat{H}_J$, to which the term of the

Heisenberg exchange interaction and the term of the dipole–dipole interaction contribute, must be included in the resulting Hamiltonian. In tensor form it can be expressed as

$$\hat{H}_J = \hat{\vec{S}}_1^T \overline{\overline{J}} \hat{\vec{S}}_2 \tag{6.9}$$

where $\overline{\overline{J}}$ is the tensor of an exchange interaction that generally does not have to be symmetrical. $\hat{\vec{S}}_1$ and $\hat{\vec{S}}_2$ are the operators of the spin momenta of unpaired electrons of individual paramagnetic particles.

The Hamiltonian of diamagnetic interactions $\hat{H}_D$ shifts all the energy levels equidistantly, therefore, it does not occur in ESR spectroscopy.

The energy values of the spin system can be obtained by solving the Schrödinger equation:

$$\hat{H}_S \Psi_i = E_i \Psi_i \tag{6.10}$$

where Ψ_i is the wave function (the eigen function of the spin Hamiltonian) and can be expressed as a linear combination of the product spin function ϕ_n:

$$\Psi_i = \sum_n c_n \phi_n \tag{6.11}$$

$$\phi_n = \left| M_{S_k}^n \; M_{I_k}^n \right> \tag{6.12}$$

where $M_{s_k}^n$ is the projection of an electron spin quantum number and $M_{I_k}^n$ is the projection of a spin quantum number of the kth nucleus with a nonzero magnetic momentum. The complete system of such functions contains

$$\left(2S_k + 1\right) \; \left(2I_k + 1\right) \tag{6.13}$$

product functions.

After substituting such product functions into the Schrödinger equation and using simple rearrangement, the following system of homogeneous linear equations results:

$$\sum_n c_n \left(H_{mn} - E_n \delta_{mn}\right) = 0 \tag{6.14}$$

where H_{mn} are the elements of the spin Hamiltonian matrix in the basis of the product wave functions:

$$H_{mn} = \left< \phi_m \middle| \hat{H}_S \middle| \phi_n \right> \tag{6.15}$$

The energies and the eigenspin functions can be determined by solving the system of the homogeneous linear equations (Equation 6.14). Because the obtained energies and spin functions are functions of the induction and orientation of an external magnetic field, the energy values are obtained as functions of these parameters:

$$E_i = E_i(B, \theta, \phi) \qquad \textbf{(6.16)}$$

where B is the induction of an external magnetic field and θ and ϕ are the spherical angles defining the mutual orientation of the system in the external magnetic field.

Due to the high number of various spin states and the high dimensions of the spin Hamiltonian matrix, perturbation methods are often used to determine the energies and functional values.

In the next step, it is necessary to determine the values of the external magnetic field induction B_0, for which the resonance condition is fulfilled, i.e., for which the following equation is valid:

$$\hbar\omega = E_i - E_j = f(B_0, \theta, \phi) \qquad \textbf{(6.17)}$$

The calculation of probable transitions proportional to the intensities of corresponding bands in ESR spectra is the next step. Because transitions are dipole transitions, the probability P_{ij} of a transition between states with the energy E_i and E_j is proportional to the square matrix element of a magnetic dipole momentum operator:

$$P_{ij} \approx \sum_k \left| < \Psi_i \left| \hat{\mu}_k \right| \Psi_j > \right|^2 \qquad k = x, y, z \qquad \textbf{(6.18)}$$

The magnetic dipole momentum operator has the following form:

$$\hat{\mu}_k = -\beta g \hat{S}_k \qquad \text{for } k = x, y, z \qquad \textbf{(6.19)}$$

where the Bohr magneton β and g-factor of g are the constant values for the given system, so for the transition probability

$$P_{ij} \approx \sum_k \left| < \Psi_i \left| \hat{S}_k \right| \Psi_j > \right|^2 \qquad \textbf{(6.20)}$$

is valid. The operator $\vec{S}_z$ has the nonzero matrix elements only between equal states, i.e., between states where $i = j$, where radiation is not absorbed. Therefore, for the probability of transition the following equation is defined:

$$P_{ij} \approx \left| < \Psi_i \left| \hat{S}_x^2 + \hat{S}_y^2 \right| \Psi_j > \right| \qquad \textbf{(6.21)}$$

Generally, individual spin functions Ψ_i are the linear combinations of various spin states:

$$\Psi_i = \sum_m c_m^i \phi_m \qquad \textbf{(6.22)}$$

and for the transition probability

$$P_{ij} \approx \sum_{m,n} \left| c^i_m \, c^i_n < \phi_m \left| \hat{S}^2_x + \hat{S}^2_y \right| \phi_n > \right| \tag{6.23}$$

is valid.

Therefore, if the states are the "pure spin states" (the spin Hamiltonian matrix has the diagonal form), where the coefficients of a linear combination C_i^m only have the value of 1 or 0, the wave function consists of only those functions describing the spin state. Hence, only transitions of the first order or "allowed" transitions are discussed. For those transitions, the choice rules of $|\Delta M_S| = 1$ and $|\Delta M_I| = 0$ are given.

In the case of "the mixed spin states", which appear when the spin Hamiltonian matrix is nondiagonal, the individual spin quantum numbers lose their original physical meaning and the spin quantum numbers during the spectral transitions can be changed differently (depending on the form of the spin Hamiltonian matrix). Therefore, transitions of the second order, or "forbidden" transitions are discussed.

For the cases where, according to complicated magnetic interactions, the mixing of nuclear spin states appears, the choice rule $|\Delta M_I| = 0$ is no longer valid.

Such systems therefore usually show very complicated ESR spectra and their analysis cannot be formulated generally; it is necessary to examine every example individually.

SHAPES OF SPECTRAL LINES IN ESR SPECTROSCOPY

In ESR spectroscopy, the resonance condition is fulfilled, as a rule, by changing the induction of an external magnetic field at constant frequency. Therefore, electromagnetic radiation in which vector $\vec{B}_x$ is perpendicular to the external magnetic field $\vec{B}$ is applied. Then for absorption of electromagnetic radiation by a spin system, the following equation can be written:[7]

$$A(\omega) = -\vec{M} \cdot \frac{d\vec{B}_x}{dt} = \frac{1}{2} B_1^2 \omega \chi''(\omega) \tag{6.24}$$

where

$$B_x = B_1 \cos(\omega t) \tag{6.25}$$

describes induction of electromagnetic radiation, $\chi''(\omega)$ is an imaginary component of microwave magnetic susceptibility, and ω and B_1 are the angular frequency and amplitude of a microwave magnetic field.

In ESR spectroscopy, the detection of radiation absorption is based on a relative change of input voltage. The relative change is defined by the change of the system Q factor and the measured signal:

$$S(\omega) = \frac{C\Delta Q}{4\pi Q_0} = C\eta Q_0 \chi''(\omega) = C_1 \frac{A(\omega)}{\omega} \quad \textbf{(6.26)}$$

where η is a factor of sample shape and C and C_1 are constants that have values dependent on the detection system and the microwave circuit in a bridge.[8]

Generally, $\chi''(\omega)$ is the function of a microwave frequency as well as the induction of the external magnetic field.[9,10] For the universal time-dependent spin Hamiltonian of the spin system, the following is valid:

$$\hat{H}(t) = \hat{H}_0 + \hat{H}' \quad \textbf{(6.27)}$$

where $\hat{H}_0$ is a defect caused by the microwave magnetic field (Equation 6.25) and

$$\hat{H}' = \hat{H}_1 + \hat{H}_2 \quad \textbf{(6.28)}$$

where $\hat{H}_1$ is a static part in which coupling coefficients are constant. This part of the Hamiltonian is used to calculate the resonance condition (see Equation 6.2). $\hat{H}_2$ represents a stochastic part of the Hamiltonian, formed because of the random motion and coupling mechanism. Let eigenvalues and eigenstates of Hamiltonians $\hat{H}_0$ and $\hat{H}_1$ be designated $E_a, |a>$; E_b, $|b>$;... and E_α, $|\alpha>$; E_β, $|\beta>$....

Hamiltonian $\hat{H}_2$ gives physical meaning for the interpretation of spectral line spreading, although it is not possible to obtain precise solutions for the states $|a>$ and $|b>$ in a representation similar to $\hat{H}_1$. It is obvious that the states $|a>$ and $|b>$ respresent many unresolved states around $|\alpha>$ and $|\beta>$.

Therefore, in the time-dependent pertubation method, for the probability of transition between the states $|a>$ and $|b>$, the following is valid:[10]

$$W = B_1^2 \sum_{a,b} \left(N_a - N_b\right) \; |<a|\hat{\mu}_x|b>|^2 \; \delta\left(\omega - \omega_{ba}\right) \quad \textbf{(6.29)}$$

where

$$\omega_{ba} = \frac{E_b - E_a}{\hbar} \quad \textbf{(6.30)}$$

N_a and N_b represent densities of corresponding spin states, and $\hat{\mu}_x$ is the operator of magnetic dipole momentum (see Introduction to Electron Spin Resonance Spectroscopy, above). As was shown by Slichter[10] for formulating the problem, it is more advantageous to write the probability of transition by means of the δ-function than in the integral form involving state density, in order to keep track of the quantum numbers of individual states. It is evident that ω_{ba} creates a set of random variables representing more ESR transitions. The summation through quasicontinuum energy states around

Table 6.1 Different Forms of the Boltzman Factor in Magnetic Resonance Theory

Form	Reference
$\frac{N_0}{2I+1}\ \frac{\hbar\omega}{kT}$	Bloemberger et al.[54]
$\hbar\omega/kT$	Pake and Purcell[55] and Castner[56]
$\exp(-E_a/kT) - \exp(-E_b/kT)$, $E_a - E_b = \hbar\omega$	Kambe and Usui[57] and Pryce and Stevents[58]
$\exp(-\hbar\omega/kT)$	Portis[59]
$\exp(-E_a/kT) - \exp(-E_b/kT)$, $E_a - E_b = \hbar\omega_0$	Coffman[60]
$n\hbar\omega/ZkT$	Pilbrow,[7] Zhong and Pilbrow,[9] and Slichter[10]

$|a>$ and $|b>$ involves the combined effect of homogeneous spreading, as well as all sources of nonhomogeneous line spreading and the time spreading leading to the natural shape of a line.

Different forms of the Boltzman factor in magnetic resonance theory have been outlined in Table 6.1.

In Equation 6.29 the Boltzman factor can be written in the form[9]

$$N_b - N_a = \left(\frac{N}{Z}\right)\exp\left(\frac{-E_b}{kT}\right)\left[1-\exp\left(\frac{\hbar\omega}{kT}\right)\right] \tag{6.31}$$

where N is a spin density, Z is the partition function, and the resonance condition (see Figure 6.1)

$$\hbar\omega = E_b - E_a \tag{6.32}$$

is valid. For sufficiently high temperatures

$$1 - \exp\left(\frac{\hbar\omega}{kT}\right) \approx \frac{\hbar\omega}{kT} \tag{6.33}$$

and

$$\exp\left(\frac{-E_b}{kT}\right) \approx 1 \tag{6.34}$$

so for the Boltzman factor the approximate relationship

$$N_a - N_b = \frac{N\hbar\omega}{ZkT} \tag{6.35}$$

is valid.[9] A coefficient for energy absorption can be written in the form[10]

$$A(\omega) = \hbar\omega W \tag{6.36}$$

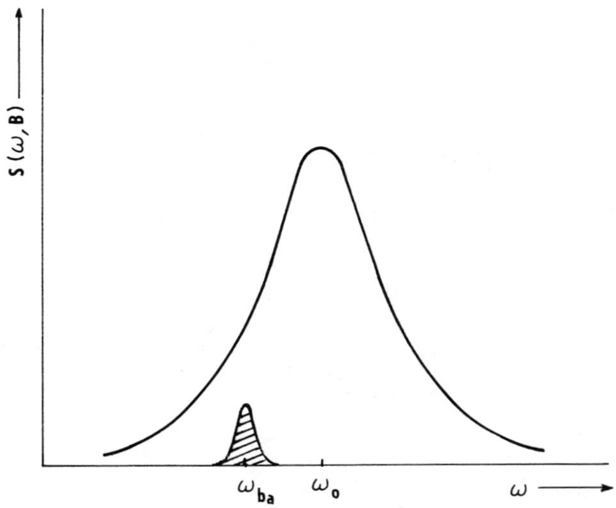

Figure 6.1 Illustation of the ESR absorption envelope showing an individual spin packet located at $\omega = \omega_{ba}$. The resonance condition $\omega = (E_b - E_a)$ holds around the resonance center $\omega_0 = (E_\alpha - E_\beta)$. (From Zhong, Y. C. and Pilbrow, J. R., *J. Magn. Reson.*, 93, 447, 1991. With permission.)

so for the susceptibility

$$\chi''(\omega) = \frac{2\hbar N\omega}{ZkT} \sum_{a,b} |<a|\vec{\mu}_x|b>|^2 \; \delta(\omega - \omega_{ba})$$
$$= C_1 \omega J(\omega) \tag{6.37}$$

where $J(\omega)$ is the spectral density.[11] If the conventional shape of function F, which is the function of ω and B, is introduced, the following relationship is obtained for the magnetic susceptibility,

$$\chi''(B, \omega) = \frac{2N\hbar\omega}{ZkT} \sum_{a,b} |<a|\vec{\mu}_x|\beta>|^2 \; F[\omega - \omega_0(B), \Omega] \tag{6.38}$$

where Ω is a solid angle, and for a resonance centrum

$$\omega_0 = \frac{E_\beta - E_\alpha}{\hbar} = \omega_0(B, \omega) \tag{6.39}$$

is valid. It is obvious from Equation 6.38 that the Boltzman factor is frequency dependent, as was emphasized in Equation 6.35. A simple illustration of this aspect is given in Figure 6.2.

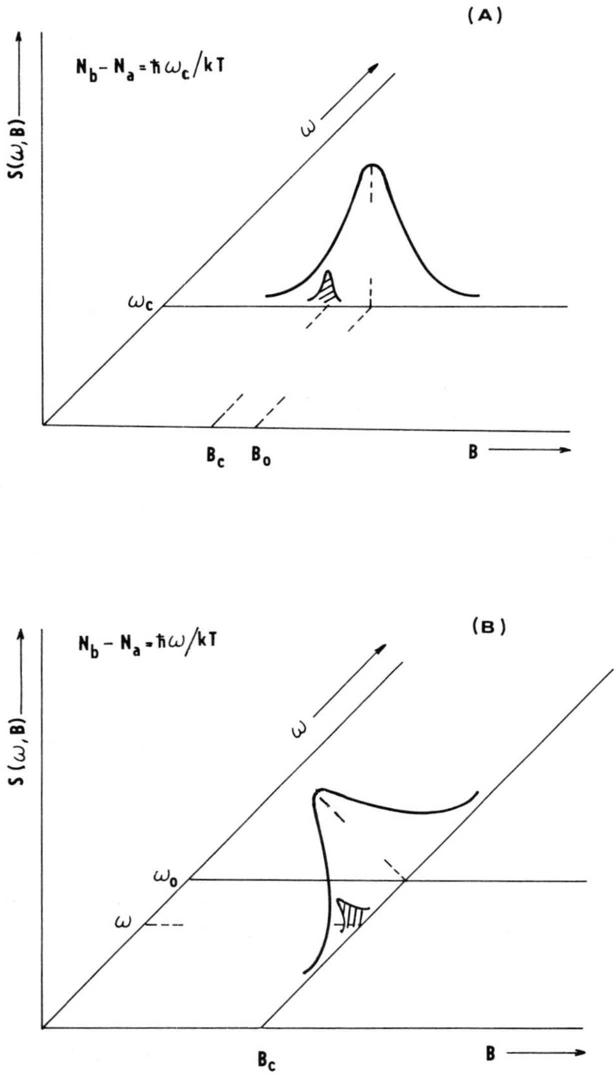

Figure 6.2 Illustration of the ESR absorption envelope in field-sweep **(A)** and frequency-sweep **(B)** experiments. The spacings determined by ω and $\omega_0(B, \Omega)$ are illustrated, where ω_0 and B_c stand for the constant frequency in field-swept EPR and the fixed magnetic field in frequency-swept EPR, respectively. (From Zhong, Y. C. and Pilbrow, J. R., *J. Magn. Reson.*, 93, 447, 1991. With permission.)

For ESR absorption by a single crystal the following is obtained:

$$S(B,\omega) = C\eta Q_0 \frac{2N\hbar\omega}{ZkT} \sum_{a,b} |<a|\vec{\mu}_x|\beta>|^2 \; F[\omega - \omega_0(B), \Omega] \quad \textbf{(6.40)}$$

For ESR absorption of randomly oriented radicals in solid phase (semicrystalline materials, powdered materials, frozen solutions, glasses, etc.), the absorption can be obtained by integrating $S(B,\omega)$ through all angle space

$$I(B,\omega) = \int_{(\Omega)} S(B,\omega,\Omega)\, d\Omega \quad \textbf{(6.41)}$$

where

$$d\Omega = \sin\theta \; d\theta \; d\phi \quad \textbf{(6.42)}$$

and where θ and ϕ are spherical coordinates.

ISOTROPIC ESR SPECTRA — RADICALS IN THE LIQUID PHASE

Free radicals in a liquid phase are, from the point of view of practical applications, the most frequent and simplest group of chemical compounds investigated by ESR spectroscopy. Due to rapid movement in such systems, the individual terms of the spin Hamiltonian are averaged to scalar quantities and most of the terms have negligible intensity. Therefore, for interpretation of the ESR spectra of such systems it is enough to use the simple scalar Hamiltonian

$$\hat{H}_S = \beta g B \hat{S}_z + \sum_k A_k \hat{S}_z \hat{I}_{z_k} \quad \textbf{(6.43)}$$

Subsequently, the individual spin levels acquire the following energies:

$$E_i = \frac{\beta g B}{2} + \frac{1}{2} \sum A_k M_k \quad \textbf{(6.44)}$$

where M_k is the projection of the spin quantum number of the kth nucleus and g is known as the g-factor, which is the mean value of the main g-tensor component (see Introduction to Electron Spin Resonance Spectroscopy, above). The selection rules for the spectral transitions are $|\Delta M_I| = 0$, so for the resonance condition

$$\hbar\omega = \Delta E = \beta g B_0 + \sum_k A_k M_k \quad \textbf{(6.45)}$$

is defined, where ω is the frequency of the clystron, β is the Bohr magneton, $\hbar$ is the Planck constant divided by 2π ($\hbar = h/2\pi$), B_0 is the induction of an

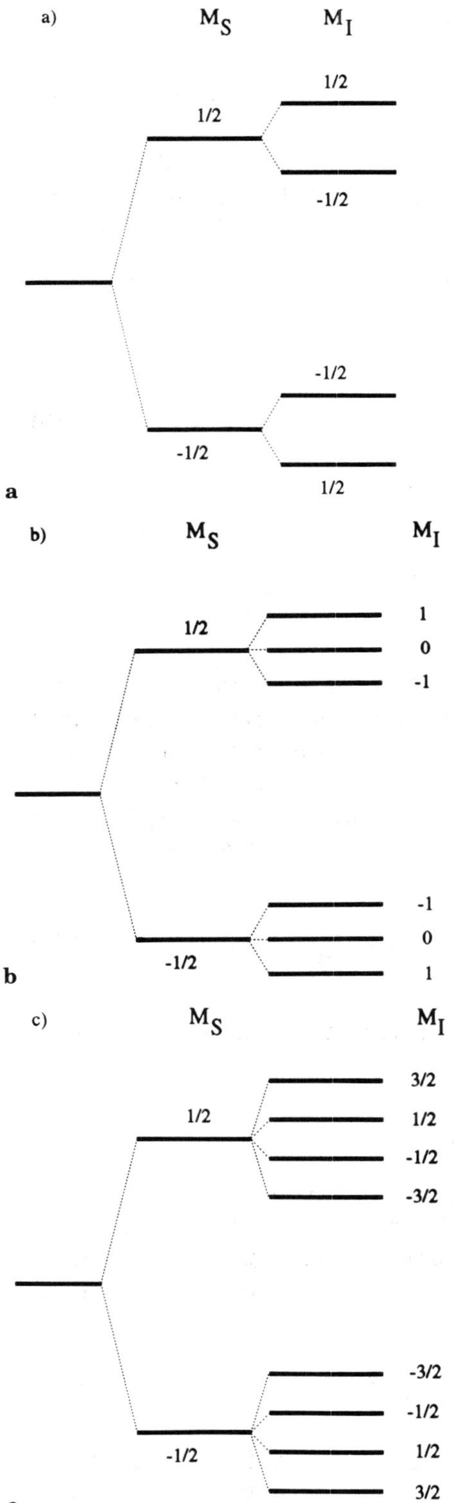

external magnetic field, g is the mean value of the g-tensor, A_k is the constant of hyperfine interaction, and M_k is the projection of the magnetic spin quantum number of the kth nucleus in the direction of an external magnetic field.

As a rule, spectra are measured so that the induction of an external magnetic field is changed at the constant frequency ω. Then the spectral transition appears for an induction

$$B_0 = \frac{\hbar\omega}{\beta g} - \sum_k a_k M_k \qquad \textbf{(6.46)}$$

where

$$a_k = \frac{A_k}{\beta g} \qquad \textbf{(6.47)}$$

One transition determining the g-factor value is observed as a rule in these types of spectra. The hyperfine structure is usually transposed through this transition (to which induction in the spectral midpoint corresponds). This hyperfine splitting gives much information about the surroundings of unpaired electrons and about molecular structure.

Each nucleus with the nonzero magnetic spin number I causes splitting to $2I + 1$ individual bands of the same intensity (Figure 6.3). Practically, the most common nucleus with a nonzero magnetic momentum is the hydrogen nucleus, which has spin $I = 1/2$. If a system contains N_i of such equivalent nuclei (due to the symmetry of molecular structure), the spectrum consists of $N_i + 1$ equidistant bands

Figure 6.3 The scheme of energy level splitting for nuclei with different magnetic quantum numbers I.

with distances that are approximately the value of a hyperfine interaction constant of the corresponding nucleus. The intensity ratio of such bands (Figure 6.4) is equal to the ratio of the coefficients of the functional series of $(1+x)^{N_i}$ — the binomial Pascal's triangle:

Relative band intensity													N_i
						1							0
					1		1						1
				1		2		1					2
			1		3		3		1				3
		1		4		6		4		1			4
	1		5		10		10		5		1		5
1		6		15		20		15		6		1	6

Generally, in the case of N_i equivalent nuclei with a spin quantum number I, the number of bands is equal to $2IN_i + 1$. The individual band intensities are given by the ratio of possible combinations of the spin quantum number projections of individual nuclei. For example, for two equivalent nuclei with spin $I = 1$, five bands with the intensity ratio of 1:2:3:2:1 are observed. Seven bands with the intensity ratio of 1:3:6:7:6:3:1 (Figure 6.5) are observed for three equivalent nuclei with spin $I = 1$, etc. The simple geometrical representation of the splitting constants is useful in determining the number of bands and their relative intensities (Figure 6.6).

For many systems containing more magnetically active nuclei, the hyperfine interaction is complicated and spectra often may be weakly resolved and deformed by spectral noise. Therefore, many numerical methods enabling their interpretation have been developed:

1. Up to now, simulation of ESR spectra[1-5,12,13] using the trial and error method is the most commonly used procedure; however, it is extraordinarily time-consuming and is not always successful for more complicated and weakly resolved spectra.
2. The iterative methods of least squares, using different techniques from nonlinear regression analysis, are also used.[14-18]
3. For the determination of hyperfine interaction constants, Fourier analysis methods,[19,20] auto-correlation functions,[21-24] and the method called cepstral analysis[25] are often used.
4. Methods leading to simplification of spectra or their combination with other methods are also known.[26-28]

Simulation and Optimization of Isotropic ESR Spectra Parameters

A model function can be formed to describe the ESR spectra of radicals in solutions as a linear superposition of individual bands. If N_e is the number

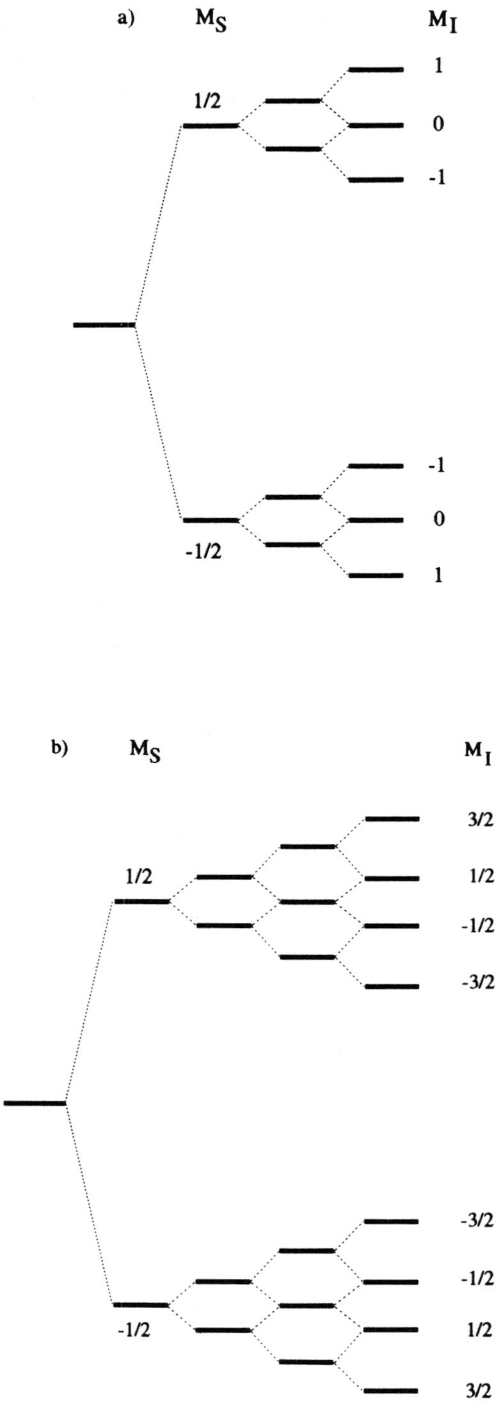

Figure 6.4 The scheme of energy level splitting for nuclei with the magnetic quantum number $I = 1/2$: one nucleus **(a)**, two equivalent nuclei **(b)**, and three equivalent nuclei **(c)**.

a)

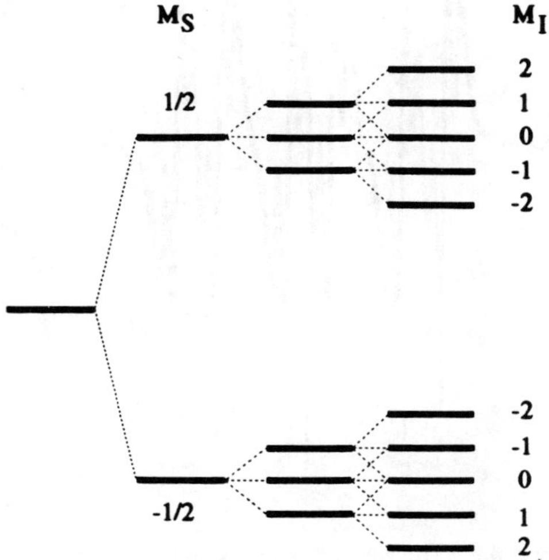

b)

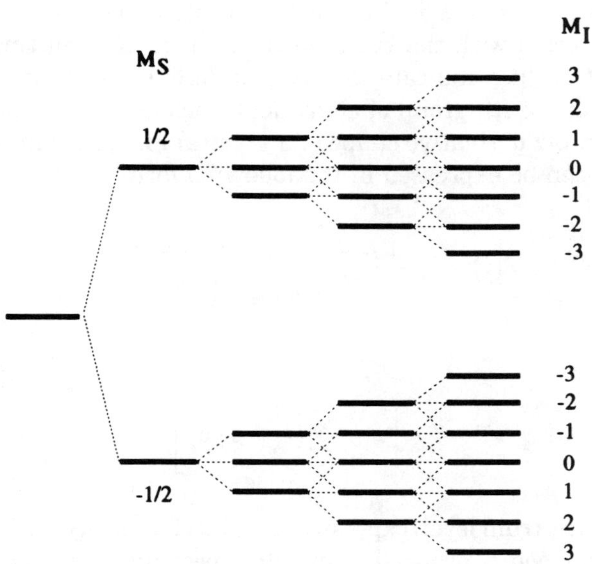

Figure 6.5 The energy level splitting scheme for nuclei with the magnetic quantum number $I = 1$: one nucleus **(a)**, two equivalent nuclei **(b)**, and three equivalent nuclei **(c)**.

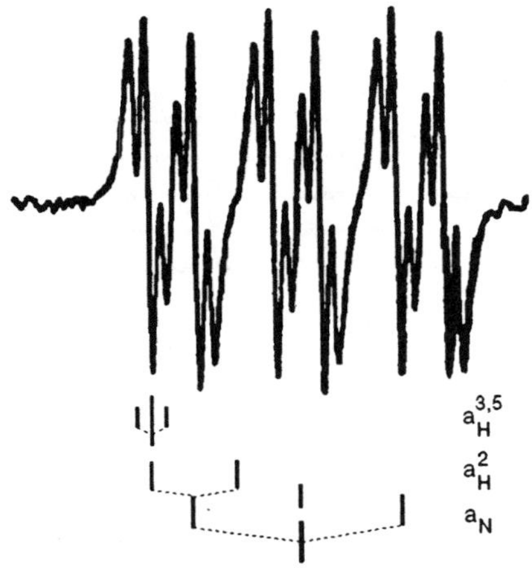

Figure 6.6 The energy level splitting scheme for nuclei with a magnetic quantum number $I = 1$ (the splitting constant a_1), one nucleus with spin $I = 1/2$ (the splitting constant a_2), and two equivalent nuclei with spin $I = 1/2$ (the splitting constant a_3).

of groups of equivalent nuclei (i.e., the number of different splitting constants) and N_i ($i = 1, 2,\ldots N_e$) is the number of bands for individual splitting constants a_i, then $N_1, N_2,\ldots N_e$ different bands can be found in the given spectrum. Each single band is represented by the well-defined group of N_e integer values $k_1, k_2,\ldots k_{N_e}$, in which k_i determines the given band $k_1, k_2,\ldots k_{N_e}$ connected with the kth component of the ith splitting and $k_i = 1, 2,\ldots N_i$. The k_i values indicate mutually different resulting projections of nuclear spin of the ith group of equivalent nuclei.

If the Lorentz derivative shape of a spectral line is assumed, the model function $Y(i)$ can be expressed in the following form:[14]

$$Y(i) = Y_0 - \frac{NW}{D^2} \sum_{k_1,k_2,\ldots k_{N_e}} \frac{M^{(k_1,k_2,\ldots k_{N_e})}\, T}{\left(3W^2 + T^2\right)} \qquad \textbf{(6.48)}$$

where

$$T = I - \left[F + \sum_{i=1}^{N_e} \left(k_i - 1\right) a_i \right] \qquad \textbf{(6.49)}$$

Y_0 is the zero spectrum level (expressed in units of intensity), D is the distance of two adjacent points of an experimental spectrum (expressed in units of magnetic field intensity), W is half of the halfwidth of individual lines (usually considered the same for all spectral lines and defined as the distance between the maximum and midpoint of the corresponding derivative shape, and F is the position of the middle point of the first independent line. $M^{(k_1,k_2,\ldots k_{N_e})}$ is the multiplicative factor for individual bands of a spectrum; it is formed as a product of the simple multiplication of the kth component of the ith splitting ($i = 1, 2,\ldots N_e$). N is a normalization constant.

The model function defined above neglects possible changes of a halfwidth of individual bands (if relaxation effects are dependent on external magnetic induction). The model function (Equation 6.48) can be defined so that the halfwidths are dependent on the magnetic field induction, e.g.,

$$W = a + bB + cB^2 \tag{6.50}$$

where a, b, and c are adjustable parameters. Another possibility is based on applying different estimates for the halfwidths of individual bands.

Another possible route to ESR spectra simulation in solution is the use of a single line shape function to describe the whole spectrum. The splitting of a hyperfine structure into individual bands is guaranteed by applying the resonance condition for individual bands to the line shape function. Then for the total shape of a spectral line

$$Y(i) = N \sum_{k_1,k_2,\ldots k_{N_e}} M^{(k_1,k_2,\ldots k_{N_e})}\, G\left(i; k_1, k_2,\ldots k_{N_e}\right) \tag{6.51}$$

is defined, where N is the normalization constant and $G(i; k_1, k_2\ldots k_{N_e})$ is the function of a line shape. The Gaussian or Lorentz form of the derivative function is used as a rule:

$$G^G\left(i; k_1, k_2,\ldots k_{N_e}\right) = -\frac{B_i - B_0\left(k_1, k_2,\ldots k_{N_e}\right)}{\Delta B^2} \cdot \exp\left\{-2\frac{\left[B_i - B_0\left(k_1, k_2,\ldots k_{N_e}\right)\right]^2}{\Delta B^2}\right\} \tag{6.52}$$

$$G^L\left(i; k_1, k_2,\ldots k_{N_e}\right) = -\frac{B_i - B_0\left(k_1, k_2,\ldots k_{N_e}\right)}{\Delta B^2} \cdot \left\{1 + \frac{4}{3}\frac{\left[B_i - B_0\left(k_1, k_2,\ldots k_{N_e}\right)\right]^2}{\Delta B^2}\right\}^{-2} \tag{6.53}$$

where ΔB is the total halfwidth of a spectral line and B_i is the induction of a magnetic field of the ith spectral point. $B_0(k_1, k_2\ldots k_N)$ is the magnetic field induction at which the resonance condition is fulfilled:

$$B_0\left(k_1, k_2, \ldots k_{N_e}\right) = -\frac{\hbar\omega}{\beta g} - \sum_j M_j \, \mathrm{a}_j \quad \textbf{(6.54)}$$

where M_j is the projection of a spin quantum number of the ith group of equivalent nuclei and a_j is the corresponding splitting constant.

For this type of model function, the variability of width of individual bands of a hyperfine structure can be guaranteed by the dependence of the halfwidth ΔB on the magnetic field induction, e.g.,

$$\Delta B = a + bB + cB^2 \quad \textbf{(6.55)}$$

An example of experimental and simulated ESR spectra of radicals in solutions is shown in Figures 6.7 and 6.8.

When the constant value of hyperfine interaction has a high value, it also is necessary to consider the effects of the second order. The spin Hamiltonian (Equation 6.43) can be written in the form[29]

$$\hat{H} = \beta g B \hat{S}_z + \sum_j A_j \sum_{i=x,y,z} \hat{S}_i \hat{I}_{j,i} \quad \textbf{(6.56)}$$

Subsequently, magnetic field induction fulfilling the resonance condition:

$$B_0\left(k_1, k_2, \ldots k_{N_e}\right) = -\frac{\hbar\omega}{\beta g} - \sum_j \mathrm{a}_j M_j + \frac{\beta g}{2\hbar\omega} \sum_j \mathrm{a}_j^2 \cdot \left[I_j\left(I_j + 1\right) - M_j^2\right] \quad \textbf{(6.57)}$$

is defined.

These types of spectra can be simulated by Equation 6.51, where the resonance condition in accordance with Equation 6.57 is replaced with the function of a line shape (Equations 6.52, 6.53). The Figure 6.9 compares the ESR spectrum of ethyl radicals simulated to the first and second order.[30]

The interpretation of ESR spectra using visual comparison of experimental and simulated model spectra is in many cases insufficient. This is so because:

1. The subjectivity arises from visual comparison of experimental and simulated spectra.
2. Often the strong influence of spectral noise appears and can obscure a hyperfine structure, especially on a spectral edge.
3. Insufficient resolution is achieved due to the strong overlap of individual bands.

Therefore, for the acceptable interpretation of weakly resolved and noisy ESR spectra, methods of nonlinear regression analysis are often used.[14-18]

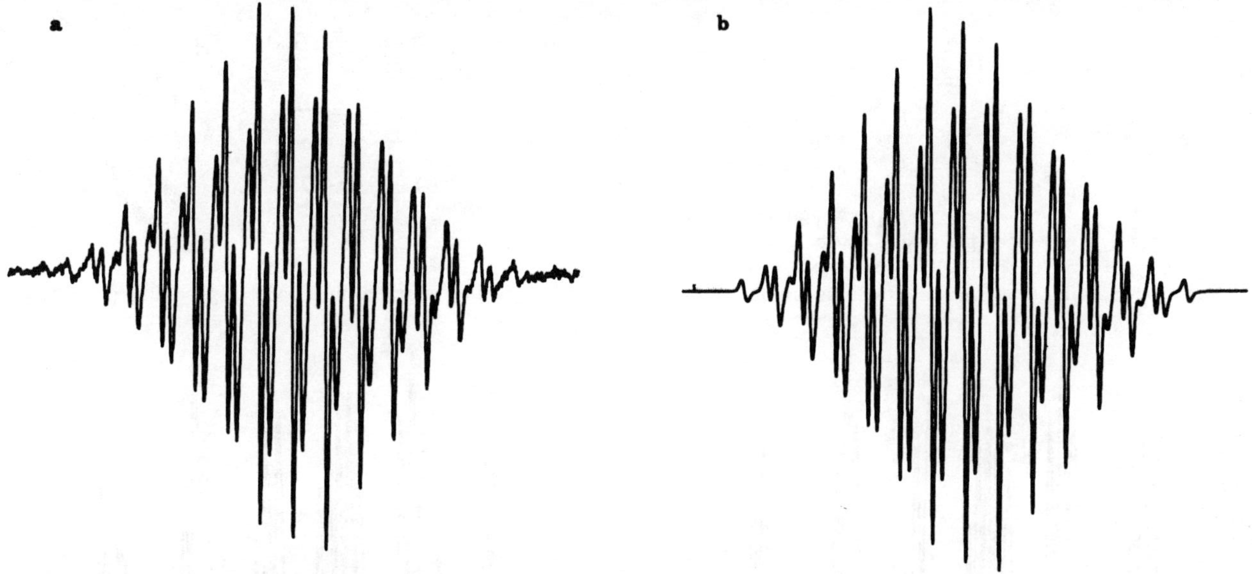

Figure 6.7 The ESR spectrum of radical

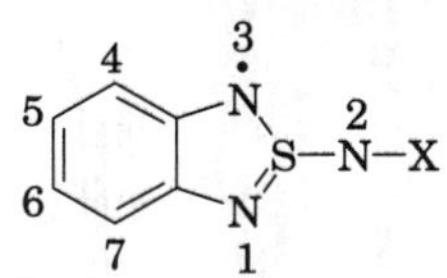

The experimental spectrum **(a)** and the simulated spectrum with the splitting constants a_{N1} = 0.424 mT, a_{N2} = 0.705 mT, a_{N3} = 0.424 mT, $a_{H4,H7}$ = 0.215 mT, and $a_{H5,\,H6}$ = 0.158 mT **(b)**.

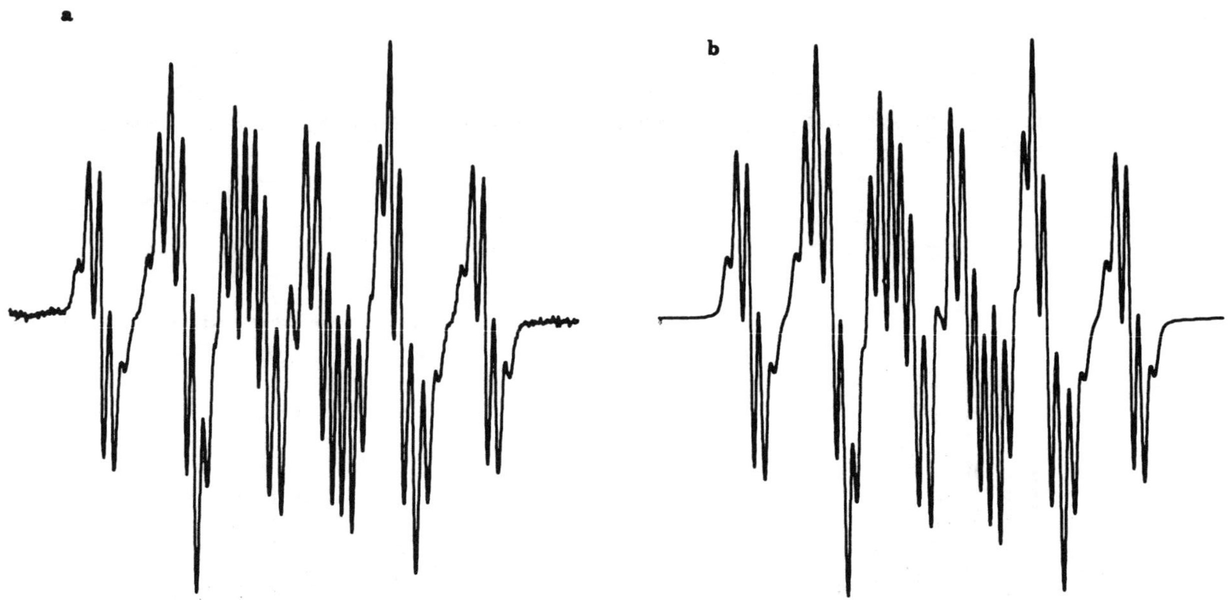

Figure 6.8 The ESR spectrum of anion radical

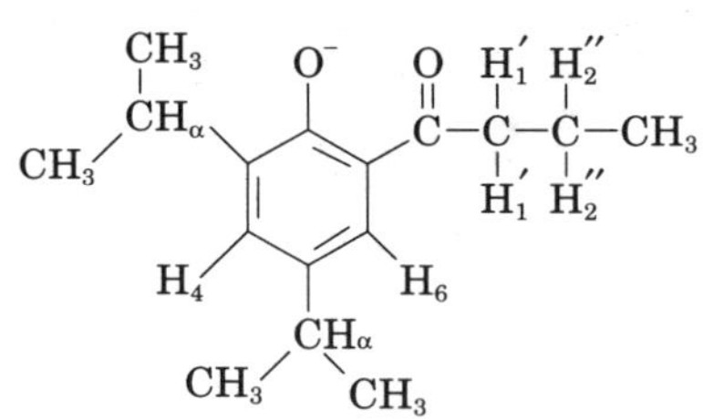

The experimental spectrum **(a)** and the simulated spectrum with the splitting constants $a_{H\alpha} = 0.064$ mT, $a_{H4} = 0.525$ mT, $a_{H6} = 0.428$ mT, $a_{H1'} = 0.828$ mT, $a_{H1''} = 0.575$ mT, and $a_{H2'} = a_{H2''} = 0.064$ mT **(b)**.

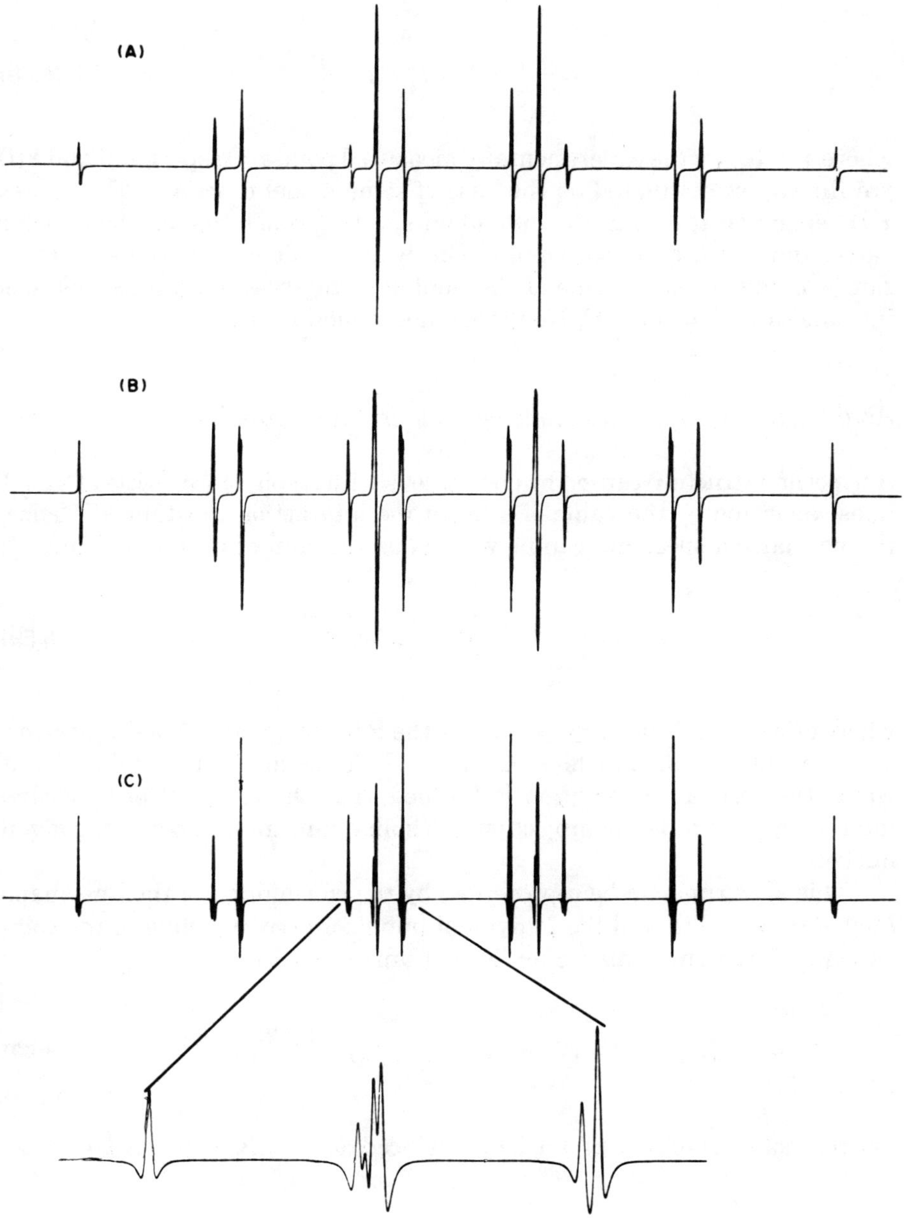

Figure 6.9 Simulation of the ethyl radical ESR spectrum with the splitting constants $a_1 = 2.238$ mT (two nuclei) and $a_2 = 2.687$ mT (three nuclei) with the Lorentz derivative curve halfwidth $\Delta B = 0.017$ mT: the derivative spectrum of the first order **(A)**, the derivative spectrum of the second order **(B)**, and the second order derivative of the second order spectrum **(C)**. (From Joula, H., *Comput. Chem.*, 12, 189, 1988. With permission.)

For this purpose, the following function is built:

$$Q = \sum_{i} \left[Y^{\exp}(i) - Y(i) \right]^2 \qquad \textbf{(6.58)}$$

where $Y^{\exp}(i)$ are the experimentally measured points of a spectrum and $Y(i)$ are the values calculated on the base of some model functions. The values $Y(i)$ depend on the values of individual splitting constants, on the g-factor value, and on the band halfwidths. The optimal values of these parameters can be obtained using some of the nonlinear regression analysis methods introduced in Appendix C, Nonlinear Regression Analysis.

Analysis of Hyperfine Structure by Fourier Transformation

A hyperfine structure caused by one nucleus with a spin of I consists of $2I + 1$ lines separated by the value of a hyperfine interaction constant A. Hence, the normalized spectrum can be written as the sum of normalized lines:[20]

$$R(\omega) = (2I + 1)^{-1} \sum_{m=-I}^{+I} F\left(\omega - \frac{\beta g B_0}{\hbar} - \frac{mA}{\hbar} \right) \qquad \textbf{(6.59)}$$

where ω is the radiation frequency, β is the Bohr magneton, g is the g-factor, $\hbar$ is the Planck constant divided by 2π, B_0 is magnetic field induction at which the resonance condition is fulfilled, A is the constant of hyperfine interaction, and m is the projection of a spin quantum number of the given nucleus.

This spectrum can be represented by the convolution of the line shape $F(\omega)$ of finite width and the individual impulses corresponding to the components of hyperfine interaction $\delta(\omega)$ of unit intensity

$$R(\omega) = (2I + 1)^{-1} F(\omega) * \sum_{m=-I}^{+I} \delta\left(\omega - \frac{\beta g B_0}{\hbar} - \frac{mA}{\hbar} \right) \qquad \textbf{(6.60)}$$

Let the notation of $N = 2I + 1$ be introduced, $\omega_0 = \beta g B_0/\hbar$, $a = A/\hbar$ and

$$S_N(\omega) = \sum_{k=0}^{N-1} \delta\left(\omega - \frac{N-1}{2+k} \right) \qquad \textbf{(6.61)}$$

then for the shape of ESR spectra

$$R(\omega) = \left(\frac{1}{Na} \right) F(\omega) * \delta(\omega - \omega_0)\, S_N\left(\frac{\omega}{\omega_0} \right) \qquad \textbf{(6.62)}$$

can be written. In the case of K different nuclei with hyperfine structures, the constant A_k is defined:

$$R(\omega) = F(\omega) * \delta(\omega - \omega_0) * G(\omega) \tag{6.63}$$

where

$$G(\omega) = \left[\prod_{k=1}^{K} \left(N_k \, \mathrm{a}_k\right)^{-1}\right] S_{N_1}\left(\frac{\omega}{\mathrm{a}_1}\right) * S_{N_2}\left(\frac{\omega}{\mathrm{a}_2}\right) * \ldots S_{N_K}\left(\frac{\omega}{\mathrm{a}_K}\right) \tag{6.64}$$

Let the Fourier transformation of a spectral line (Equation 6.63) be considered. According to the convolution theorem (see Appendix D), the Fourier transformation of the convolution product is equal to the product of the corresponding Fourier transformations:

$$\mathcal{F}\{R(\omega)\} = \mathcal{F}\{F(\omega)\} \, \mathcal{F}\{G(\omega)\} \, \mathcal{F}\{\delta(\omega - \omega_0)\} \tag{6.65}$$

For the individual terms of the product the following is given:

$$\mathcal{F}\{\delta(\omega - \omega_0)\} = \exp\left(i2\pi\omega_0 \, t\right) \tag{6.66}$$

$$\mathcal{F}\{G(\omega)\} = g(t) = \prod_{k=1}^{K} \left[\frac{\mathcal{F}\left\{S_{N_k}\left(\frac{\omega}{\mathrm{a}_k}\right)\right\}}{N_k \, \mathrm{a}_k}\right] \tag{6.67}$$

$$\mathcal{F}\left\{\frac{S_N\left(\frac{\omega}{\mathrm{a}_k}\right)}{\left(N_k \, \mathrm{a}_k\right)}\right\} = \frac{\sin\left(N_k \, \pi \, \mathrm{a}_k \, t\right)}{N_k \sin\left(\pi \, \mathrm{a}_k \, t\right)} \tag{6.68}$$

Due to symmetry, different nuclei have the same constant value of hyperfine interaction. Equation 6.48 can be written in the form

$$\mathcal{F}\{G(\omega)\} = g(t) = \prod_{k=1}^{K_e} \left[\frac{\mathcal{F}\left\{S_{N_k}\left(\frac{\omega}{\mathrm{a}_k}\right)\right\}}{N_k \, \mathrm{a}_k}\right]^{N_k} \tag{6.69}$$

where the product is carried through K_e equivalent sets of nuclei.

If the Fourier transformation of the line-shape function is labeled as $F(\omega)$

$$f(t) = \mathcal{F}\{F(\omega)\} \tag{6.70}$$

Then for the total Fourier transformation of a spectral line, the shape

$$\mathcal{F}\{R(\omega)\} = r(t) = f(t) \prod_{k=1}^{K_e} \left[\frac{\sin(N_k \pi a_k t)}{N_k \sin(\pi a_k t)} \right]^{N_k} \exp(i 2 \pi \omega_0 t) \tag{6.71}$$

results.

Figure 6.10 shows the normalized Fourier transformation of components of a hyperfine structure (Equation 6.69), which corresponds to nuclei with spin I = 1/2 (Figure 6.10a) and I = 1 (Figure 6.10b) for one to four equivalent nuclei (N_k = 1, 2, 3, 4).

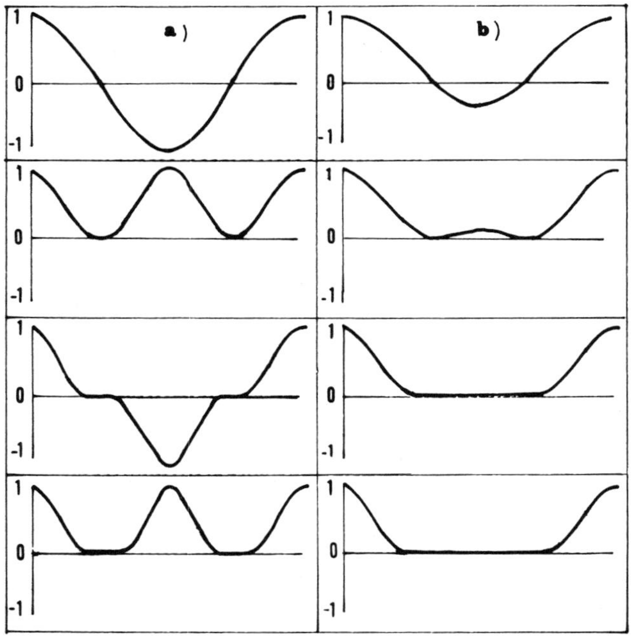

Figure 6.10 The Fourier transformation of the hyperfine structure components corresponding to one, two, three, and four equivalent nuclei: the nuclei with spin I = 1/2 **(a)** and the nuclei with spin I = 1 **(b)**. (From Dunham, W. R., Fee, J. A., Harding L. J., and Grande, H. J., *J. Magn. Reson.*, 49, 351, 1980. With Permission.)

As an example of the Fourier transformation of an experimental ESR spectrum, the ESR spectrum of methyl-viologene (Figure 6.11a) dissolved in a buffering phosphate solution after irradiation by visible light can be introduced. Fourier analysis of this spectrum (Fig. 6.11b) shows the presence of the following splitting constants:

$a_N = 0.423$ mT (2 equivalent nuclei)
$a_N = 0.157$ mT (4 equivalent nuclei)
$a_N = 0.133$ mT (4 equivalent nuclei)
$a_N = 0.399$ mT (6 equivalent nuclei)

Cepstral Analysis

Cepstral analysis uses some of the properties of Fourier transformation of ESR spectra.[25] As introduced in the previous section, the Fourier transformation of ESR spectra can be written in the form

$$\mathcal{F}\{R(\omega)\} = r(t) = f(t) \prod_{k=1}^{K_e} \left[\frac{\sin(N_k \pi a_k t)}{N_k \sin(\pi a_k t)} \right]^{N_k} \exp(i2\pi\omega_0 t) \qquad \textbf{(6.72)}$$

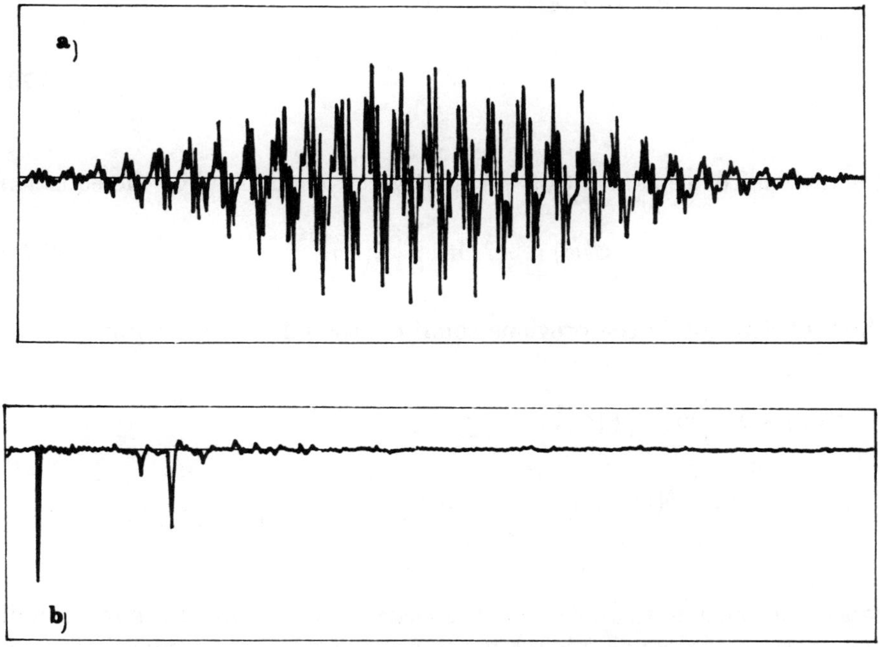

Figure 6.11 Fourier analysis of the methyl-viologen ESR spectrum: the experimental spectrum **(a)** and the Fourier transformation of the experimental spectrum **(b)**. (From Dunham, W. R., Fee, J. A., Harding, L. J., and Grande, H. J., *J. Magn. Reson.*, 40, 351, 1980. With permission.)

where $f(t)$ is the Fourier mapping of a line shape without hyperfine interaction:

$$f(t) = \mathcal{F}\{F(\omega)\} \tag{6.73}$$

The single terms of the product function (Equation 6.53) correspond to the peaks of hyperfine interactions for K_e sets of equivalent nuclei. The natural logarithm of the absolute value of $r(t)$ can be written as follows:

$$\ln|r(t)| = \ln|f(t)| + \sum_{k=1}^{K_e} N_k \left[\ln \left| \frac{\sin(N_k \pi a_k t)}{N_k \sin(\pi a_k t)} \right| \right] \tag{6.74}$$

Because the Fourier series of an absolute value of a natural logarithm of a sinusoidal function (Equation 6.75) is defined as

$$\ln|\sin(\pi a t)| = -\ln 2 - \sum_{j=1}^{\infty} \frac{cox(2j\pi a t)}{j} \tag{6.75}$$

so that

$$\ln\left[\mathcal{F}\left\{\frac{\sin(N\pi a t)}{N\sin(\pi a t)}\right\}\right] = \sum_{j=1}^{\infty} \frac{\cos(2\pi j a t)}{j} - \sum_{k=1}^{\infty} \frac{\cos(2\pi k N a t)}{k} - \ln N \tag{6.76}$$

is obtained. The cepstrum $C(\omega)$ for ESR spectra is defined by the equation

$$C(\omega) = \mathcal{F}^{-1}\left\{\ln\left|\mathcal{F}\{R(\omega)\}\right|^2\right\} \tag{6.77}$$

After substituting in the previous equation, the following results:

$$C(\omega) = \mathcal{F}^{-1}\left\{\ln\left|\mathcal{F}\{F(\omega)\}\right|^2\right\} + \sum_{k=1}^{N_e} N_k \cdot \left[\sum_{i=1}^{\infty} \frac{\delta(\omega \pm i a_k)}{i} - \sum_{j=1}^{\infty} \frac{\delta(\omega \pm j N_k a_k)}{j} - 2(\ln N_k)\,\delta(\omega)\right] \tag{6.78}$$

Hence, the cepstrum of a resonance spectrum is the sum of the individual line cepstra corresponding to a hyperfine structure of single nuclei and to a complete line shape.

An ESR spectrum of a radical with one nucleus with spin $I = 1/2$ and with a hyperfine interaction constant value a = 1 is shown in Figure 6.12.

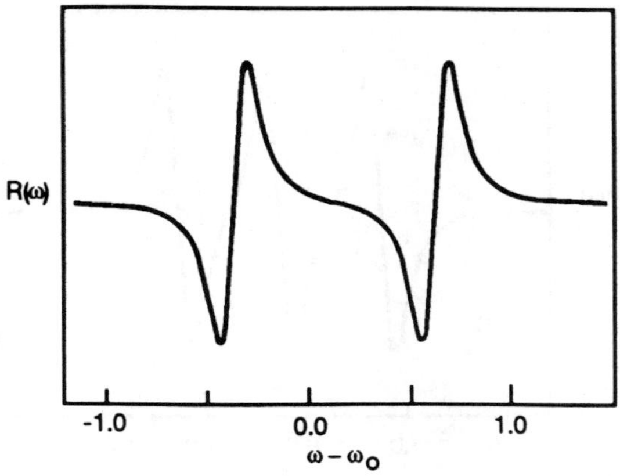

Figure 6.12 The hyperfine structure corresponding to one nucleus with spin I = 1/2 and with the hyperfine interaction constant a = 1. (From Kirmse, D. W., *J. Magn. Reson.*, 11, 1, 1971. With permission.)

A corresponding spectrum is shown in Figure 6.13. The cepstral lines change sign and decrease, as is indicated by Equation 6.78. The value of the hyperfine interaction constant can be determined from the position of the first cepstral band or from the difference in position of two successive peaks.

An ESR spectrum of radicals with one nucleus of spin I = 1 is shown in Figure 6.14. Its cepstrum is shown in Figure 6.15. Again, the value of a hyperfine interaction constant can be determined from the position of the first peak of a cepstrum or from the difference in position of two successive peaks. The fact that every third peak of a cepstrum has a negative sign is indicative that the spin I = 1.

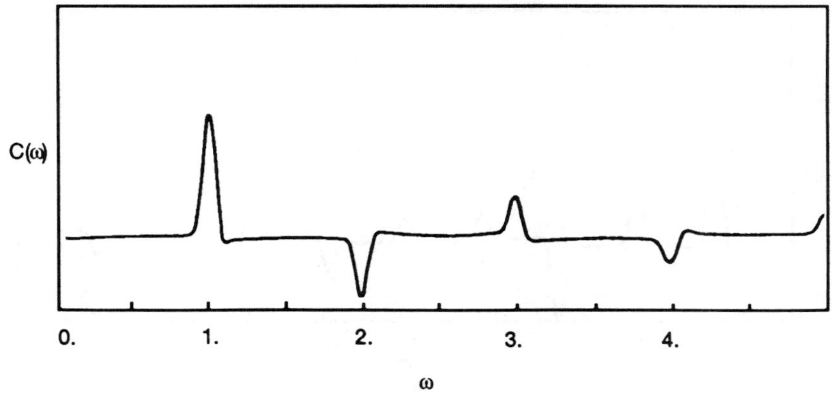

Figure 6.13 The cepstrum of the ESR spectrum corresponding to one nucleus with spin I = 1/2 with the hyperfine interaction constant a = 1. (From Kirmse, D. W., *J. Magn. Reson.*, 11, 1, 1971. With permission.)

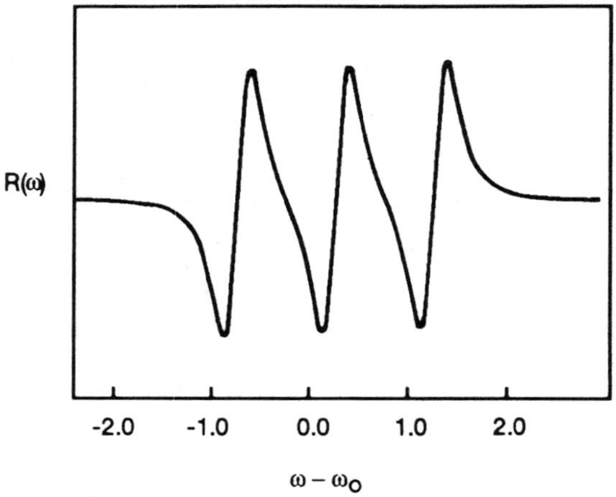

Figure 6.14 The hyperfine structure corresponding to one nucleus with spin $I = 1$ with the hyperfine interaction constant a = 1. (From Kirmse, D. W., *J. Magn. Reson.,* 11, 1, 1971. With permission.)

The spectrum in Figure 6.16 represents a hyperfine structure of 10 nuclei with spin $I = 1/2$, which can be divided into three groups of equivalent nuclei. The cepstrum of this spectrum (Figure 6.17) shows three main lines corresponding to the nuclei with the following values for a hyperfine interaction constant: a = 0.25 mT (a pair of nuclei), a = 0.3 mT (a group of four nuclei), and a = 0.45 mT (a group of four nuclei). The bands at 0.5 mT and 0.6 mT correspond to the first of the two bands, but the negative band corresponding to double the third hyperfine interaction constant (0.9 mT) is overlapped by the positive cepstral peak corresponding to triple the splitting constant.

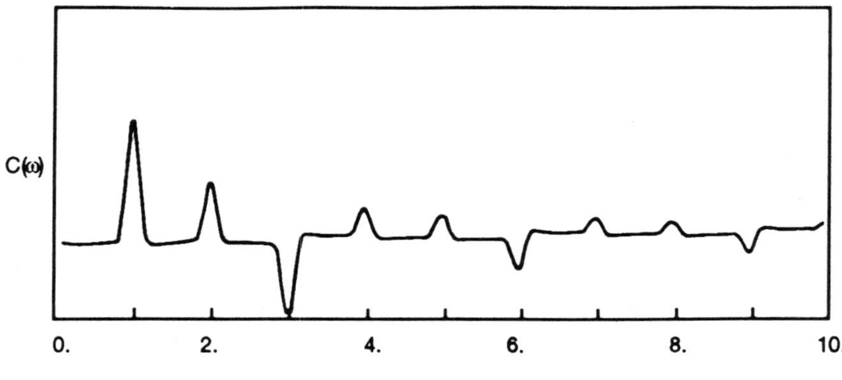

Figure 6.15 The cepstrum of the ESR spectrum corresponding to one nucleus with spin $I = 1$ with the hyperfine interaction constant a = 1. (From Kirmse, D. W., *J. Magn. Reson.,* 11, 1, 1971. With permission.)

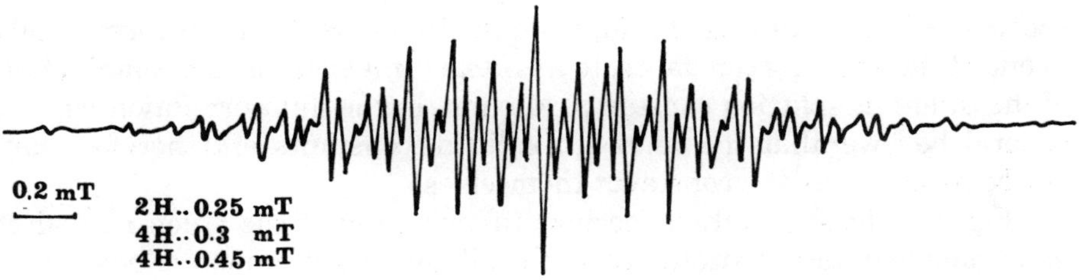

Figure 6.16 The ESR spectrum corresponding to 10 nuclei with spin I = 1/2 with the hyperfine interaction constants a_1 = 0.25 mT (2×), a_2 = 0.30 mT (4×), and a_3 = 0.45 mT (4×). (From Kirmse, D. W., *J Magn. Reson.*, 11, 1, 1971. With permission.)

Application of Correlation Functions for Analysis of Symmetric ESR Spectra

Application of correlation functions is one way to determine the splitting constants of symmetric ESR spectra.[21-24]

The autocorrelation function $A(\Delta B)$ is defined by the equation

$$A(\Delta B) = \int_{-\infty}^{+\infty} Y(B)\, Y(B - \Delta B)\, dB \qquad \textbf{(6.79)}$$

where $Y(B)$ is the intensity of ESR spectra as a function of external magnetic field induction B. The autocorrelation function, defined above, shows local maxima when ΔB is equal to the distance between a pair of lines in a

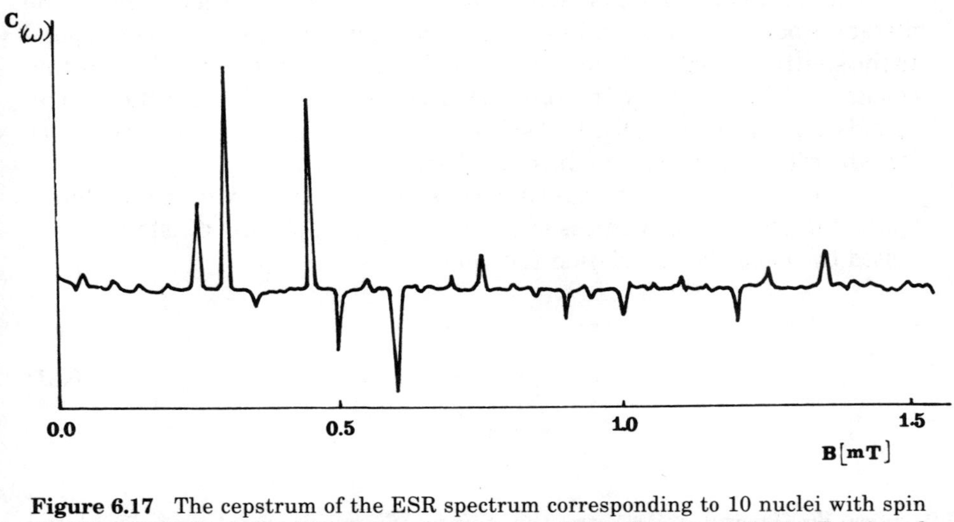

Figure 6.17 The cepstrum of the ESR spectrum corresponding to 10 nuclei with spin I = 1/2 with the hyperfine interaction constants a_1 = 0.25 mT (2×), a_2 = 0.30 mT (4×), and a_3 = 0.45 mT (4×). (From Kirmse, D. W., *J. Magn. Reson.*, 11, 1, 1971. With permission.)

spectrum. The maximum of a function $A(\Delta B)$ occurs when ΔB corresponds to one of the splitting constants, to its integer multiple, or to a combination of the different splitting constants. Maxima in the autocorrelation will, in general, be fewer than the number of splitting constants, and therefore may not be as useful as the constants themselves.

Figure 6.18 shows the experimental spectrum of the anion radical of naphthalene in tetrahydrofurene in the alkaline environment of potassium hydroxide. The autocorrelation function (Figure 6.19) shows two non-equivalent protons ($I = 1/2$) with different values of splitting constants.

Motten and Schreiber[24] proposed an autocorrelation function convenient for analyzing digitized ESR spectra. Let $Y(i)$ be an experimental spectrum, $i = 1, N$, in which N is the number of digitized spectral points. The normalized autocorrelation function $A(n)$ (n is the shift of spectral points) subsequently can be defined by

$$A(n) = \frac{\left\{ \sum_{i=1}^{N-n} Y(i)\,Y(i+n) - \frac{\sum_{i=1}^{N-n} Y(i) \cdot \sum_{i=1}^{N-n} Y(i+n)}{N-n} \right\}}{\left\{ \left(\sum_{i=1}^{N-n} Y(i)^2 - \frac{\left[\sum_{i=1}^{N-n} Y(i) \right]^2}{N-n} \right)^{\frac{1}{2}} \left(\sum_{i=1}^{N-n} Y(i+n)^2 - \frac{\left[\sum_{i=1}^{N-n} Y(i+n) \right]^2}{N-n} \right)^{\frac{1}{2}} \right\}} \tag{6.80}$$

The values of a function $A(n)$ are from the interval between 0 and 1. The distances between local maxima of these functions and the origin correspond to the splitting constant — or to a sum, difference, or multiple of the splitting constants. The position of the highest maximum of an $A(n)$ function corresponds, usually, to an individual splitting constant; however, some examples for which this intuitive rule has failed are already known.[24]

After evaluation of the splitting constant, it is necessary to simulate a spectrum, and the correctness of determining the splitting constant can be tested by using the correlation function:

$$P = \frac{\sum_{i=1}^{N} Y(i)^{\exp}\, Y(i)^{sim}}{\left[\sum_{i=1}^{N} \left(Y(i)^{sim} \right)^2 \right]^{\frac{1}{2}}} \tag{6.81}$$

where $Y(i)^{\exp}$ and $Y(i)^{sim}$ are the points of experimental and simulated spectra, respectively. The maximal value of a correlation function P shows the most efficient choice of hyperfine interaction constants.

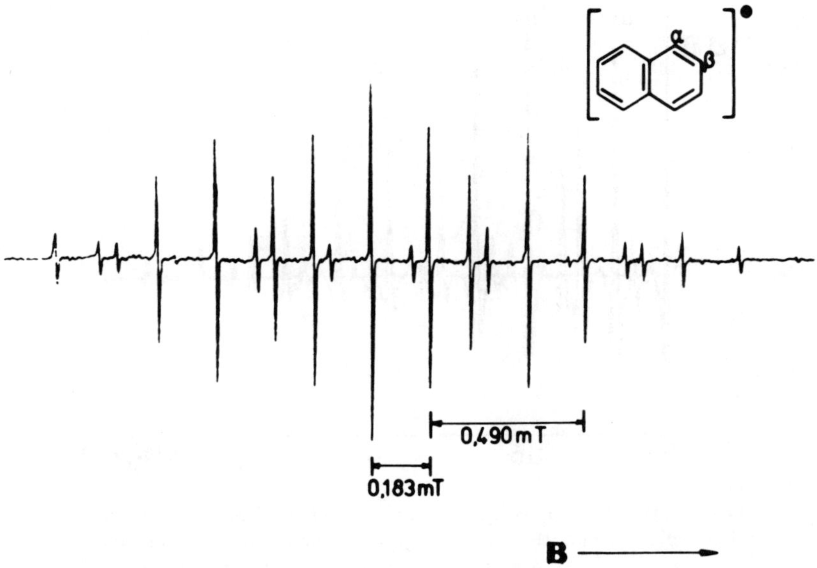

Figure 6.18 The ESR spectrum of the anion radical of naphthalene in tetrahydrofuran in an alkaline environment (KOH). The splitting constants are 0.183 mT for β-protons and 0.490 mT for α-protons. (From Grampp, G. and Schiller, C. A., *Anal. Chem.*, 53, 560, 1984. With permission.)

Subsequent Determination and Elimination of Splitting Constants

This method is based on the subsequent determination and elimination of the individual splitting constants from an original spectrum. This process is repeated until only a simple spectrum without hyperfine interaction is obtained.[26]

Let us consider a spectrum Y (Figure 6.20A) consisting of two subspectra y_A and y_B (Figure 6.20B), which are identical, except for the fact that they are mutually shifted by the hyperfine interaction constant a (for spin $I = 1/2$). Let the following function $S(i)$ be defined:

$$S(i) = \sum_{j=j_0}^{j_l} (-i)^j \; Y(i + jx) \tag{6.82}$$

where i is any chosen point of a spectrum

$$Y(i) = y_A(i) + y_B(i) \tag{6.83}$$

and j is the index of summation through individual points of a spectrum. Single subspectra are identical, but they are shifted by a corresponding splitting constant:

$$y_A(i) = y_B(i + \mathrm{a}) \tag{6.84}$$

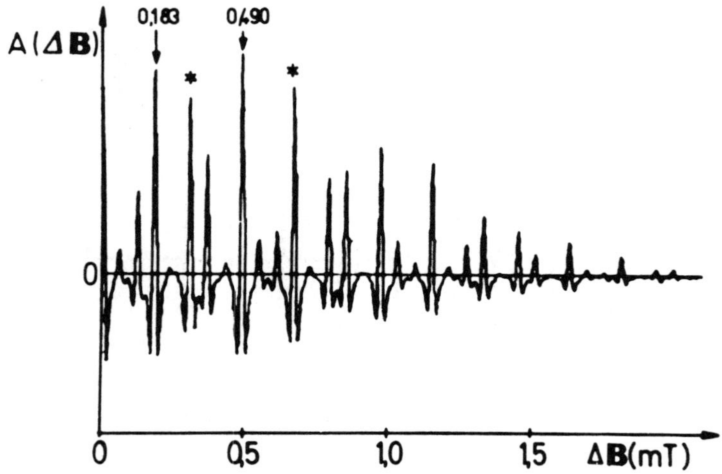

Figure 6.19 The autocorrelation function of the ESR spectrum of the anion radical of naphthalene. The splitting constants are indicated and their sums and differences are labeled by a star. (From Grampp, G. and Schiller, C. A., *Anal. Chem.*, 53, 560, 1984. With permission.)

If the summation is carried out over the whole spectrum, then in the case of an identical spectrum without spectral noise, the function $S(i)$ will acquire the null value at x = a.

Placing Equation 6.83 into Equation 6.63 the following results:

$$S(i) = \sum_{j=j_0}^{j_l} (-1)^j \; y_A(i + jx) + \sum_{j=j_0}^{j_l} (-1)^j \; y_B(i + jx) \tag{6.85}$$

Using Equation 6.84 for $S(i)$:

$$S(i) = \sum_{j=j_0}^{j_l} (-1)^j \; y_A(i + jx) + \sum_{j=j_0}^{j_l} (-1)^j \; y_A(i + jx - \mathrm{a}) \tag{6.86}$$

respective to

$$S(i) = \sum_{j=j_0-1}^{j_l-1} (-1)^{j+1} \; y_A\left[i + (j+1)x\right] + \sum_{j=j_0}^{j_l} (-1)^j \; y_A(i + jx - \mathrm{a}) \tag{6.87}$$

can be written and for x = a:

$$\begin{aligned} S(i) &= \sum_{j=j_0-1}^{j_l-1} (-1)^{j+1} \; y_A(i + j\mathrm{a} + \mathrm{a}) + \sum_{j=j_0}^{j_l} (-1)^j \; y_A(i + j\mathrm{a} - \mathrm{a}) \\ &= (-1)^{j_0} \; y_A(i + j_0\mathrm{a}) + (-1)^{j_l} \; y_A(i + j_l\mathrm{a} - \mathrm{a}) \end{aligned} \tag{6.88}$$

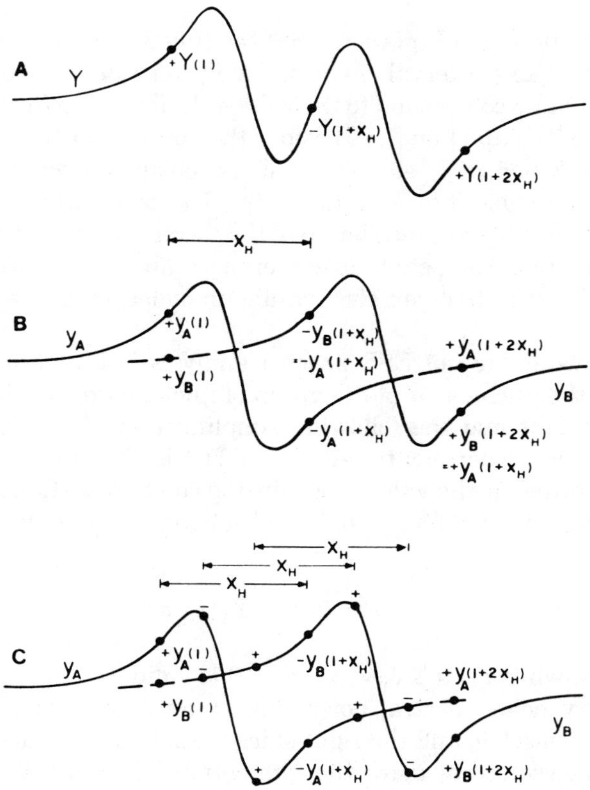

Figure 6.20 The model spectra describing the relationship between the original spectrum and the subspectra: the original spectrum **(A)**, the subspectra shifted by the hyperfine interaction constant a = x_H **(B)**, and the function $\mathcal{S}(x)$ for x = a/3 **(C)**. (From Motten, A. G., Duling, D. R., Schreiber, J., *J. Magn. Reson.*, 71, 34, 1987. With permission.)

is valid. If x = a, all the terms except one are eliminated from every summation. If j_0 and j_i are chosen close to the end points of a spectrum, the two remaining terms will be very low. The contribution of these two members causes the summation to be realized through the whole spectrum (and not through only one part), obtaining the zero value for $S(i)$. Therefore, $i - j_0x$ must be the first point of a spectrum and $i + j_0x$ must be the last.

The function $S(i)$ uses information from only a small number of spectral points. Let us form the following function including the maximal available information:

$$\mathcal{S}(x) = \sum_{i=1}^{x-1} \left| S(i) \right| = \sum_{i=1}^{x-1} \left| \sum_{j_0=0}^{j_l} (-1)^j \, Y(i+jx) \right| \tag{6.89}$$

We can calculate (from Equation 6.89) the function $\mathcal{S}(x)$ for all values of x from the reasonable interval of the splitting constants. Then the values of splitting constants correspond to the values of x for which $\mathcal{S}(x)$ is minimized.

Minima of the function $\mathcal{S}(x)$, besides the points x = a, also appear in the points x = a/3, a/5, a/7, etc. This can be easily proven by placing the corresponding values into Equation 6.89. The odd parts of splitting constants cause the mutual cancellation of the corresponding contributions, but the even parts (a/2, 3a/4, etc.) cause their accumulation. In weakly resolved spectra, the function $\mathcal{S}(x)$ can also contain many local minima, which will be very shallow.

The following step of ESR spectra analysis is the elimination of the influence of the nucleus, or of the group of nuclei with the given value of a splitting constant and the following simplification of a spectrum. Let us suppose that an experimental spectrum $Y(i)$ is the sum of two identical subspectra shifted at the value of a splitting constant a (Equation 6.83). By connecting Equations 6.83 and 6.84 and solving for $y_A(i)$, the following will be yielded:

$$Y_A(i) = Y(i) - Y_A(i - \mathrm{a}) \tag{6.90}$$

but for cases, when i – a ≤ 0, $y_A(i - \mathrm{a}) = 0$ is valid.

With very noisy spectra, noise can accumulate during generation of individual subspectra, and the right sides of spectra are much noisier. This problem can be solved by using the symmetry of the ESR spectrum. Therefore, noise can be uniformly distributed through the whole generated spectrum. If the spectrum $Y(i)$ is centered at the midpoint, the symmetry requirement is that

$$Y_A(i) = -Y_B(N - i) \tag{6.91}$$

where N is the total number of spectral points. Using Equations 6.90 and 6.91, and after solving for $y_A(i)$ in the terms of the original spectrum, the recurrent formula can be derived:

$$2y_A(i) = Y(i) + \sum_{j_0=1}^{j_l} (-1)^j \, Y(i - j\mathrm{a}) + \sum_{k=0}^{k_l} (-1)^k \, Y(N - i - k\mathrm{a}) \tag{6.92}$$

where j_j = INT (i/a), k = INT [($N - i$)/a], and the function INT gives the closest lower integer value for an argument. In the final subspectrum y_A, the terms from both sides of a spectrum $Y(i)$ contribute to each point i, and spectral noise is uniformly distributed though the whole subspectrum y_A.

The decomposition of one part of a nitrogen triplet (a_N = 2.938 mT) corresponding to the lowest triplet component of anion radical of 1-oxo-1,2-dihydronaphthalene of 4-sulpho-iminoxyl is introduced as an example of the application of this analysis.[26] An ESR spectrum of this radical contains the splitting of various protons, but the spectral resolution is very weak (Figure 6.21A). The function $\mathcal{S}(x)$ (Figure 6.21C) shows a well-defined minimum at the value of $x \approx 0.0419$ mT, three minima in the interval of 0.012 to 0.018 mT, and many minima below 0.008 mT, which can represent the false minima corresponding to the odd parts of the splitting constants (see above). If the part corresponding to the splitting with a hyperfine interaction constant a = 0.0419 mT is subtracted from the experimental spectrum, the spectrum in Figure 6.21D will be yielded. A subspectrum with the value of a splitting constant a = 0.0176 mT (Figure 6.21E) and another with splitting constant a = 0.007 mT (Figure 6.21F) were also subtracted from the same experimental spectrum. One component spectrum (Figure 6.21G), which proves that the values of all hyperfine interaction constants were already determined, was obtained after subtracting the corresponding subspectra. The splitting constants calculated by this procedure were used to simulate a theoretical spectrum (Figure 6.21B).

The method is sufficiently reliable for weakly resolved spectra, but it is relatively sensitive to spectral noise. Therefore, it is advantageous to apply some methods to filter out spectral noise before application of this method (see Chapter 1). With extremely weakly resolved spectra, it is convenient, first, to differentiate the experimental spectrum (i.e., to yield a spectrum corresponding to the second derivative of an absorption profile) and only afterward apply the introduced procedure.

Numerical Decoupling Analysis

Numerical decoupling analysis of isotropic ESR spectra is an efficient method for complicated or weakly resolved spectra. The application of this analysis can be split into two steps:[27]

1. Some constants of hyperfine interaction can be determined using the least squares method, and any part of an original spectrum corresponding to the splitting is eliminated from it.
2. The final numerical decoupling is evaluated by examination of the special index function. From this examination, whether or not the given value is the real splitting constant can be determined.

Analysis of Nuclei with Spin I = *1/2.* If a nucleus with spin $I = 1/2$ is given, the value of a hyperfine interaction constant a, contributing to an isotropic spectrum with a constant halfwidth, gives the function of a line shape $R(\omega)$:

$$R(\omega) = F\left(\omega + \frac{\mathrm{a}}{2}\right) + F\left(\omega - \frac{\mathrm{a}}{2}\right) \quad \textbf{(6.93)}$$

where $F(\omega)$ is an ESR spectrum with no contribution from hyperfine splitting. This spectrum is called a subspectrum. $R(\omega)$ is represented by the

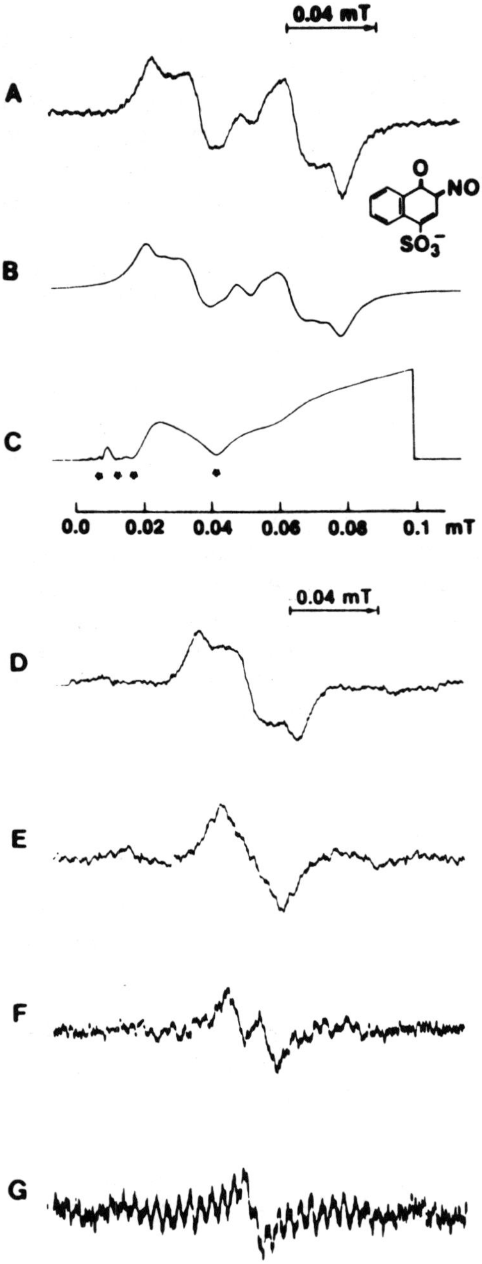

Figure 6.21 The application of the method of subsequent determination and elimination of splitting constants: one component of sodium triplet of the ESR spectrum of the anion radical of 1-oxo-1,2-dihydronaphthalene-4-sulpho-iminoxyl **(A)**, the computer simulation with the constants of hyperfine interaction 0.0425, 0.0175, 0.0126, and 0.007 mT **(B)**, the function $\mathcal{S}(x)$ (the stars label the hyperfine interaction constants) **(C)**, the spectrum obtained after the elimination of the hyperfine structure with the splitting constant a = 0.0419 mT **(D)**, the spectrum after elimination of the hyperfine structure with the splitting constants 0.0419 and 0.0176 mT **(E)**, the spectrum after elimination of the hyperfine structure with the splitting constants 0.0419, 0.0176, and 0.007 mT **(F)**, and the final one-line spectrum after subtracting the hyperfine structure caused by all hyperfine interaction constants **(G)**. (From Motten, A. G., Duling, D. R., and Schreiber, J., *J. Magn. Reson.*, 71, 34, 1987. With permission.)

series of experimental data for values of a magnetic field increasing by a constant increment.

Let $S(\omega)$ be any function for which

$$R(\omega) = S\left(\omega + \frac{x}{2}\right) + S\left(\omega - \frac{x}{2}\right) \tag{6.94}$$

is valid for at least one value of x. This x value can be determined by minimizing the function:

$$d(x) = \sum_{i=\frac{1+x}{2}}^{\frac{N-x}{2}} \left[R(i) - S\left(i + \frac{x}{2}\right) - S\left(i - \frac{x}{2}\right)\right]^2 \tag{6.95}$$

where N is the total number of spectral points and $x/2$ takes on all integer values in the interval of $(0, N)$.

A minimization of $d(x)$ leads to the system of linear equations:

$$\begin{aligned}
S(1) + 2S(1+x) + S(1+2x) &= R\left(1 + \frac{x}{2}\right) + R\left(1 + \frac{3x}{2}\right) \\
S(2) + 2S(2+x) + S(2+2x) &= R\left(2 + \frac{x}{2}\right) + R\left(2 + \frac{3x}{2}\right) \\
\ldots \qquad\qquad & \ldots \qquad\qquad \ldots \\
S(N-2x) + 2S(N-x) + S(N) &= R\left(N - \frac{3x}{2}\right) + R\left(N - \frac{x}{2}\right)
\end{aligned} \tag{6.96}$$

If the fact that the intensities on spectral edges have practically zero value is taken into consideration, then

$$S(1) = S(2) = \ldots = S(x) = 0 \tag{6.97}$$

is valid. The function d(x) obtains the minimal values ($d[x] = 0$) for values x = a, but also for a/3, a/5, a/7, etc.:

$$S(N - x + 1) = \ldots = S(N) = 0 \tag{6.98}$$

$$d(x) = 0 \tag{6.99}$$

for

$$x_n = \frac{\mathrm{a}}{2n+1} \qquad n = 0, 1, 2, 3, \ldots \tag{6.100}$$

Values $S_n(\omega)$ corresponding to the values x_n can be expressed in the following form:

$$S_n(\omega) = \sum_{j=-n}^{n} (-1)^{j+n} F(\omega - jx_n) \tag{6.101}$$

These values x_n (besides $n = 0$) are not real constants of hyperfine interaction; they can be called "false" constants of hyperfine interaction. Therefore, for ESR spectra that contain more hyperfine interaction constants, a function $d(x)$ cannot be used to evaluate correct values for hyperfine interaction constants.

For the differentiation of correct values, instead of false values for hyperfine interaction constants, the following index function can be defined:

$$I_2(x) \equiv \left\{ \frac{\int_{-\infty}^{+\infty} \left| R(\omega) - R(\omega - \delta) \right| d\omega}{\sum_{j=1}^{2} \int_{-\infty}^{+\infty} \left| R(\omega) - S\left(\omega - \frac{x}{2}\right) - S\left[\omega + \frac{x}{2} + (-1)^j \delta\right] \right| d\omega} \right\}^2 \tag{6.102}$$

where δ has a low value (usually $\delta = 1$ or $\delta = 2$, which are very low values compared to those of N). Labeling the index function $I_2(x)$ takes into consideration the nuclear spin $I = 1/2$, with the multiplicity of $2I + 1 = 2$. The numerator of Equation 6.102 guarantees the normalization of the function $I_2(x)$. Because the denominator of $I_2(x)$ is also the error function, it comes to zero for $x = \mathrm{a}$. So for this value $I_2(x)$ takes on the maximal value. Using Equations 6.93, 6.100, and 6.102, $I_2(x)$ can be rewritten as

$$I_2(\mathrm{a}) = \left[\frac{\int_{-\infty}^{+\infty} \left| F'\left(\omega + \frac{\mathrm{a}}{2}\right) - F'\left(\omega - \frac{\mathrm{a}}{2}\right) \right| d\omega}{2 \int_{-\infty}^{+\infty} \left| F'\left(\omega + \frac{\mathrm{a}}{2}\right) \right| d\omega} \right]^2 \tag{6.103}$$

where $F'(\omega)$ represents the approximate numerical derivative of the function $F(\omega)$:

$$F'(\omega) = \frac{F(\omega) - F(\omega + \delta)}{\delta} \tag{6.104}$$

or

$$F'(\omega) = \frac{F(\omega) - F(\omega - \delta)}{\delta} \tag{6.105}$$

If $F'(\omega + a/2)$ and $F'(\omega - a/2)$ are well resolved (not overlapped) functions, the numerator of Equation 6.102 can be written:

$$\int_{-\infty}^{+\infty} \left| F'\left(\omega + \frac{a}{2}\right) + F'\left(\omega - \frac{a}{2}\right) \right| d\omega \approx \int_{-\infty}^{+\infty} \left| F'\left(\omega + \frac{a}{2}\right) \right| d\omega$$

$$+ \int_{-\infty}^{+\infty} \left| F'\left(\omega - \frac{a}{2}\right) \right| d\omega = \int_{-\infty}^{+\infty} \left| F'(\omega) \right| d\omega \qquad \textbf{(6.106)}$$

and $I_2(a)$ is close to 1. Because most measured bands are more or less overlapped, $I_2(a)$ is usually lower than 1. It is necessary to note that the shape of the spectral line $R(\omega)$ is not specified. Analysis can be improved by using spectral line derivatives of higher orders, which are especially convenient for weakly resolved spectra (Figure 6.22).

It can be proved that the ratio

$$\frac{I_2\left(\frac{a}{3}\right)}{I_2(a)} = \left[\frac{\int_{-\infty}^{+\infty} \left| F'\left(\omega + \frac{a}{2}\right) \right| d\omega}{\int_{-\infty}^{+\infty} \left| F'\left(\omega + \frac{a}{2}\right) - F'\left(\omega + \frac{a}{6}\right) + F'\left(\omega - \frac{a}{6}\right) \right| d\omega} \right]^2$$

$$= \left[\frac{\int_{-\infty}^{+\infty} \left| F'(\omega) \right| d\omega}{\int_{-\infty}^{+\infty} \left| F'\left(\omega + \frac{a}{3}\right) - F'(\omega) + F'\left(\omega - \frac{a}{3}\right) \right| d\omega} \right]^2 \qquad \textbf{(6.107)}$$

can be changed within the interval (1/9, 1) depending on the overlap of functions $F'(\omega + a/3)$ and $F'(\omega - a/3)$. If the overlap of these functions is small (i.e., the spectrum is well resolved), this ratio is equal only to 1/9. This value increases with decreasing spectral solution, but it is always lower than 1. Similar consideration is valid also for the values a/5, a/7, etc. Figures 6.23 and 6.24 show the relationship between an original spectrum and its subspectra. The dependence of the index function $I_2(x)$ on the band width (the degree of their overlap) is shown in Figure 6.22.

Analysis of Nuclei with Spin I = *1.* For nuclei with spin $I = 1$, Equations 6.93 and 6.96 can be written in as

$$R(\omega) = F(\omega + a) + F(\omega - a) \qquad \textbf{(6.108)}$$

and

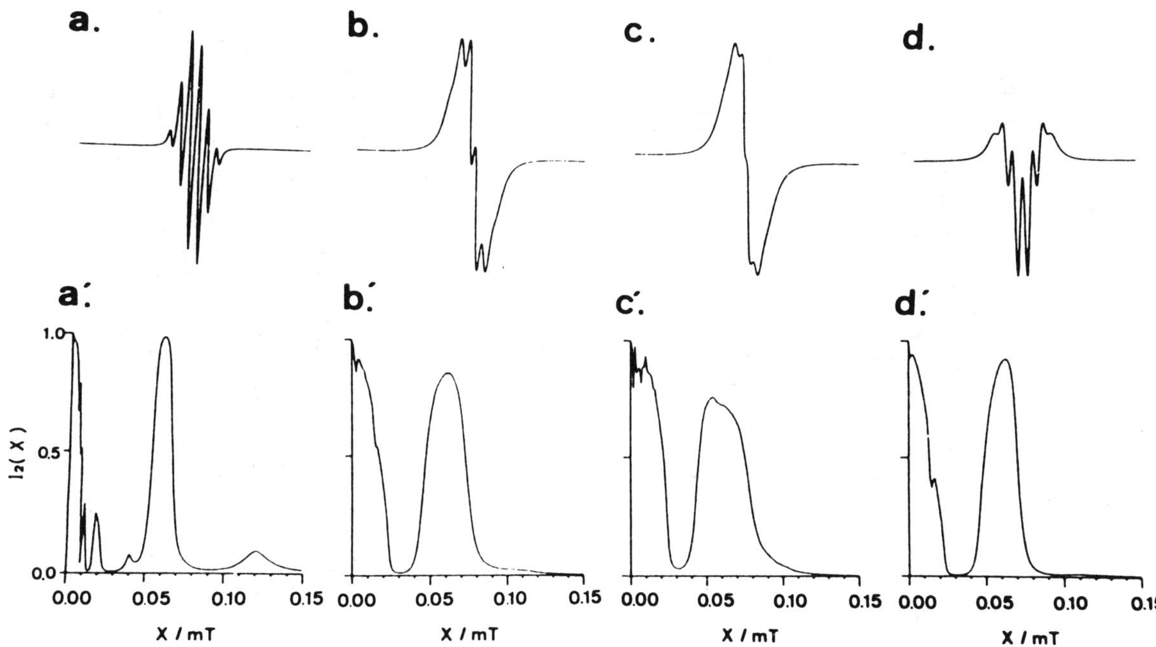

Figure 6.22 The dependence of the spectral resolution on the band half-widths for the spectrum of five equivalent nuclei with spin I = 1/2 and with the constant of hyperfine interaction a = 0.06 mT: the halfwidth of 0.02 mT **(a)**, the halfwidth of 0.05 mT **(b)**, the halfwidth of 0.06 mT **(c)**, the second order derivative with the halfwidth 0.02 mT **(d)**, and the dependence of the function $I_2(x)$ for the parameter value $\delta = 1$ (a', b', c', d'). (From Nozaki, K., Naito, A., Hatano, H., and Okazaki, S., *Bull. Chem. Soc., Jpn.*, 62, 1715, 1989. With permission.)

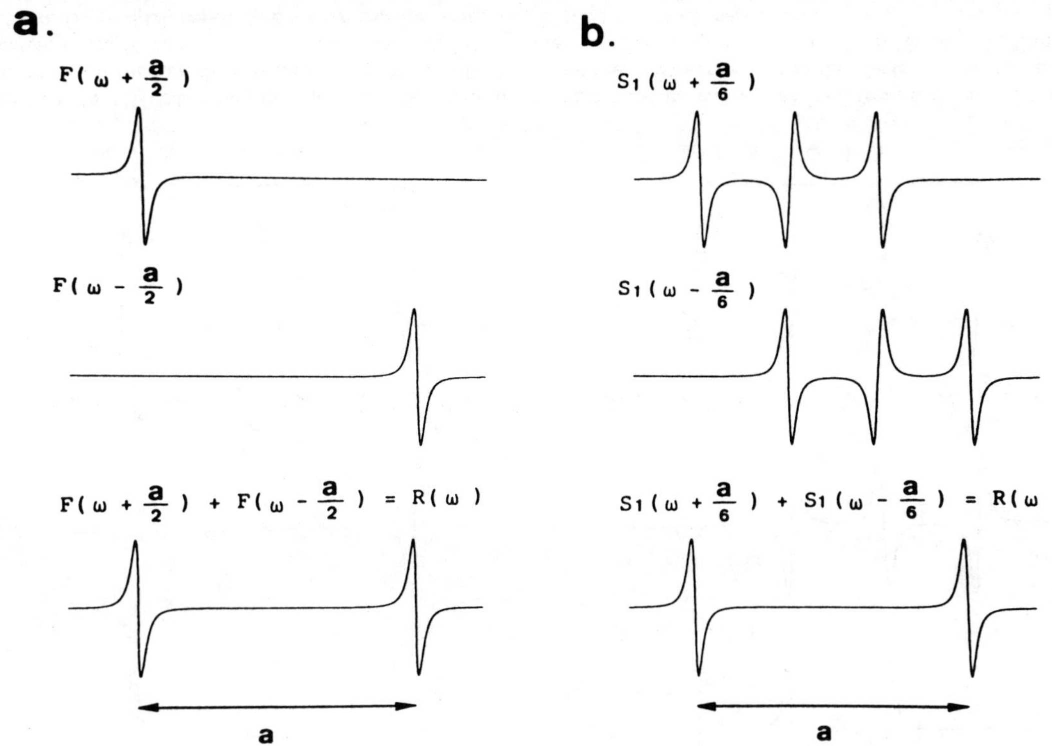

Figure 6.23 The relationship between the subspectrum and the original spectrum: the real constant of hyperfine interaction a **(a)** and the "false" constant of hyperfine interaction a/3 **(b)**. (From Nozaki, K., Naito, A., Hatano, H., and Okazaki, S., *Bull. Chem. Soc. Jpn.,* 62, 1715, 1989. With permission.)

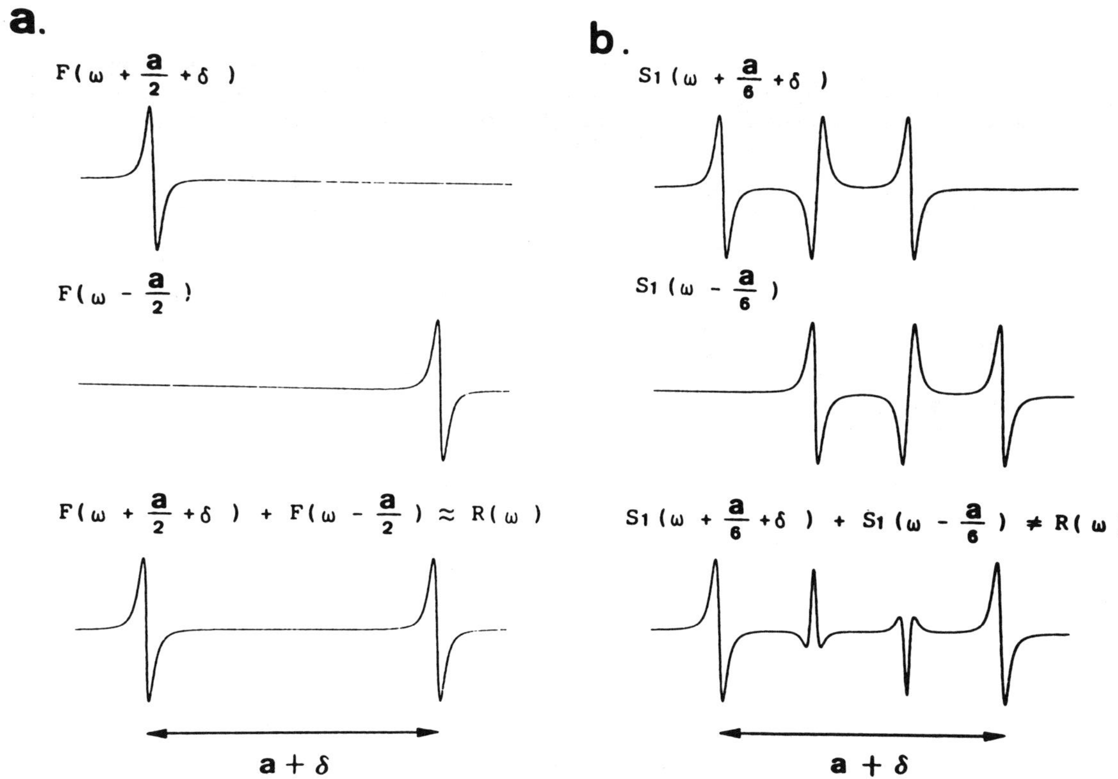

Figure 6.24 By introduction of a small distortion d the false constant can be resolved from the real constant of hyperfine interaction: the synthesized spectrum $F(\omega)$ is almost identical with the original spectrum $R(\omega)$ **(a)** and the spectrum synthesized by the function $S(\omega)$ is different from the original spectrum **(b)**. (From Nozaki, K., Naito, A., Hatano, H., and Okazaki, S., *Bull Chem. Soc. Jpn.*, 62, 1715, 1989. With permission.)

$$\begin{aligned}
&S(1) + 2S(1+x) + 3S(1+2x) + 2S(1+3x) + S(1+4x) \\
&\quad = R(1+x) + R(1+2x) + R(1+3x) \\
&S(2) + S(2+x) + 3S(2+2x) + 2S(2+3x) + S(2+4x) \\
&\quad = R(2+x) + R(2+2x) + R(2+3x) \\
&\qquad \cdots \qquad \cdots \qquad \cdots \\
&S(N-4x) + 2S(N-3x) + 3S(N-2x) + 2S(N-x) + S(N) \\
&\quad = R(N-3x) + R(N-2x) + R(N-x)
\end{aligned} \tag{6.109}$$

Then false hyperfine interaction constants will appear for values

$$x_n = \frac{a}{3n+1} \qquad \text{for } n = 1, 2, \ldots \tag{6.110}$$

$$x_m = \frac{a}{3m+2} \qquad \text{for } m = 0, 1, 2, \ldots \tag{6.111}$$

For functions $S_n(\omega)$ and $S_m(\omega)$ corresponding to values x_n respective to x_m:

$$S_n = \sum_{i=-3n}^{3n} c(|i|)\, F(\omega - ix_n) \tag{6.112}$$

$$S_m = \sum_{i=3m-1}^{3m+1} c(|i|+2)\, F(\omega - ix_m) \tag{6.113}$$

are defined where

$$\begin{aligned}
c(k) &= 1 \quad \text{for} \quad k = 3p \\
&= 0 \quad \text{for} \quad k = 3p+1 \\
&= -1 \quad \text{for} \quad k = 3p+2 \qquad p = 0, 1, 2, \ldots
\end{aligned} \tag{6.114}$$

The index function $I_3(x)$ for spin $I = 1$ of nuclei and multiplicity $2I + 1 = 3$ is defined by the equation

$$I_3(x) = \left[\frac{2 \int_{-\infty}^{+\infty} \left| R(\omega) - R(\omega - \delta) \right| d\omega}{3 \sum_{j=1}^{2} \int_{-\infty}^{+\infty} \left| R(\omega) - S(\omega + x) - S\left(\omega + [-1]^j \delta\right) - S(\omega - x) \right| d\omega} \right]^2 \tag{6.115}$$

The maximal value of this function is close to 1 at x = a. For the ratio $I_3(a/2)/I_3(a)$

$$\frac{I_3\left(\frac{a}{2}\right)}{I_3(a)} = \left[\frac{\int_{-\infty}^{+\infty} \left| F'(\omega) \right| d\omega}{\int_{-\infty}^{+\infty} \left| F'\left(\omega + \frac{a}{2}\right) - F'(\omega) + F'\left(\omega - \frac{a}{2}\right) \right| d\omega}\right]^2 \quad \textbf{(6.116)}$$

is valid. It can be proved that this ratio changes from the value 1/9 to 1 according to the degree of overlap of functions $F'(\omega + a/2)$, $F'(\omega)$, and $F'(\omega - a/2)$. The application of the index function for ESR spectra of different types of nuclei is shown in Figure 6.25. Figure 6.26 shows the numerical decoupling analysis method for the spectrum consisting of 10 equivalent nuclei and 1 nonequivalent nucleus with spin I = 1/2.

SIMULATION AND OPTIMIZATION OF ESR SPECTRA PARAMETERS OF RANDOMLY ORIENTED RADICALS IN THE SOLID PHASE

The most precise and complete information about radicals in solid phase can be obtained from measurements on single crystals. Such measurements pertain only to a limited set of compounds.

ESR spectra of single crystals have a character similar to those of spectra in liquid phase. Because of this, the values of various magnetic interactions are not averaged; they usually have a more complicated shape. The parameters of the spin Hamiltonian are unambiguously dependent on the single crystal orientation according to an external magnetic field. From the angle dependence of these parameters it is possible to evaluate not only the main values of individual tensors included in the spin Hamiltonian, but also their direction to the crystallographic axes of the measured crystal. In analyzing these spectra, the numerical problems connected with transforming individual coordinate systems have specific characteristics. They are more complicated and their explanation would be time consuming. For readers interested in this problem, special literature can be recommended.[1-6,12,13,31-33]

If it is not possible to prepare a particular compound as a single crystal, ESR spectra can be investigated in a polycrystalline state as powdered material, glass, or frozen solutions. In all such cases, these are systems in which individual radicals are randomly oriented in space.

The range and type of information which can be obtained from ESR measurements of randomly oriented radicals in the solid phase are dependent on individual contributions of the spin Hamiltonian. Usually, neither quadruple interaction nor nuclear Zeeman interaction are observed. Similarly, the influence of a crystal field is shown only in specific cases of high spin complexes. Therefore, during the interpretation of ESR spectra of randomly oriented radicals in solid phase, these terms in the Hamiltonian can be eliminated.

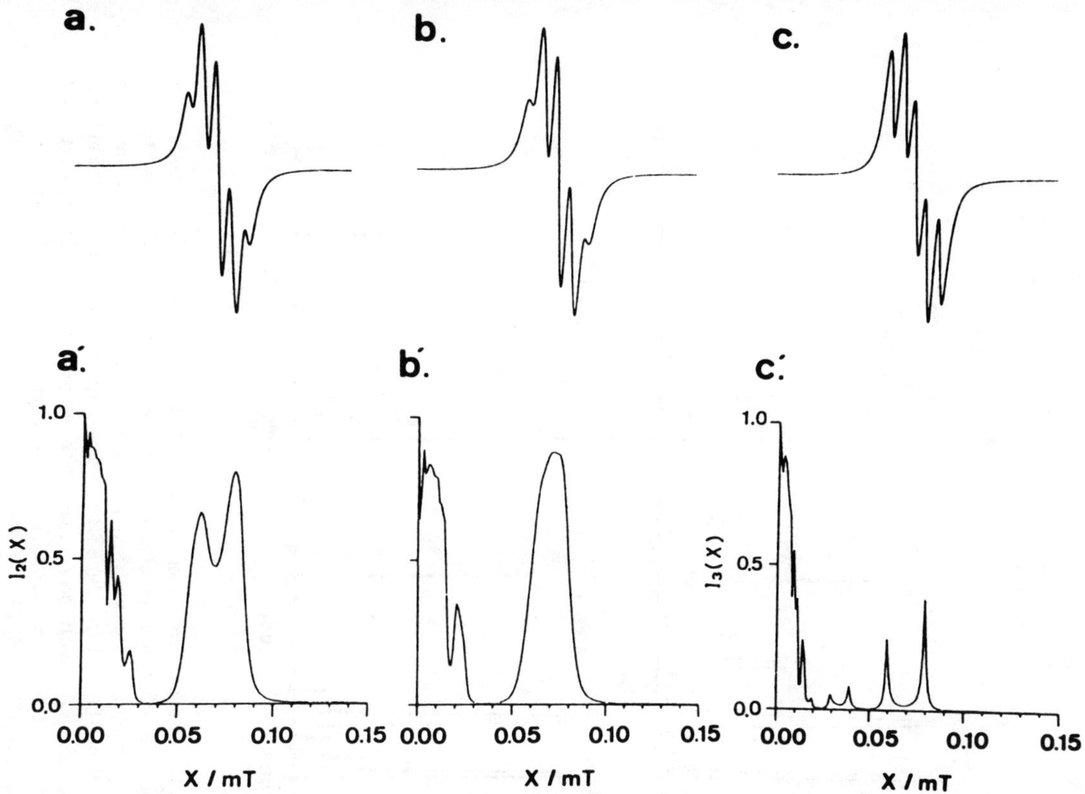

Figure 6.25 Using the index function for the ESR spectra corresponding to nuclei of different types: the ESR spectrum for two pairs of equivalent nuclei with spin $I = 1/2$ ($a_1 = 0.06$ mT, $a_2 = 0.08$ mT), with the halfwidth 0.04 mT **(a)**, the ESR spectrum for two nonequivalent nuclei with spin $I = 1/2$ (a = 0.07 mT), with the halfwidth 0.045 mT **(b)**, the ESR spectrum for two nonequivalent nuclei with spin $I = 1$ ($a_1 = 0.06$ mT, $a_2 = 0.08$ mT), with the halfwidth 0.04 mT **(c)**, and the corresponding index functions for the value $\delta = 1$ (*a′*, *b′*, *c′*, *d′*). (From Nozaki, K., Naito, A., Hatano, H., and Okazaki, S., *Bull. Chem. Soc. Jpn.*, 62, 1715, 1989. With permission.)

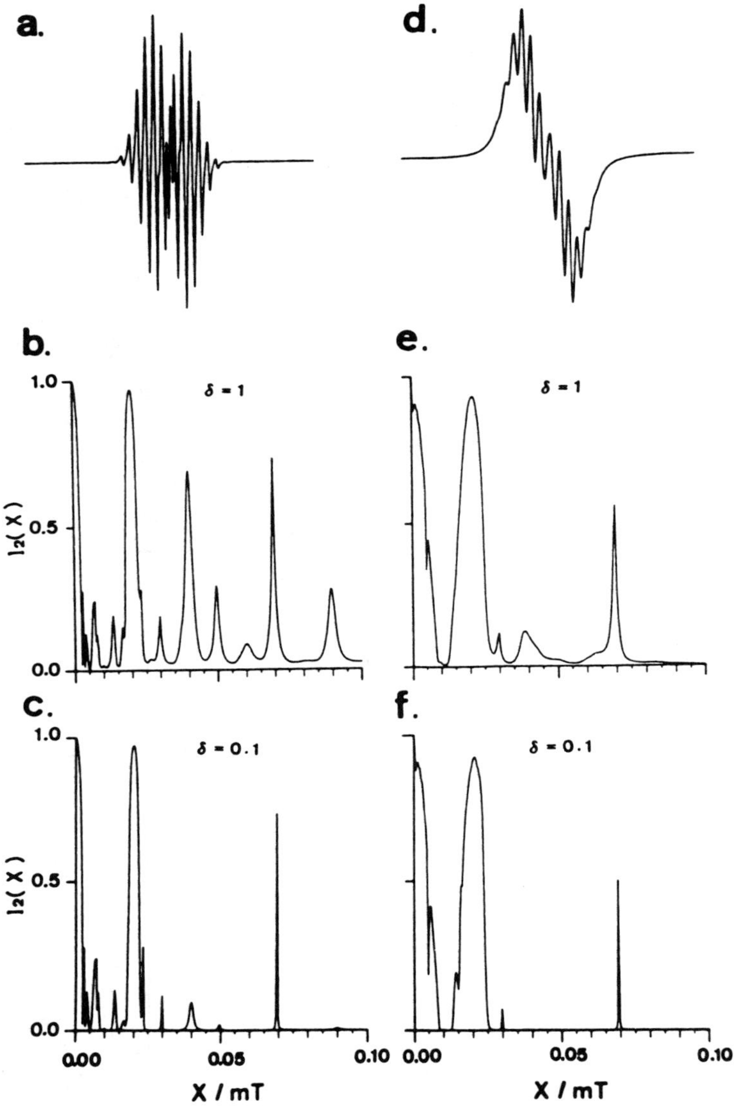

Figure 6.26 The application of the numerical decoupling analysis method for interpreting the ESR spectra of 10 equivalent nuclei with spin $I = 1/2$ ($a_1 = 0.02$ mT) and 1 nonequivalent nucleus with spin $I = 1/2$ ($a_2 = 0.07$ mT): the ESR spectrum with the halfwidth of 0.005 mT **(a)**, the ESR spectrum with the halfwidth of 0.015 mT **(d)**, the index function $I(x)$ with the value $\delta = 1$ **(b, e)**, and the index function $I(x)$ with the value $\delta = 0.1$ **(c, f)**. (From Nozaki, K., Naito, A., Hatano, H., and Okazaki, S., *Bull. Chem. Soc. Jpn.*, 62, 1715, 1989. With permission.)

If individual terms in the spin Hamiltonian are isotropic, the same information will be produced during ESR study of randomly oriented radicals in solid phase as is obtained during the measurements of single crystal compounds. In the case of strong anisotropy, however, the range of obtainable information is usually decreased. Recent work[34-39] reports algorithms to optimize the ESR spectra of randomly oriented radicals in solid phase. By efficient computer processing the interpretation can reach that available from single crystals.

In the case of single crystals, the spectral line intensity in an external magnetic field with induction B, the following can be expressed as[1,34,35]

$$Y(B,\theta,\phi,i) = N\,P(\theta,\phi,i)\,G(B,\theta,\phi,i) \quad \textbf{(6.117)}$$

where i determines the type of given spectral transition, $P(\theta, \phi, i)$ is the probability of the ith spectral transition, $G(B, \theta, \phi, i)$ is the function of the spectral line shape, N is the normalization constant, and θ and ϕ are the spherical coordinates, that show the orientation of an external magnetic field to the coordinate system of a measured paramagnetic particle.

For derivative ESR spectra, either the Lorentz or Gaussian shapes of the derivative function are used:

$$G^L(B,\theta,\phi,i) = -\frac{B-B_{0,i}}{\Delta B_i^2}\left[1+\frac{4}{3}\frac{\left(B-B_{0,i}\right)^2}{\Delta B_i^2}\right]^{-2} \quad \textbf{(6.118)}$$

$$G^G(B,\theta,\phi,i) = -\frac{B-B_{0,i}}{\Delta B_i^2}\exp\left[-2\frac{\left(B-B_{0,i}\right)^2}{\Delta B_i^2}\right] \quad \textbf{(6.119)}$$

where $B_{0,i} = B_{0,i}(\theta, \phi)$ is the magnetic field induction at which the resonance condition is satisfied and $\Delta B_i = \Delta B_i(\theta, \phi)$ is the spectral line width.

The total intensity of a spectral line can be obtained by summing all the possible spectral transitions:

$$Y(B,\Theta,\phi) = \sum_i Y(B,\theta,\phi,i) = N\sum_i P(\theta,\phi,i)\,G(B,\theta,\phi,i) \quad \textbf{(6.120)}$$

Because they are randomly oriented in solid phase, it is necessary to average the line function over the elemental angle volume

$$d\Omega = \sin\theta\, d\theta\, d\phi \quad \textbf{(6.121)}$$

For the total line intensity of randomly oriented radicals in solid phase, therefore, the following is obtained:

$$Y_t(B) = N\sum_i \int_\theta \int_\phi P(\theta,\phi,i)\,G(B,\theta,\phi,i)\sin\theta\, d\theta\, d\phi \quad \textbf{(6.122)}$$

The solution of the integral through the angular variables θ and ϕ can be carried out explicitly in some easier cases,[40-42] which makes the whole optimization procedure simpler and faster.

Johnston and Hecht[39] showed that, due to the dependence on the orientation of relaxation effects, spectral line widths are also dependent on the orientation. Detailed theoretical studies of such phenomena are still lacking, and this effect is neglected in most research works. Such negligence can lead to deformation of the values of the spin Hamiltonian parameters.

Because the Zeeman term (whose anisotropy is expressed by a g-tensor) is the most dominant term of the spin Hamiltonian, it is logical that relaxation effects will have the same angular dependence as this term in the spin Hamiltonian. Therefore, it is reasonable to choose the following dependence of a spectral line halfwidth on angular variables:

$$\Delta B_i(\theta,\phi) = \left(\Delta B_{x,i}^2 \sin^2\theta \cos^2\phi + \Delta B_{y,i}^2 \sin^2\theta \sin^2\phi + \Delta B_{z,i}^2 \cos^2\theta\right) \quad \mathbf{(6.123)}$$

where $\Delta B_{x,i}$, $\Delta B_{y,i}$, and $\Delta B_{z,i}$ characterize the halfwidth in the direction of the main axes of a g-tensor. In the case of axially symmetric systems

$$\Delta B_{x,i} = \Delta B_{y,i} = \Delta B_{\perp,i} \quad \mathbf{(6.124)}$$

$$\Delta B_{z,i} = \Delta B_{\parallel,i} \neq \Delta B_{\perp,i} \quad \mathbf{(6.125)}$$

are defined; therefore, Equation 6.123 is simplified to

$$\Delta B_i(\theta,\phi) = \left(\Delta B_{\perp,i} \sin^2\theta + \Delta B_{\parallel,i} \cos^2\theta\right)^{\frac{1}{2}} \quad \mathbf{(6.126)}$$

The transition probabilities $P(\theta, \phi, i)$ in the case of paramagnetic systems with anisotropic spin Hamiltonian also show anisotropy, so it is necessary to place into Equation 6.122 the relationship obtained for the corresponding type of magnetic interactions. In the simplest case, if the Zeeman contribution is the dominant term in the spin Hamiltonian, the transition probability:

$$P(\theta,\phi,i) \approx \frac{g_x^2 g_y^2 \sin^2\theta + g_y^2 g_z^2\left(\sin^2\phi + \cos^2\theta\cos^2\phi\right)}{g_{ef}^2} + \frac{g_z^2 g_x^2\left(\cos^2\phi + \cos^2\theta \sin^2\phi\right)}{g_{ef}^2} \quad \mathbf{(6.127)}$$

can be written, where g_x, g_y, and g_z are the main components of the g-tensor and g_{ef} is defined by the following equation:

$$g_{ef} = \left(g_x^2 \sin^2\theta\cos^2\phi + g_y^2 \sin^2\theta \sin^2\phi + g_z^2 \cos^2\phi\right)^{\frac{1}{2}} \quad \mathbf{(6.128)}$$

In the case of axially symmetrical systems, Equations 6.127 and 6.128 are simplified to the form

$$P(\theta, \phi, i) \approx g_\perp^2 \frac{\left[g_\perp^2 \sin^2\theta + g_\parallel^2 \left(1 + \cos^2\theta\right)\right]}{g_{ef}^2} \tag{6.129}$$

$$g_{ef} = \left(g_\perp^2 \sin^2\theta + g_\parallel^2 \cos^2\theta\right)^{\frac{1}{2}} \tag{6.130}$$

Equation 6.122 can be put into a more convenient form for computer elaboration. The integration over the angular variables can be changed to a summation, so that Equation 6.122 becomes

$$Y_t(B) = N \sum_i \sum_k \sum_l P(\theta_k, \phi_l, i)\, G(B, \theta_k, \phi_l, i) \cdot \sin\theta_k\, \Delta\theta_k\, \Delta\phi_l \tag{6.131}$$

where $\Delta\theta_k$ and $\Delta\theta_i$ are the steps with which the values of the angle variables, i.e., θ_k and q_l, are changed. It is suitable during practical application to use the substitution:

$$\sin\theta\, d\theta = -d(\cos\theta) \tag{6.132}$$

or

$$\sin\theta_k\, \Delta\theta_k = -\Delta(\cos\theta_k) \tag{6.133}$$

Then Equation 6.131 becomes

$$Y_t(B) = -N \sum_i \sum_k \sum_l P(\theta_k, \phi_l, i)\, G(B, \theta_k, \phi_l, i) \cdot \Delta(\cos\theta_k)\, \Delta\phi_l \tag{6.134}$$

If the constant intervals $\Delta(\cos\theta_k)$ and $\Delta\phi_l$ are chosen, their values can be included in the normalization constant and the spectral line shape is expressed by the relationship

$$Y_t(B) = -N' \sum_i \sum_k \sum_l P(\theta_k, \phi_l, i)\, G(B, \theta_k, \phi_l, i) \tag{6.135}$$

This equation can be used directly for computer simulation of the ESR spectra of randomly oriented radicals in solid phase.

Another method of pairs (θ_n, ϕ_m) selection involves an equatorially centered "regular" grid.[43] The confluence of meridians toward the poles is balanced here with an increasing angular width of the zones. Provided that the number of chosen orientations is L, then with m, $n = 1, 2, \ldots \sqrt{L}$:

$$\phi_m = \frac{2\pi\left(m - \frac{1}{2}\right)}{L} \tag{6.136}$$

$$\theta_n = \arccos\left(\frac{2n-1}{L}\right) \tag{6.137}$$

A third scheme employs a polar-centered grid in which the confluence of meridians toward the poles is compensated for with decreasing occupation of zones with representatives.[42] Provided that the total number of zones is $N - 1$, then with the number M_n of surface elements in the nth zone

$$M_n = \frac{\cos\left(\frac{n\pi}{2N}\right) - \cos\left[\frac{(n+1)\pi}{2N}\right]}{1 - \cos\left(\frac{\pi}{2N}\right)} \tag{6.138}$$

and set of (ϕ_{m_n}, θ_n) pairs is obtained with

$$\theta_n = \frac{(2n+1)\pi}{4N} \qquad n = 1, 2, \ldots M_n \tag{6.139}$$

$$\phi_{m_n} = \frac{(2m_n - 1)\pi}{M_n} \qquad m_n = 1, 2, \ldots M_n \tag{6.140}$$

Finally, a stochastic distribution of $\{\theta, \phi\}_n$ pairs can be chosen with $\theta = \arccos(x_n)$, $\phi = 2\pi y_n$, and x_n, $y_n \in (0, 1)$ random numbers.

Algorithms and descriptions of programs for simulation of EPR spectra have been published by many authors.[35-37,43-52]

Before applying Equation 6.135, it is usually necessary to include a large number of orientations. Therefore, the calculation of a spectral line becomes extraordinarily difficult. For obtaining a reasonable ratio of signal/noise, usually about 5000 orientations are necessary. Calculating the large number of points for different values of magnetic field induction is simplified by efficient interpolation functions.

Such a function must fulfill two fundamental requirements:

1. It can contain discontinuities only for the derivatives of second or higher order of magnetic induction B.

2. The number of extremes of an interpolation function inside the defined interval must be the same as those of the numbers of extremes of the experimental dependence.

According to these requirements all types of polynomials can be eliminated, because polynomials of higher degree have a tendency to oscillate and polynomials of lower degree are not flexible enough. The interpolation function developed by Lagerlöv[53] gives a simple solution, and it can be written in the following form

$$f(x) = \ln\left\{1 + \exp\left[b(x-a)\right]\right\} \tag{6.141}$$

This type of interpolation function seems to be sufficiently flexible, but without the disadvantage of oscillations.

For the analysis of ESR spectra of randomly oriented radicals in solid phase, it is necessary to have considerable experience. It is not possible to formulate the simple rules of analysis by their interpretation, as it is in the case of isotropic spectra in solutions, because of anisotropy of magnetic interaction; often because of inconvenient relaxation effects, their shape is strongly deformed. Figures 6.27 to 6.31 show examples of simple spectra of those types; the problem of their interpretation is already apparent. The spin Hamiltonian parameters visually subtracted from the spectra are only approximate values, due to often strong overlap of individual bands and their extensive anisotropy. Therefore, it is necessary to use computer techniques for "tolerable" interpretation of ESR spectra of randomly oriented radicals in solid phase.

Methods of nonlinear numerical analysis enable adequate interpretation of spectra with such characteristics. Optimization of the spin Hamiltonian parameters can be realized by minimization of the function:

$$\mathcal{F}(B) = \sum_{B} \left[Y^{\exp}(B) - Y_t(B)\right]^2 \tag{6.142}$$

where $Y^{\exp}(B)$ are the experimentally measured points of spectra and $Y_t(B)$ are the points calculated by the model method and Equation 6.135 with respect to Equation 6.134. Nonlinear regression analysis methods can be used for this purpose (see Appendix C, Nonlinear Regression Analysis).

In this context, it is necessary to note that, mostly in well equipped laboratories, the spin Hamiltonian parameters of randomly oriented radicals in solid phase are obtained only by simulation programs (software), using only visual subtraction of the values. This can cause many inaccuracies, because of the subjectivity of visual comparison of the experimental and model spectra. Suitably efficient personal computers can run the optimization programs (software) convenient for interpretation of such spectra. The results of analysis of a simple ESR spectrum of randomly oriented radicals in solid phase using visual subtraction of the spin Hamiltonian parameters and their optimization are shown in Figure 6.30.

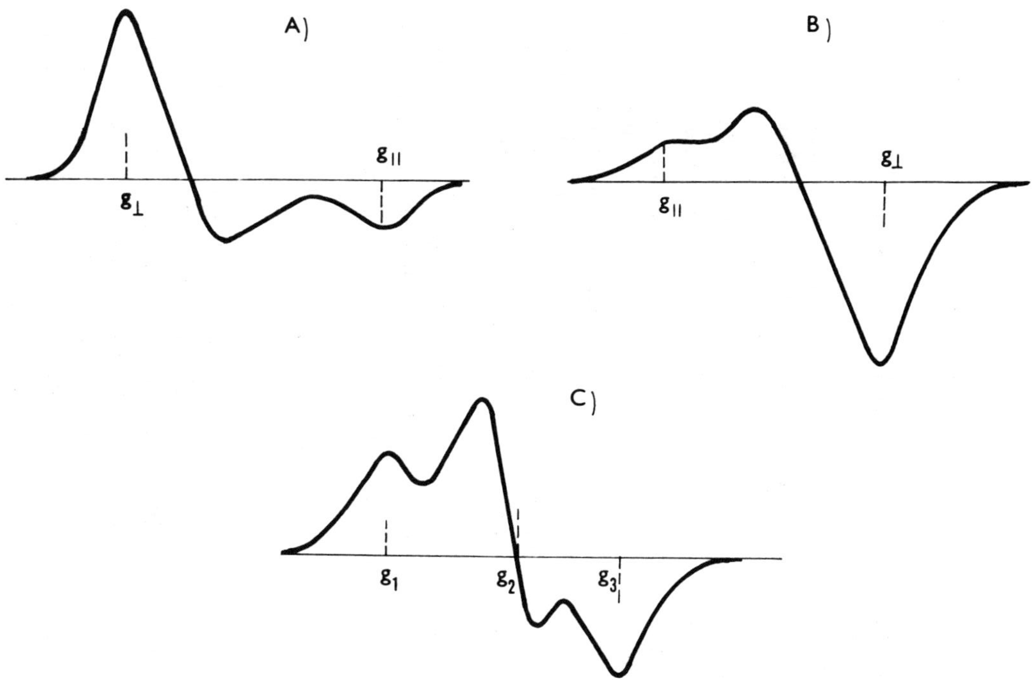

Figure 6.27 The ESR spectra of randomly oriented radicals in solid phase with spin $S = 1/2$ without hyperfine interaction: the axially symmetrical spectrum ($g_\perp > g_\parallel$) **(A)**, the axially symmetrical spectrum ($g_\perp < g_\parallel$) **(B)**, and the orthorhombic spectrum ($g_1 > g_2 > g_3$) **(C)**.

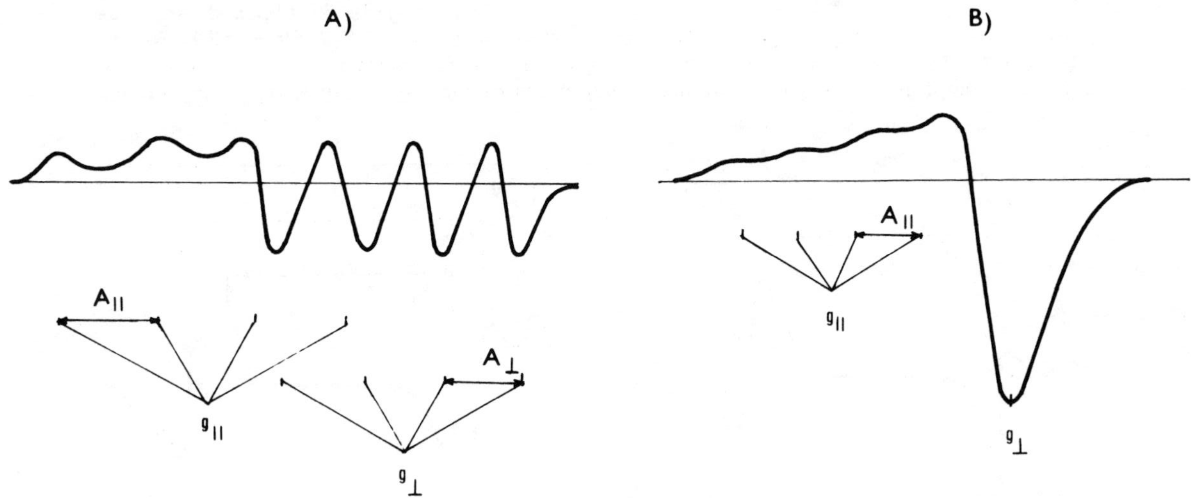

Figure 6.28 The shape of the ESR spectra of randomly oriented radicals in solid phase with spin $S = 1/2$ and with one nucleus with spin $I = 3/2$: the axially symmetrical spectrum $g_1 < g_{\|}$, $A_1 < A_{\|}$ (a part of hyperfine interaction is fade down in the line width) **(A)** and the axially symmetrical spectrum $g_{\|} > g_{\|}$, $A_{\|} \neq 0$, $A_{\perp} = 0$ **(B)**.

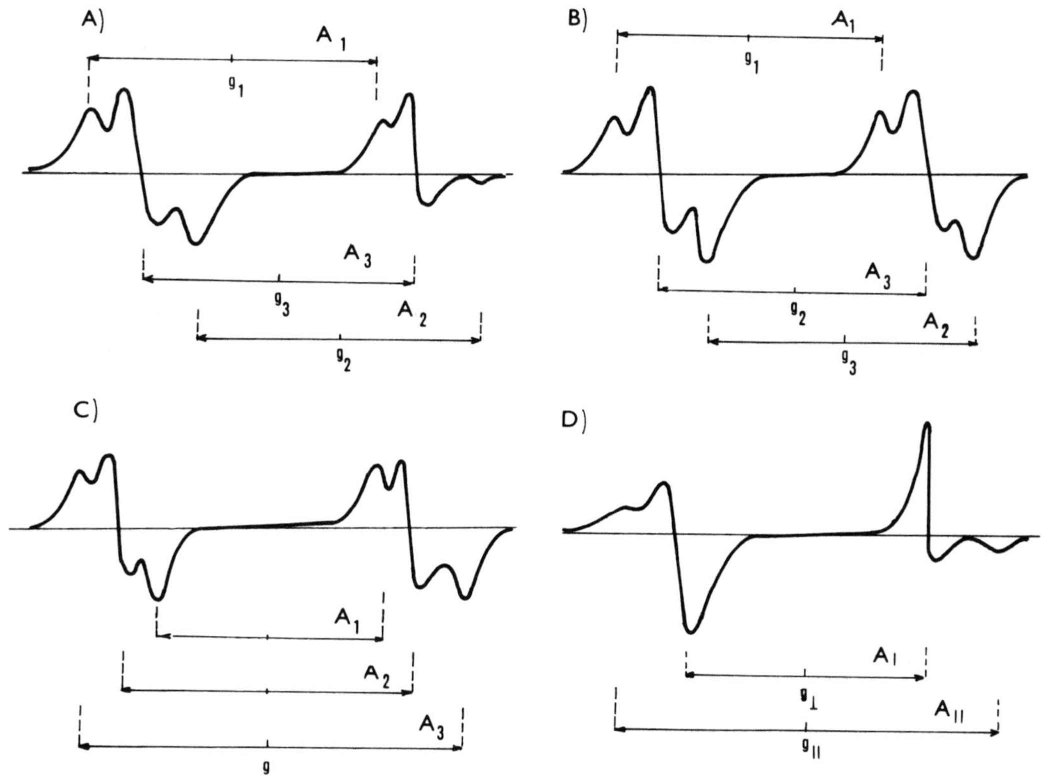

Figure 6.29 The typical ESR spectra of randomly oriented radicals in solid phase with spin $S = 1/2$ and with one nucleus with spin $I = 1/2$: $g_1 > g_2 > g_3$, $A_1 > A_2$, $A_2 < A_2$, $A_2 > A_3$ (**A**); $g_1 > g_2 > g_3$; $A_1 = A_2 = A_3 = A$ (**B**); $g_1 = g_2 = g_3 = g$, $A_1 < A_2 < A_3$ (**C**), and the axially symmetrical spectrum $g_\perp > g_\parallel$, $A_\perp > A_\parallel$ (**D**).

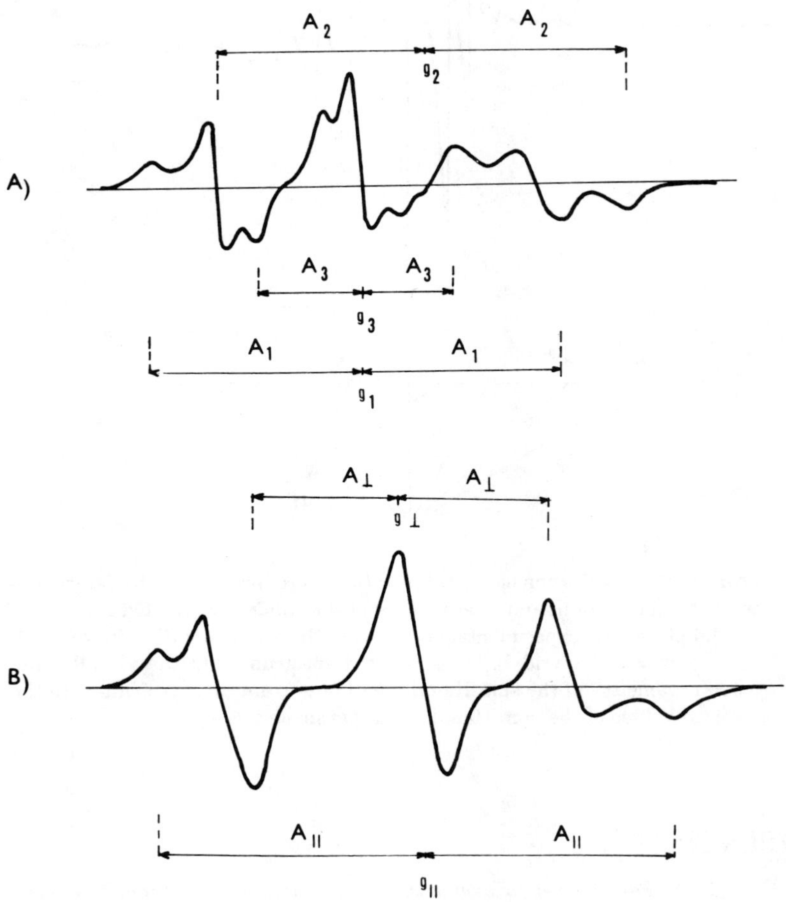

Figure 6.30 The axially symmetric spectrum of the vanadyl cation (VO_2^+) with spin S = 1/2 and with nuclear spin I = 7/2.

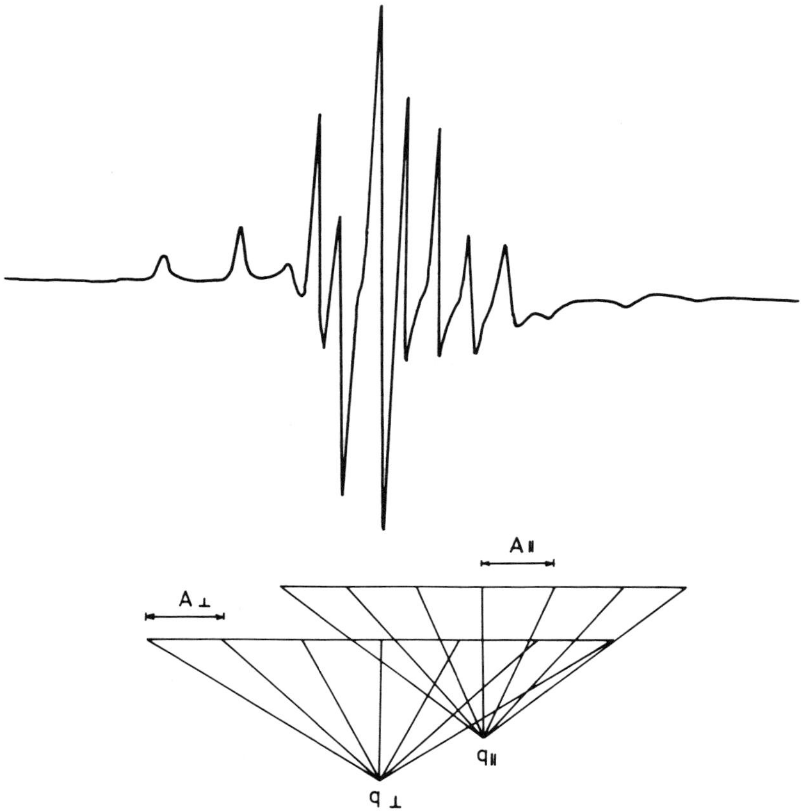

Figure 6.31 Application of the optimization procedure during the determination of the spin Hamiltonian parameters of the randomly oriented radicals in the solid phase: the experimental spectrum of the compound $[Cu(bipy)_2I]ClO_4$ (– – –) (bipy = 2,2'bipyridyl), the simulated spectrum with the visually subtracted parameters of the spin Hamiltonian (——), and the spectrum with the optimized values of the spin Hamiltonian parameters (– • – • –).

REFERENCES

1. Pake, G. A., *Paramagnetic Resonance, an Introductory Monograph,* New York, 1968.
2. Gordy, W., *Theory and Applications of Electron Spin Resonance Spectroscopy,* John Wiley & Sons, New York, 1980.
3. Gerson, F., *High Resolution Electron Spin Resonance Spectroscopy,* John Wiley & Sons, New York, 1970.
4. Abragam, A. and Bleaney, B., *Electron Paramagnetic Resonance of Transition Ions,* Clarendon Press, Oxford, 1970.
5. Poole, C. P., Jr. and Farach, H. A., *The Theory of Magnetic Resonance,* John Wiley & Sons, New York, 1972.
6. McGavin, D. G., *J. Magn. Reson.,* 74, 19, 1987.
7. Pilbrow, J. R., *J. Magn. Reson.,* 58, 186, 1984.

8. Feher, G., *Bull. Syst. Tech. J.,* 36, 449, 1957.

9. Zhong, Y C. and Pilbrow, J. R., *J. Magn. Reson.,* 93, 447, 1991.

10. Slichter, C. P., *Principles of Magnetic Resonance,* Harper & Row, New York, 1963.

11. Kubo, R. and Tomita, K., *J. Phys. Soc. Jpn.,* 9, 888, 1954.

12. Atkins, P. W. and Symons, M. C. R., The structure of inorganic radicals, in *An Application of Electron Spin Resonans to the Study of Molecular Structure,* 1st ed., Amsterdam, 1967.

13. Marquardt, D. W., Bennett, R. G., and Burrell, E. J., *J. Molec. Spectrosc.,* 7, 269, 1961.

14. Marquardt, D. W., Bennett, R. G., and Burrell. E. J., *J. Molec. Spectrosc.,* 7, 269, 1961.

15. Bauder, A. and Myers, R. J., *J. Molec. Spectrosc.,* 27, 110, 1968.

16. Fischer, L., *J. Molec. Spectrosc.,* 40, 414, 1971.

17. Barzaghi, M. and Simonetta, *M., J. Magn. Reson.,* 51, 175, 1983.

18. Barzaghi, M., Gamba, A., Morosi, G., and Simonetta, M., *J. Phys. Chem.,* 82, 2105, 1978.

19. Dunham, W. R., Fee, J. A., Harding, L. J., and Grande, H. J., *J. Magn. Reson.,* 40, 351, 1980.

20. Silsbee, R. H., *J. Chem. Phys.,* 45, 1710, 1966.

21. Ziegler, E. and Hoffmann, E. G., *Fresenius' Z. Anal. Chem.,* 240, 145, 1966.

22. Bieber, K. D. and Gough, T. E., *J. Magn. Reson.,* 21, 285, 1976.

23. Grampp, G. and Schiller, C. A., *Anal. Chem.,* 53, 560, 1981.

24. Motten, A. and Schreiber J., *J. Magn. Reson.,* 67, 42, 1986.

25. Kirmse, D. W., *J. Magn. Reson.,* 11, 1, 1971.

26. Motten, A. G., Duling, D. R., and Schreiber, J., *J. Magn. Reson.,* 71, 34, 1987.

27. Nozaki, K., Naito, A., Hatano, H., and Okazaki, S., *Bull. Chem. Soc. Jpn.,* 62, 1715, 1989.

28. Dračka, O., *J. Magn. Reson.,* 65, 187, 1985.

29. Harriman, J. E., Theoretical Foundations of Electron Spin Resonance, Loebl, E. M., Ed., Academic Press, New York, 1978.

30. Joela, H., *Comput. Chem.,* 12, 189, 1988.

31. Schonland, D. S., *Proc. Phys. Soc.,* 73, 788, 1959.

32. Waller, W. G. and Rogers, M. T., *J. Magn. Reson.,* 9, 92, 1973.

33. Waller, W. G. and Rogers, M. T., *J. Magn. Reson.,* 13, 53, 1974.

34. Byrn, M. P. and Strouse, Ch. E., *J. Magn. Reson.,* 53, 32, 1983.

35. Pelikán, P., Valko, M., Biskupič, S., and Staško, A., *J. Comput. Chem.,* in press.

36. Weil, J. A. and Hecht, H. G., *J. Chem. Phys.,* 38, 281, 1963.

37. Rieger, P. H., *J. Magn. Reson.,* 50, 485, 1982.

38. van Ween, G., *J. Magn. Reson.,* 30, 91, 1978.

39. Johnston, T. S. and Hecht, H. G., *J. Molec. Spectrosc.,* 17, 98, 1965.

40. Bertrán-López, V. and Castro-Tello, J., *J. Magn. Reson.,* 39, 437, 1980.

41. Bertrán-López, V. and Castro-Tello, J., *J. Magn. Reson.,* 47, 19, 1982.

42. Gonzáles-Tovany, L. and Bertrán-López, V., *J. Magn. Reson.,* 89, 227, 1990.

43. Kreiter, A. and Huettermann, J., *J. Magn. Reson.,* 93, 12, 1991.

44. Hoffman, B. M., Martinsen, J., and Venters, R. A., *J. Magn. Reson.,* 59, 110, 1984.

45. Hoffman B. M., Venters, R. A., and Martinsen, J., *J. Magn. Reson.,* 62, 537, 1985.

46. Keijzers, C. P., Reijerse, E. J., Stam, P., Dumont, M. F., and Gribnau, M. C. M., *J. Chem. Soc. Faraday Trans.,* 183, 3493, 1987.

47. Kopp, M. and Mackey, J., *QCPE Program 150, MAGNSPEC3.*

48. Griscom, D. L. and Griscom, R. E., *J. Chem. Phys.,* 47, 2711, 1967.

49. Reynolds, R. W., Boatner, L. A., Finch, C. B., Chatelain, A., and Abraham, M. M., *J. Chem. Phys.,* 56, 5607, 1972.

50. Pace, M. D. and Masumura, S. M., *Gov. Rep. Announce,* 89, 1989.

51. Jimenez-Dominquez, H. and Cabral-Prieto, A., *J. Magn. Reson.,* 93, 178, 1991.

52. Lassmann, G., Wessel, R., Grasshoff, P., Jakobs, C., and Schwarz, D., *Exp. Tech. Phys.,* 38, 249, 1990.

53. Lagerlöv, R. Q., *Comm. ACM,* 17, 476, 1974

54. Bloemberger, N., Purcell, E. M., and Pound, R. V., *Phys. Rev.,* 73, 679, 1948.

55. Pake, G. E. and Purcell, E. M., *Phys. Rev.,* 74, 1184, 1948.

56. Castner, T. G., *Phys. Rev.,* 115, 1506, 1955.

57. Kambe, K. and Usui, T., *Prog. Theor. Phys.,* 8, 302, 1952.

58. Pryce, M. H. L. and Stevents, K. W., *Proc. Phys. Soc. A,* 63, 36, 1955.

59. Portis, A. M., *Phys. Rev.,* 91, 1071, 1953.

60. Coffman, R. E., *J. Phys. Chem.,* 79, 1129, 1975.

Chapter 7

Numerical Methods in NMR Spectroscopy

FUNDAMENTALS OF NMR SPECTROSCOPY

The existence of directly measurable resonance in the magnetized sample is a characteristic property of the method of nuclear magnetic resonance (NMR) spectroscopy. Unlike the classical absorption methods, which just produce a frequency spectrum, NMR also produces a spectrum in the time domain.

The application of coherent electromagnetic radiation, i.e., a monochromatic radiation with high phase coherence, is the basic condition for generating macroscopically measurable sample resonance. For the magnetic component of coherent radiation in a laboratory coordinate system (Figure 7.1), the following is valid:

$$\vec{B}_1 = 2\,B_1 \cos(\omega t + \alpha)\,\vec{i} \tag{7.1}$$

where ω is the frequency, α is the phase shift, and $\vec{i}$ is the unit vector perpendicular to the direction of radiation propagation. This linearly polarized radiation can be divided into the sum of two opposite, rotating, circularly-polarized components

$$\vec{B}_1 = B_1\left[\cos(\omega t + \alpha)\,\vec{i} - \sin(\omega t + \alpha)\,\vec{j}\right] \tag{7.2}$$

where $\vec{j}$ is the unit vector perpendicular to the vector $\vec{i}$ as well as to the direction of radiation propagation. A phase sensitive detector with reference frequency identical to the frequency $\vec{B}$ is used in NMR spectroscopy. Therefore, the resonance experiment is best described in a rotating coordinate system $x', y', z' \equiv z$, which is identical to a laboratory coordinate system at time zero. A dashed coordinate system rotates with the angle frequency

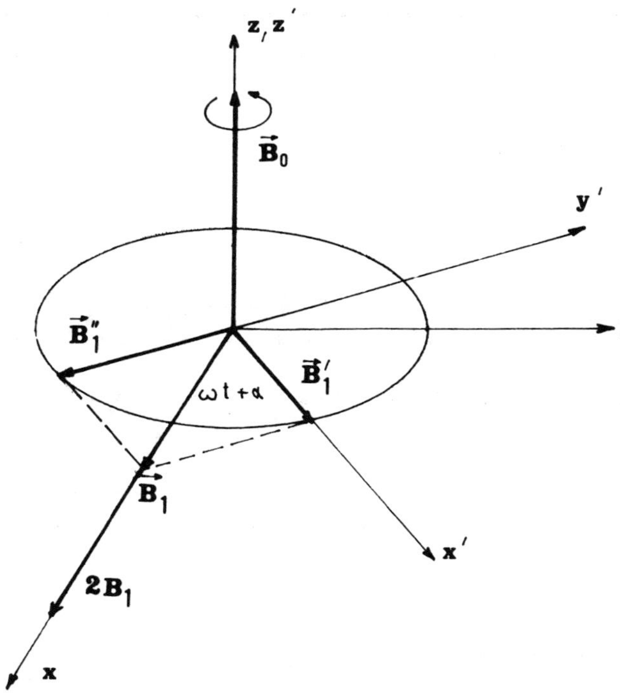

Figure 7.1 The definition of a laboratory and rotating coordinate system and the decomposition of linearly polarized radiofrequency radiation ($\vec{B}_1$) into two circularly polarized components ($\vec{B}_1'$, $\vec{B}_1''$).

ω, so that the coordinates of the vector $\vec{B}$ are not changed with time in this coordinate system:

$$\vec{B}_1 = B_1 \left\{ \cos\alpha \, \vec{i} - \sin\alpha \, \vec{j} \right\} \tag{7.3}$$

The total magnetic moment of a sample, called the magnetization $\vec{M}$, depends on the magnitude and orientation of the magnetic moments of individual nuclear spins $\vec{\mu}_i$,

$$\vec{M} = \sum \vec{\mu}_i \tag{7.4}$$

Magnetization is the measurable quantity in NMR spectroscopy.

The magnetization is not equal to zero only if a sample has at least partially unarranged spins. Populations of various spin energy levels determined by the Boltzmann distribution are in equilibrium. In an external magnetic field $\vec{B}_0$, the energy of individual spin states depends on their orientation (θ_i) according to the external field:

$$E_i = -\vec{\mu}_i \cdot \vec{B}_0 = -\mu_i B_0 \cos\theta_i \tag{7.5}$$

A spin state can be changed only by the effect of magnetic fields. These fields can be external ($\vec{B}_0$ and $\vec{B}_1$) or internal, created by thermal motions of various electrical charges, magnetic dipoles, and so on. An external field $\vec{B}_1$ is coherent and therefore causes precession of $\vec{M}$ in the direction of the static magnetic field $\vec{B}_0$. Internal fields have incidental character and break the arrangement of magnetic moments, causing the relaxation of magnetization.

A sample's equilibrium arrangement is reached through energy exchange between spins and a lattice (other degrees of freedom of a system). This exchange is realized by fluctuating internal magnetic fields created by thermal movement. The time dependence of such movement has an exponential character, as a rule, and is described by the Bloch equations for the components of magnetization:

$$\frac{dM_x}{dt} = \gamma\left(M_y B_0 + M_z B_1 \sin\omega t\right) - \frac{M_x}{T_2} \quad \textbf{(7.6)}$$

$$\frac{dM_y}{dt} = \gamma\left(M_z B_1 \cos\omega t - M_x B_0\right) - \frac{M_y}{T_2} \quad \textbf{(7.7)}$$

$$\frac{dM_z}{dt} = -\gamma\left(M_x B_1 \sin\omega t + M_y B_1 \cos\omega t\right) - \frac{M_z - M_0}{T_1} \quad \textbf{(7.8)}$$

where M_0 is the equilibrium magnetization caused by a static field $\vec{B}_0$, T_1 and T_2 are called the spin-lattice and spin-spin relaxation time, and γ is the gyromagnetic ratio of a nucleus (a constant for the given nucleus). Due to the nonhomogeneity of a magnetic field in a sample, a resonance frequency scattering contributes to the extinction of components M_x and M_y. Therefore, it is necessary to introduce an effective relaxation time, T_2^*, instead of the relaxation time T_2

$$\frac{1}{T_2^*} = \frac{1}{T_2} + \frac{\Delta B_0}{2\gamma} \quad \textbf{(7.9)}$$

where ΔB_0 is the scattering of an external magnetic field induction caused by nonhomogeneities of a sample.

If only the external magnetic field $\vec{B}_0$ affects the sample, this field causes a free precession of spins in the direction $\vec{B}_0$, with a frequency known as the Lamor frequency:

$$\vec{\omega}_0 = -\gamma \vec{B}_0 \quad \textbf{(7.10)}$$

However, this precession is attenuated due to relaxation effects, so the precession rate is decreased to the value

$$\vec{\Omega}_0 = \vec{\omega}_0 - \vec{\omega} \quad \textbf{(7.11)}$$

The effective induction of an external static field:

$$\vec{B}_0^{ef} = \frac{\vec{\omega}_0 - \vec{\omega}}{\gamma} \tag{7.12}$$

corresponds to this value. The magnetic component of a radiofrequency field $\vec{B}_1'$ (Equation 7.3) also affects the precession, so that a real effective magnetic field affecting the given spin has the following value:

$$\vec{B}^{ef} = \vec{B}_0^{ef} + \vec{B}_1' \tag{7.13}$$

Relaxation processes can be neglected only in short-duration applications of the radiofrequency field. In other cases these relaxation effects must be put into an equation of the sample motion, so that

$$\frac{d\vec{M}}{dt} = \gamma\, \vec{M} \times \vec{B}^{ef} + \left(\frac{d\vec{M}}{dt}\right)_{\text{relax}} \tag{7.14}$$

Nonequilibrium magnetization of a sample changes with time. This dynamic form of magnetization is the only measurable resonance in a set of nuclear spins. Its component in an area x, y ($\vec{M}_{xy}$) induces an electromotive voltage in a receiver coil, which is expressed, after treatment, in the form of a frequency or time spectrum.

$\vec{M}_0$ and $\vec{M}(t)$ are not the only possible forms of arrangement of the spin system. Specific arrangements of interacting spins, known as coherencies, can also be created by affecting external fields in a system of mutually interacting spins. In time-based experiments, a sample passes through various types of coherency that characterize the sample memory, i.e., the period over which the given spin arrangement will attenuate. In NMR spectroscopy this period is relatively long (10^{-1} to 10 s). Together with the application of coherent radiation it causes the development of impulse NMR techniques. At the present time two basic methods of recording NMR spectra, the CW (continuous wave) method and impulse methods, are applied.[1-4]

CW METHODS

The stable response of a sample ($d\vec{M}/dt = 0$) of a radiofrequency field $\vec{B}_1$ effect is measured during application of continuous wave methods. A radiofrequency field disturbs a magnetization $\vec{M}_0$, while relaxation processes cause its reverse formation. In a short time, a dynamic equilibrium appears. Its state depends on $\vec{B}^{ef}$, $\vec{B}_1$, and the relaxation rate. As resonance $\vec{B}^{ef} \doteq \vec{B}_1$, its effect is minimal. Therefore, a strong component of magnetization in a plane x, y ($\vec{M}_{xy}$) (measurable quantity) is created by affecting $\vec{B}_1$. Simultaneously, the shape of spectral curves depends on the amplitude $\vec{B}_1$, a relaxation character, and a detector β phase (Figure 7.2).

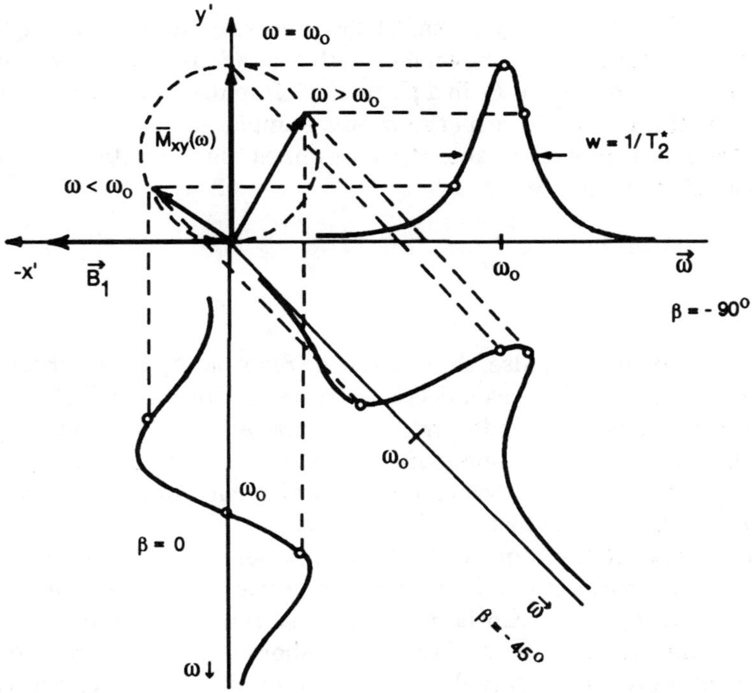

Figure 7.2 Stable response $M_{xy}(\omega)$ of a sample and its record during the different phases of detection.

The Lorentz curves are obtained at exponential relaxation (Equations 7.6 to 7.8) and at the low value of the amplitude $\vec{B}_1$ ($\gamma^2 B_1^2 T_1 T_2 \ll 1$). If a detector phase is shifted $-\pi/2$ according to a phase $\vec{B}_1$, the absorption Lorentz curve is obtained (see Chapter 2). Because absorption curves have shorter tails than those of their first derivation, it is more advantageous to record an absorption spectrum given spectra with multiple absorption peaks. The data obtained in this way can be numerically analyzed using any of the methods explained in Chapter 2.

IMPULSE METHODS

A sample is subjected to the effect of one or more radiofrequency impulses during application of impulse methods. A sample response, i.e., a change of magnetization in an area x, y ($\vec{M}_{xy}$), is recorded. A response induced by these impulses is attenuated continuously over time, and is called the FID (Free Induction Decay).[1-4]

The duration of a radiofrequency impulse is usually very short (1 to 100 μs), and relaxation effects can be neglected over this span. $\vec{M}$ clearly precesses around $\vec{B}^{ef}$ during an impulse. The effect of the impulse is frequency dependent (it is known as the "off-resonance" effect). A radiofrequency impulse is characterized by means of the precession angle of a magnetization component, which is in resonance with the radiofrequency field and the orientation α of the vector $\vec{B}_1$ in a plane x', y'. Impulses with $\alpha = \pi/2$ produce the maximal measurable response from a sample.

Pulse power (proportional to B_1) determines the characteristic period of duration of the response:

$$t_{\frac{\pi}{2}} = \frac{\frac{\pi}{2}}{\gamma B_1} \tag{7.15}$$

The more powerful the pulse, the shorter the response ($t_{\pi/2}$). Simultaneously, an increase in power causes a decrease in the "off-resonance" effect, and a pulse becomes frequency independent — nonselective. Composite impulses consisting of a series of successive impulses are used to increase the nonselectivity. These composite impulses also can compensate considerably for any nonhomogeneity of $\vec{B}_1$.

Some applications require the frequency selective pulses that can be realized by decreases of power or by prolonging impulse time. For example, impulses with $t_{\pi/2} \approx 1$ s can be applied for selective excitation of a single spectral component. A series of successive short impulses, which are separated by free precession periods, can also be used for selective excitation.

NUMERICAL METHODS FOR FID PROCESSING

Detection of FID

The detection system of an impulse spectrometer consists of one or two (quadrature detection) phase sensitive detectors. The reference frequency of a detector is set to a frequency B_1. With quadrature detection, the second detector has its reference phase shifted about $\pi/2$ from that of the first detector.

The same spectral parameters as from the frequency spectrum can be determined from simple FID (Figure 7.3). However, the sign of the relative resonance frequency can be determined only using quadrature detection. Therefore, simple detection requires that the reference frequency be at one end of the measured spectral region.

If a sample contains more frequency components, the FID is a complicated interferogram from which spectral parameters can be obtained only by extensive mathematical analysis.

Digitization is the first step of FID processing. This operation is realized using AD (analog-digital) converters in regular time intervals. The digitization rate ($1/t_d$) and an acquisition time of FID – A_T determine the basic parameters of the resulting frequency spectrum

$$S_W = \frac{1}{2t_d} \tag{7.16}$$

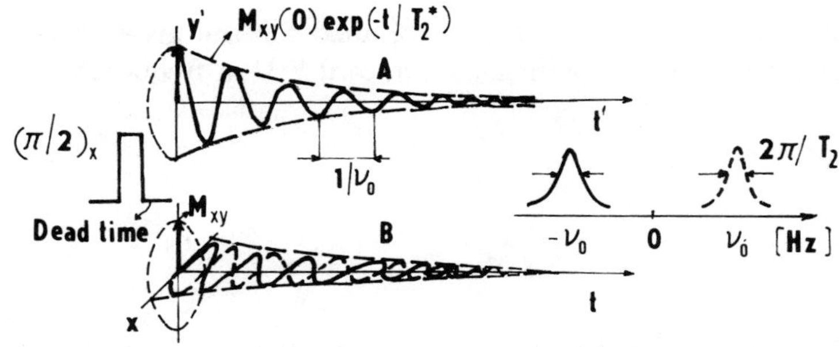

Figure 7.3 The scheme of the simple impulse experiment. With simple detection it is not possible to determine the sign of the precession $\vec{M}_{xy}$. From the time course of the simple FID, it is possible to determine the same spectral parameters as from the frequency spectrum.

$$N_P = \frac{A_T}{t_d} \quad \textbf{(7.17)}$$

where S_W (spectral width) is the width of the measured spectral region and N_p is the number of points representing the digitized FID. During simple detection it is necessary to digitize FID at frequency (Figure 7.4)

$$\frac{1}{t_d} = 2S_W \quad \textbf{(7.18)}$$

with

$$N_P = 2S_W A_T \quad \textbf{(7.19)}$$

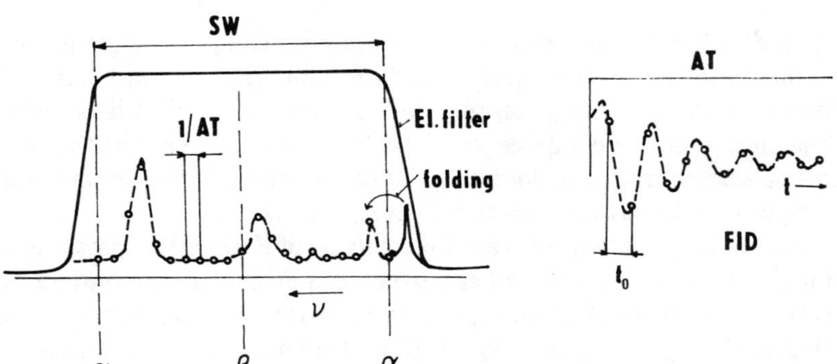

Figure 7.4 The influence of the FID digitization procedure on parameters of a frequency spectrum after digitization: simple detection **(a)** and quadrature detection **(b)**.

and real points $M_{y'}^{j}$, $j = 1, 2\ldots N_p$ obtained for the given FID. During quadrature detection it is sufficient to record FID at frequency

$$\frac{1}{t_d} = S_W \tag{7.20}$$

with

$$N_P = S_W\ A_T \tag{7.21}$$

and complex points $M_{y'}^{j} + i\ M_{x'}^{j}$, $j = 1, 2,\ldots N_p$ obtained. Detection by means of the TPPI (Time Proportional Phase Increment) method is the equivalent of quadrature detection. During this method, the FID is recorded at frequency

$$\frac{1}{t_d} = 2S_W \tag{7.22}$$

During each FID recording, the detector phase is incremented by $\pi/2$; this means that the resonance frequency is shifted by the value of $S_W/2$, i.e., to an edge of the spectrum.

All frequency components can be present in an FID. However, components outside the spectral area S_W mutually overlap the bands inside the spectral area (folding). This effect causes trouble with interpretation of a resulting spectrum and is undesirable. Electronic filters with appropriate frequency characteristics are used to remove folding. These filters are able to record selective parts of a spectrum. But removing noise, which implies increasing the sensitivity of the method, is the main task of these filters.

At this time, all NMR spectrometers using impulse techniques use powerful computers and software to analyze the FIDs obtained. Therefore, the problems which remain with standard software will not be explained here, but further attention will be focused on the treatment of NMR spectra, which enhances resolving capacity and unambigous interpretation.

FID Processing

A simple FID (consisting only of one frequency component) contains the same information as a CW spectrum, and therefore they are equivalent. This equivalence is stable for any complicated spectra, although it is not obvious from the course of the complicated FID. This equivalence mathematically causes a spectrum from one domain to be transformed to the second domain by a simple Fourier transformation (see Appendix D).

A scheme for complete software for analysis of FID-NMR spectra is illustrated in Figure 7.5. As a rule, all spectrometer parameters are automatically checked by the software. A subprogram for recording data repeatedly accumulates them in the given table. After repeated impulses the data recording is repeated according to the original scheme, and after finishing the record, all data are averaged. Because of the large number of FID repetitions, some data could overflow, and it is necessary to use a suitable scaling procedure.

The original data are then treated, as a rule, by three types of operations:

1. Adding zeros to the FID end points will increase the apparent resolution in the frequency domain spectrum and will improve the representation of narrow spectral curves.
2. The FID is measured over a finite time. Therefore, some slowly relaxing components can have sufficiently high intensity at the end of the

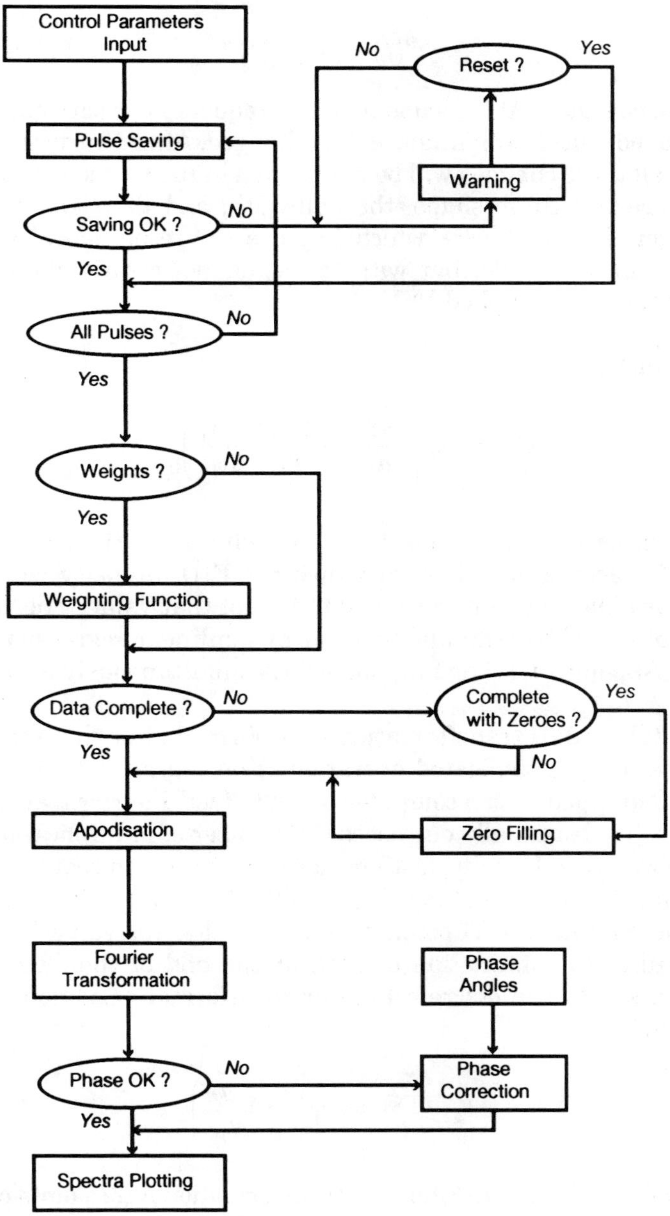

Figure 7.5 Flowchart of software for FID treatment.

FID response. After Fourier transformation the deterioration of spectral curve shape is affected by truncation of these components. Therefore, many apodizing procedures, which are based on the multiplication of the FID by a function decreasing monotonically to zero at the end of the FID, are used.

3. Multiplying the FID by an exponential weighting function ($w[t] \geq 0$), which will change the original FID $f(t)$ to a function $f'(t)$

$$f'(t) = f(t)\, w(t) \tag{7.23}$$

is the next step. At the same time the frequency of each component will be saved, but its amplitude will be changed. After Fourier transformation all these changes will be manifested in the frequency spectrum in the spectral curve shape, the halfwidth, and the signal/noise ratio. Various functions $w(t)$, which improve the resolution of a resulting spectrum as a rule (but with the signal/noise ratio simultaneously worsened), are applied.[5,6]

The function

$$w(t) = \exp\left[\frac{\ln 2}{a} t - \frac{\ln 2}{4a}\left(\frac{\Delta\omega_0}{\Delta\omega_1}\right)^2 t^2\right] \tag{7.24}$$

is one example of a function of this type, where a is the time expressed in units A_T (acquisition time) at which the FID intensity will decrease from the original function (for $t = 0$) to half of that value. The constant a may be optimized by trial and error or by nonlinear regression analysis methods (Appendix C). The function $w(t)$ is simultaneously normalized to unity.

A digital Fourier transformation is performed after these treatments. This numerically complicated operation is only practical on NMR spectrometers equipped with a computer using a "fast" Fourier transformation (FFT), which substantially decreases the number of multiplications during the transformation. Usually, fast Fourier transformation requires the number of points N_p to be equal to a power of 2. Therefore, before the transformation the number of FID points is made equal to the value $N_p = \ldots 4096, 8192, \ldots$ either by elimination of data at the end of the FID or by the addition of zeros. The discrete Fourier transformation is defined by the relationship

$$F_k = \sum_{l=0}^{N_p-1} f_l \exp\left(-i\,\frac{2\pi l k}{N_P}\right) \tag{7.25}$$

where $k = 0, 1, \ldots N_p/2 - 1$. After this transformation $N_p/2$ points of the real axes and $N_p/2$ points of the imaginary part of a frequency spectrum will be obtained.

After the Fourier transformation the spectrum obtained usually has a deformed phase, and this deformation is, as a rule, frequency dependent. The main reasons for this phase deformation are (1) incorrect setting of the detector phase (this part of the deformation is not usually frequency dependent) and (2) the appearance during some experiments of delays inherent in FID recording, known as dead time (DT).

Application of Digital Filters for Measured Spectral Region Elimination

The phase error introduced prevents the real part of a transformed spectrum $F^{\text{real}}(\omega)$ from corresponding to an absorption and the imaginary part of a spectrum $F^{\text{imag}}(\omega)$ from corresponding to a dispersion curve. Instead, the mixed spectra are obtained. Pure absorption $A(\omega)$ and dispersion $D(\omega)$ spectra can be obtained after the mathematical rearrangement

$$A(\omega) = \cos\varphi\, F^{\text{real}}(\omega) - \sin\varphi\, F^{\text{real}}(\omega) \quad \textbf{(7.26)}$$

$$D(\omega) = \sin\varphi\, F^{\text{real}}(\omega) + \cos\varphi\, F^{\text{real}}(\omega) \quad \textbf{(7.27)}$$

The phase shift φ is optimized by an interactive procedure based on the spectral shape, or in newer spectrometers by means of an automatic optimization method. Usually a linear dependence φ on $\Delta\omega$:

$$\varphi = a + b\Delta\omega \quad \textbf{(7.28)}$$

is assumed. This functional dependence compensates for phase deformation due to incorrect setting of the detector phase, and due to the delays at the beginning of the FID recording (dead time).

Application of Numerical Methods for Increasing NMR Spectral Resolution

Interpretation of NMR spectra is often limited by the strong overlap of peaks. Various numerical procedures are used to eliminate this problem.

De Marco and Wüthrich[7] used digital filtering of the FID by means of a sinusoidal "window" function, which decreases the individual curve width, while simultaneously decreasing the distortion of curve shape and partially reducing the signal/noise ratio.

During this procedure FID is multiplied by the sinusoidal function with zero phase and period equal to double the FID recording time (A_T). For this procedure, the curve shape is given by the following relationship:

$$L_{\sin}(\Delta\omega) = \int_0^{A_T} \exp\left(\frac{-t}{T_2^*}\right) \cos(\Delta\omega t) \sin\left(\frac{\pi t}{A_T}\right) dt \quad \textbf{(7.29)}$$

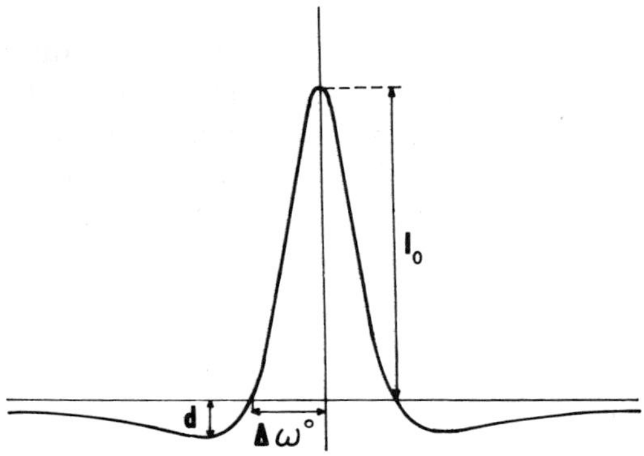

Figure 7.6 The definition of the signal intensity (I_0), curve halfwidth ($\Delta\omega^0$), and distortion (I_0/d) for the resonance curve. (From De Marco, A. and Wüthrich, K., *J. Magn. Reson.*, 24, 201, 1976. With permission.)

where T_2^* is the effective time of the spin–spin relaxation (Equation 7.9) and $\Delta\omega$ is the difference between resonance and considered frequency. As with the method of convolution difference (see Appendix D), the broad components of a spectrum are attenuated more than the narrow components. However, the deformation of spectrum end points does not appear because the response of a sample is zero at time A_T.

The defect of this method is that it does not eliminate the consequences of the phase error introduced above and, therefore, the band obtained by this method is not a pure absorption curve but a combination of absorption and dispersion curves (Figure 7.6). Moreover, this method does not decrease the signal/noise ratio.

DeMarco and Wüthrich[7] and Brereton et al.[8] proposed the application of a partially modified shape function (Equation 7.29):

$$L_{\sin}(\Delta\omega) = \int_0^{A_T} \exp\left(\frac{-t}{T_2^*}\right) \cos(\Delta\omega t) \cdot \left[K + \sin\left(\frac{\pi t}{A_T}\right)\right] dt \tag{7.30}$$

where K is an adjustable parameter ensuring a reasonable compromise between the increasing resolving ability and the decreasing signal/noise ratio (Figure 7.7).

As a rule, the signal/noise ratio increases during exponential attenuation of the FID. Multiplication by the term $e^{-t/\tau}$ corresponds to a convolution with the Lorentz curve of halfwidth:

$$\delta = (\pi\tau)^{-1} \tag{7.31}$$

The consequence of this convolution is that the narrowing of peaks with half-widths below the threshold value of δ does not appear. Gueron[9] therefore proposed an empirical scheme for optimal narrowing of the FID curve, which is based on multiplication of the FID by the function

$$\exp(-\pi\delta t)\,\sin\left[2\pi(0,6\delta)t\right] \quad \textbf{(7.32)}$$

If $0.6\delta = 1/2A_T$, Equation 7.32 reduces to Equation 7.30. This shows that Equation 7.30 is only suitable with a reasonable choice for the value A_T.

Clin et al.[10] proposed application of the convolution of the FID by means of the general sinusoidal function (Figure 7.8)

$$\mathscr{F}\{\text{FID}\cdot\sin(\alpha t)\} = \mathscr{F}\{\text{FID}\} * \mathscr{F}\{\sin(\alpha t)\} \quad \textbf{(7.33)}$$

where * indicates a convolution product (see Appendix D). The Fourier transformation of sin(αt) is represented by the difference of two δ functions: $\delta(\Delta\omega + \alpha)$ and $\delta(\Delta\omega - \alpha)$. Due to that fact, the Fourier transformation of FID filtered by the sinusoidal "window" with the period $2\pi/\alpha$ will give the transformed FID shifted $\alpha/2\pi$ and at the same time shifted $-\alpha/2\pi$. The value of the maximum position will not be changed, because for the maximum the following is valid:

$$\int_{-\infty}^{+\infty} f(x)\,\delta(x - x_0) = f(x_0) \quad \textbf{(7.34)}$$

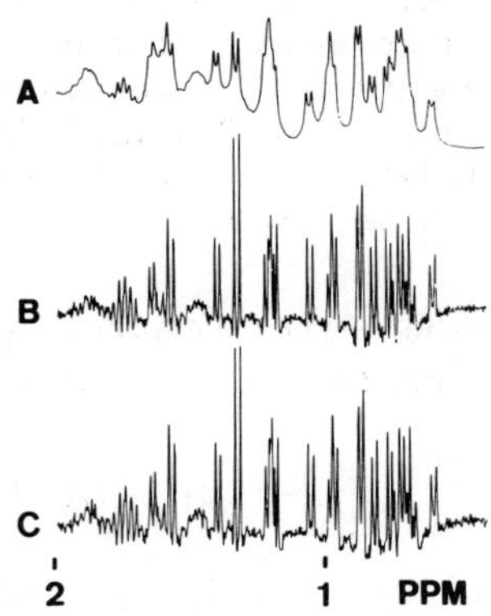

Figure 7.7 Aliphatic region of an FT-NMR spectrum ^{1}H of the base inhibitor of pancreatic tripsine obtained at 360 MHz: the measured spectrum **(A)**, the spectrum after applying Equation 7.29 **(B)**, and the spectrum after applying Equation 7.30 **(C)**. (From DeMarco, A. and Wüthrich, K., *J. Magn. Reson.*, 24, 201, 1976. With permission.)

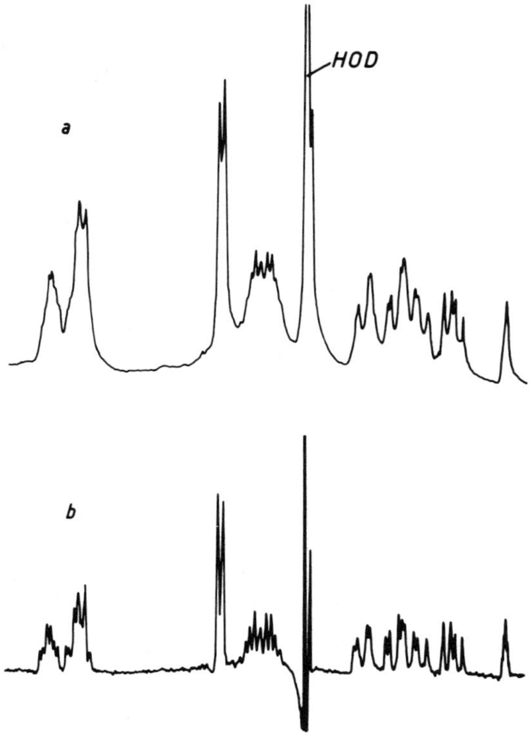

Figure 7.8 A part of the FT-NMR spectrum ^1H in polypeptides: the measured spectrum **(a)** and the spectrum after application of the convolution (Equation 7.33) **(b)**. (From Clin, B., De Bony, J., Lalanne, P., Biais, J., and Lemanceau, B., *J. Magn. Reson.*, 33, 457, 1979. With permission.)

To increase the signal/noise ratio Lohman[11] proposed multiplying the peak-shape function, after Fourier transformation, by a filter of the form

$$L(\Delta\omega) = T_2^* \frac{1+(2d)^{-2}+\left(\Delta\omega T_2^*\right)^2}{\left[1+(2d)^{-2}+\left(\Delta\omega T_2^*\right)^2\right]^2 - 4(2d)^{-2}\left(\Delta\omega T_2^*\right)^2} \quad \textbf{(7.35)}$$

where

$$d = \frac{A_T}{\pi T_2^*} \quad \textbf{(7.36)}$$

is the digital resolution, defined as the number of the spectral points above half the height of the unfiltered curve. The mean quadratic value of noise in the frequency spectrum is reduced by the factor $1/\sqrt{2}$ by the application of this filter, improving the signal/noise ratio by approximately 40%.

Delsuc et al.[12] developed another method for improving the signal/noise ratio. In this method the measured signal is expressed in the form of K exponentially attenuated sinusoidals:

$$x_n = \sum_{k=1}^{K} c_k \exp\left(-d_k n\Delta t\right) \exp\left[i\left(\omega_k n\Delta t + Q_k\right)\right] \tag{7.37}$$

which includes $4K$ spectral parameters (amplitudes c_k, frequencies ω_k, attenuating factors d_k, and phases Q_k) and $n\Delta t$ are the intervals over which the spectrum is recorded. The filtering of Equation 7.37 leads to nonlinear regression analysis. Delsuc et al.[12] proposed reducing this nonlinear problem to the model known as a predictive linear model, in which data x_n are expressed in the form of a linear combination K of the previous and following data:

$$x_n = \sum_{k=1}^{K} a_k \, x_{n-k} \tag{7.38}$$

$$x_n = \sum_{k=1}^{K} b_k \, x_{n+k} \tag{7.39}$$

The coefficients a_k and b_k can be determined by linear algebra.[13]

THE METHOD OF MAXIMAL ENTROPY IN NMR SPECTROSCOPY

NMR spectroscopy is often not very sensitive. Its restrictions are significant, especially in complicated biological systems of samples with limited size or concentration and for sample compounds that are insoluble or instable, and which do not permit accumulation of data over a long time. FT-NMR[1-5] accelerates data accumulation, strongly increases the quality of the spectra obtained, and quickly gives the FID. Treatment of these data can be improved by multiplying the FID with a suitable apodization function before Fourier transformation, applying sensitive filters, and application of other methods described in previous chapters. Simultaneously, the signal/noise ratio can always be improved to permit recognition of individual bands.

The method of maximal entropy (MEM)[14-18] is a generally applicable technique for solving data analysis problems of this type, and compared to other procedures it has many advantages. At the same time MEM does not introduce any new structural characteristics or correlations during the treatment of data and therefore gives the simplest and purest resulting spectra.

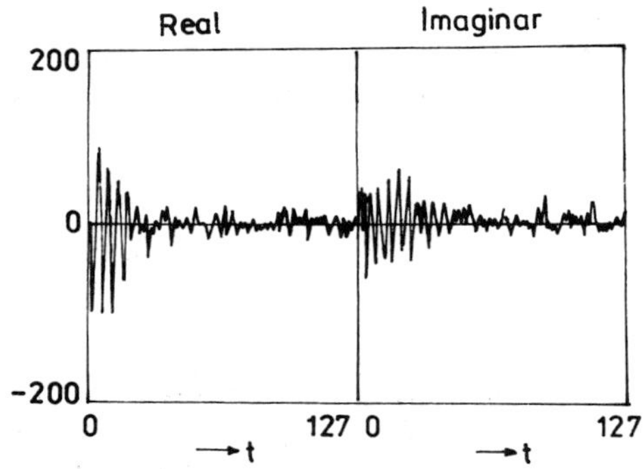

Figure 7.9 The real and imaginary parts of the typical FID. (From Lave, E. D., Skilling, J., Staunton, J., Sibisi, S., and Brereton, R., *J. Magn. Reson.*, 62, 437, 1985. With permission.)

FT-NMR spectroscopy records the time course of induction attenuation — FID, which can be expressed, for example, in the form of a linear combination of attenuated exponential dependencies:

$$M(t) = \sum_j n_j \exp\left(-k_j t\right) \exp\left(-i\, 2\pi \nu_j t\right) \qquad \textbf{(7.40)}$$

A simple example of this type of data is illustrated in Figure 7.9. These data need to be interpreted in terms of a density series n_j of different nuclei j, from which each is connected with frequency ν_j and rate constant of relaxation k_j. The inverse Fourier transformation (see Appendix D) is the simplest algorithm to obtain them, and gives

$$\mathrm{Re}\left[\mathcal{F}^{-1}\left\{M(t)\right\}\right] = \mathrm{Re}\left[\int_0^\infty M(t) \exp(i\, 2\pi\nu t)\, dt\right]$$

$$= \sum_j \frac{n_j k_j}{4\pi^2\left(\nu - \nu_j\right)^2 + t^2} \qquad \textbf{(7.41)}$$

Thus, a series of Lorentz bands centered at frequencies ν_j, with halfwidth k_j and area proportional to densities n_j, result from the Fourier transformation. The spectral data obtained are usually strongly overlapped and become confused, which is the main defect of this direct procedure. Therefore, plotting of a spectrum given by Equation 7.41 shows a broadened response to spectra with poor resolution. Experimental data include more information than Equation 7.41 can extract.

Let's go back to Equation 7.40, which defines the measuring process. Because the FID is measured only a finite number of times $t = 0, 1, 2,\ldots N - 1$ (in corresponding units), Equation 7.40 can be written in the form of a discrete sum over the scaled integer frequencies ν:

$$M_t = \sum_{\nu=0}^{N-1} n_\nu \exp\left(-k_\nu t\right) \exp\left(-i\,\frac{2\pi\nu t}{N}\right) \tag{7.42}$$

If a simplifying assumption, namely that all nuclei relax at the same rate (k = constant, see general case in Sibisi[16]), is supposed, the frequency spectrum:

$$n_\nu = \frac{1}{N} \sum_{t=0}^{N-1} M_t \exp\left(kt\right) \exp\left(i\,\frac{2\pi\nu t}{N}\right) \tag{7.43}$$

is obtained after inverse Fourier transformation. This transformation involves multiplication of the FID by the positive exponential exp(kt), which broadens individual bands (especially for higher t values). The measurement results become worse due to the fact that this function also increases the noise exponentially. This noise then produces a spectrum n_ν, which contains negative regions for certain values of ν. Because the value n_ν must be positive over the whole region ν, for obvious physical reasons, the spectrum obtained is physically incorrect (physically correct negative values n_ν are obtained only if a phase shift of signals exists — this effect appears equally with all nuclei).

These problems always arise during direct solution of Equation 7.42. The corresponding trial solutions $\bar{n}_\nu$ are shown on the left side of Figure 7.10. They can be transformed back to the corresponding trial FIDs on the right side of Figure 7.10. Many of these trial FIDs can be rejected *a priori* because they are strongly different from the real experimental data in Figure 7.9. However, it is practically impossible to choose between the FIDs in Figures 7.10e and 7.10g given the experimental data in Figure 7.9, if the contribution of indeterminacy caused by the presence of noise is considered. Therefore, some way to choose the "correct" spectrum must be made available. This is the motivation for the MEM method.

Principle of Maximal Entropy

Let this simple example be considered in more detail. Given a spectrum consisting of six bands (A, B, C, D, E, and F) at frequencies

$$\begin{array}{ll} \nu(A) = 0 & \nu(B) = 1 \\ \nu(C) = 2 & \nu(D) = 3 \\ \nu(E) = 4 & \nu(F) = 5 \end{array} \tag{7.44}$$

according to ν(A). Let us assume that the delay time of free induction is very long. Then their FIDs, measured at times t = 0, 1, 2,...5, can be written in the form

$$M_t = \sum_{\nu=0}^{5} n_\nu \exp\left(-i\,\frac{2\pi\nu t}{6}\right) \tag{7.45}$$

After application of the Fourier transformation the following results:

$$n_\nu = \frac{1}{6}\sum_{t=0}^{5} M_t \exp\left(i\,\frac{2\pi\nu t}{6}\right) \tag{7.46}$$

Because n_ν are all real numbers, it is sufficient to measure six real numbers:

$$\begin{aligned}
M_0 &= n_0 + n_1 + n_2 + n_3 + n_4 + n_5 \\
\mathrm{Re}\left[M_1\right] &= \frac{2n_0 + n_1 - n_2 - 2n_3 - n_4 + n_5}{2} = \text{unmeasurable} \\
\mathrm{Im}\left[M_1\right] &= \frac{\sqrt{3}\left(-n_1 - n_2 + n_4 + n_5\right)}{2} = \text{unmeasurable} \\
\mathrm{Re}\left[M_2\right] &= \frac{2n_0 - n_1 - n_2 + 2n_3 - n_4 - n_5}{2} = 6 \\
\mathrm{Im}\left[M_2\right] &= \frac{\sqrt{3}\left(-n_1 + n_2 - n_4 + n_5\right)}{2} = -2\sqrt{3} \\
M_3 &= n_0 - n_1 + n_2 - n_3 + n_4 - n_5 = 18
\end{aligned} \tag{7.47}$$

The right side of Equation 7.47 involves the specific measured values, in addition to the value for $t = 1$, which is not available (e.g., instrumentation limits acquisition of data occurring soon after starting the measurement); therefore, the FID data are not complete. Practically, this case often occurs due to the influence of noise and therefore, the FID does not specify a spectrum unambiguously. The set of data from Equation 7.47 can be used to give the sums of intensities:

$$\begin{aligned}
n(A) + n(D) &= n_0 + n_3 = 16 \\
n(C) + n(F) &= n_2 + n_5 = 8 \\
n(B) + n(E) &= n_1 + n_4 = 12
\end{aligned} \tag{7.48}$$

and the ratio of intensities:

$$\frac{n(A) + n(C) + n(E)}{n(B) + n(D) + n(F)} = \frac{n_0 + n_2 + n_4}{n_1 + n_3 + n_5} = 3 \tag{7.49}$$

Two degrees of freedom remain in the data set. The number of degrees of freedom will be much higher in experimental data, which involve hundreds or thousands of noise values. It is also necessary to select a spectrum from these data that adequately describes experimental measurements. There is no procedure to choose the “correct” spectrum by rigorous deduction.

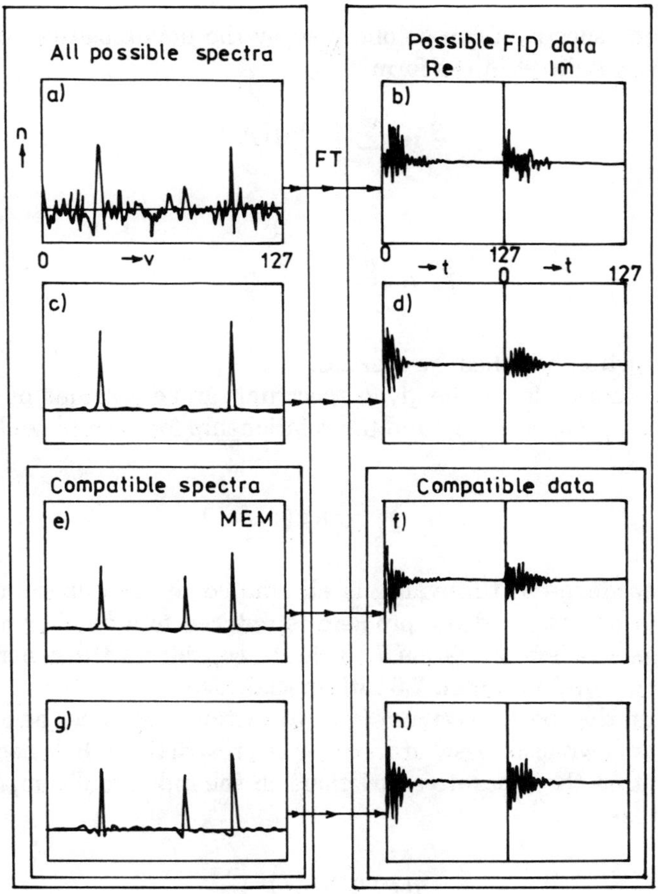

Figure 7.10 The application of the MEM method. Each from the possible resolved spectra **(a, c, e, g)** has the corresponding FID **(b, d, f, h)**. Only some of these FIDs (f, h) are compatible with the original spectrum in Figure 7.9. (From Lave, E. D., Skilling, J., Staunton, J., Sibisi, S., and Brereton, R., *J. Magn. Reson.*, 62, 437, 1985. With permission.)

The following example is overly simple, and the choice of "correct" data is also simple. This choice can be realized by introducing the equality of ratio values $n(A)/n(D)$, $n(C)/n(F)$, and $n(E)/n(B)$ to the total ratio defined by Equation 7.49. The only "correct" spectrum for the following values is obtained:

$$
\begin{aligned}
n(A) &= 12 \qquad & n(B) &= 3 \\
n(C) &= 6 \qquad & n(D) &= 4 \\
n(E) &= 9 \qquad & n(F) &= 2
\end{aligned}
\tag{7.50}
$$

No other spectrum will fulfill the condition for the total intensity ratios given by Equation 7.49.

The same spectrum can be obtained by the maximization of informational entropy defined in the form[14]

$$S = -\sum_{\nu} p_{\nu} \log\left(p_{\nu}\right) \quad \textbf{(7.51)}$$

where

$$p_{\nu} = \frac{n_{\nu}}{\sum n_{\nu}} \quad \textbf{(7.52)}$$

No other function has these properties.

A priori knowledge of the given spectrum can be obtained by means of the known model spectra m_{ν}, and the relationship for entropy will have the form[19-24]

$$S = -\sum_{\nu} p_{\nu} \log\left(p_{\nu}/m_{\nu}\right) \quad \textbf{(7.53)}$$

A priori knowledge of this value is eliminated by this operation in the optimization process, and the problem is reduced to a smaller number of optimized parameters. Equation 7.53 can be considered the general definition of entropy and Equation 7.51 its special case.

To obtain the "best" spectrum P, any function $R(P)$ must be optimized, which explains what is "best" at meeting the restrictions. It is necessary to define R so that the separate restrictions on the individual components P_{ν} can also be met:

$$R(P) = \sum_{\nu} r\left(P_{\nu}\right) \quad \textbf{(7.54)}$$

by accounting for the influence of characteristics related to one part of a spectrum on those of other parts of a spectrum. Let the spectrum P consist of M bands M_j, $j = 1, 2,\ldots M$, and each jth band is split by digitization into N-multiple of a multiplet P_{kj}, $k = 1, 2,\ldots N$, so that the spectrum consists of $N \cdot M$ individual components.

Then for the global intensity of the jth band the following is valid:

$$X_j = \sum_{k=1}^{N} P_{jk} \qquad \text{for} \quad j = 1, 2,\ldots M \quad \textbf{(7.55)}$$

Similarly, for the total intensity of the kth component of the mutliplet in the whole spectrum:

$$Y_k = \sum_{j=1}^{M} P_{jk} \qquad \text{for} \quad k = 1, 2,\ldots N \quad \textbf{(7.56)}$$

is valid. For simplicity, the minimized condition

$$\sum_{j=1}^{M} X_j = \sum_{k=1}^{N} Y_k = 1 \tag{7.57}$$

is introduced. If only data for X and Y are known, then the uncorrelated spectrum

$$P_{jk} = X_j Y_k \tag{7.58}$$

must be preferred, because there are no values showing that different bands will have different structure.

Given that values X and Y are maximized, the function $R(P)$ leads to the variation condition

$$\delta\left(R - \sum_j \lambda_j X_j - \sum_k \mu_k Y_k \right) = 0 \tag{7.59}$$

where λ_j, $j = 1, 2,\ldots M$, and μ_k, $k = 1, 2,\ldots N$, are Lagrange multiplicators for the restriction conditions. The solution of the variation problem posed in Equation 7.59 is

$$\frac{\partial R}{\partial P_{jk}} = \lambda_j + \mu_k \tag{7.60}$$

If R/P_{jk} is labeled as a function $\theta(P_{jk})$, λ_j as a function $\phi(X_j)$, and μ_k as a function $\psi(Y_k)$, then the following functional equation results:

$$\theta(xy) = \phi(x) + \psi(y) \tag{7.61}$$

The solution[19-21] is:

$$\theta(x) = A \log(x) + B \tag{7.62}$$

Therefore, for the variation function, the equation

$$R(P) = A \sum_{j,k} P_{jk} \log\left(P_{jk}\right) + (B - A) \sum_{j,k} P_{jk} + C \tag{7.63}$$

will be obtained. Equation 7.63 has the same shape as the formula for entropy if constants A, B, C, and ΣP_{jk} are not taken into consideration.

Experimental FID data are always more complicated than those in the example discussed. Practical measurements are often very noisy, and they can be written in the form

$$M_t \pm \sigma_t \tag{7.64}$$

where σ_t is the standard deviation of noise. It would not be suitable to record these data exactly, because the noise would be interpreted as true signal. The χ^2 statistic

$$\chi^2 = \frac{\sum_t \left| \tilde{M}_t - M_t \right|^2}{\sigma_t^2} \tag{7.65}$$

can be used as a measure of the difference of real data M and "error" data $\overline{M}$, which would be measured in a hypothetical testing spectrum for resolution of noise. Only the spectra for which $\chi^2 \approx N$, are acceptable.

For the MEM method, a spectrum is selected that has higher entropy as the "result" of the experiment. Such spectra can be formally obtained by maximization of entropy, fulfilling the restriction of the condition $\chi^2 \approx N$.

Numerical Problems Associated with the MEM Method

Numerical problems associated with the MEM method are a consequence of the attempted optimization of nonlinear functions (entropy) under the constraints for all possible spectra. Typical spectra are characterized by thousands of degrees of freedom for ν. Because the entropy S is a nonlinear function, an algorithm for its optimization must be iterative and must redefine its repeatable "error" spectra $\overline{n}_\nu$. At the same time entropy has only a maximum, therefore, S and χ^2 are convex functions and, for this reason, the problem of selecting the starting point is eliminated. It is possible to start, for example, from the point $\overline{n}$ = constant. Then in an iterative cycle, n_ν is modified:

$$\tilde{n}_\nu \rightarrow \tilde{n}_\nu + \delta\tilde{n}_\nu \tag{7.66}$$

with the negative gradient value:

$$\delta\tilde{n}_\nu \approx \frac{-\partial\left(\chi^2\right)}{\partial\tilde{n}_\nu} \tag{7.67}$$

The gradient is calculated by using an attenuating Fourier transformation:

$$\tilde{M}_t = \exp(-kt) \sum_{t=0}^{N-1} \tilde{n}_\nu \exp\left(-i\,\frac{2\pi\nu t}{N}\right) \tag{7.68}$$

(which simulates the measuring process), by the following calculation — weighting the residual noise:

$$\frac{\partial\left(\chi^2\right)}{\partial\tilde{M}_t} = \frac{2\left(\tilde{M}_t - M_t\right)}{\sigma_t^2} \tag{7.69}$$

and repeating the Fourier transformation, leading to the system of differential equations:

$$\frac{\partial(\chi^2)}{\partial \tilde{n}_\nu} = \sum_{t=0}^{N-1} \left[\frac{\partial(\chi^2)}{\partial \tilde{M}_t}\right] \left(\frac{\partial \tilde{M}_t}{\partial \tilde{n}_\nu}\right)$$

$$= \sum_{t=0}^{N-1} \left[\frac{\partial(\chi^2)}{\partial \tilde{M}_t}\right] \exp(-kt)\ \exp\left(-i\,\frac{2\pi\nu t}{N}\right) \qquad \textbf{(7.70)}$$

An ambiguous "error" spectrum also can be modified by the increase of its entropy:

$$\delta\tilde{n}_\nu = \frac{\partial S}{\partial \tilde{n}_\nu} = \text{constant} - \log(\tilde{n}_\nu) \qquad \textbf{(7.71)}$$

It is more advantageous to combine both of the procedures introduced above in one iterative step:

$$\tilde{n}_\nu \rightarrow \tilde{n}_\nu + u\,\frac{\partial(\chi^2)}{\partial \tilde{n}_\nu} + v\,\frac{\partial S}{\partial \tilde{n}_\nu} \qquad \textbf{(7.72)}$$

applying these changes to modify both S and χ^2.

The iterations will be finished when

$$\chi^2 = \text{true value } (=N) \qquad \textbf{(7.73)}$$

and the gradients

$$\frac{\partial(\chi^2)}{\partial \tilde{n}_\nu} \text{ and } \frac{\partial S}{\partial \tilde{n}_\nu} \qquad \textbf{(7.74)}$$

are zero.

Fulfilling these two conditions is necessary and unavoidable when obtaining the solution by the MEM method.

Figure 7.11 illustrates application of the MEM method during interpretation of the FT-NMR spectrum of 2-vinylpyridine in $CDCl_3$.

SIMULTANEOUS APPLICATION OF THE MEM METHOD AND FOURIER DECONVOLUTION ANALYSIS

Ni et al.[25] developed on effective procedure using a combination of the MEM method and spectral deconvolution (the MEFSD method), which significantly improved the spectral resolution and simultaneously increased the signal/noise ratio.

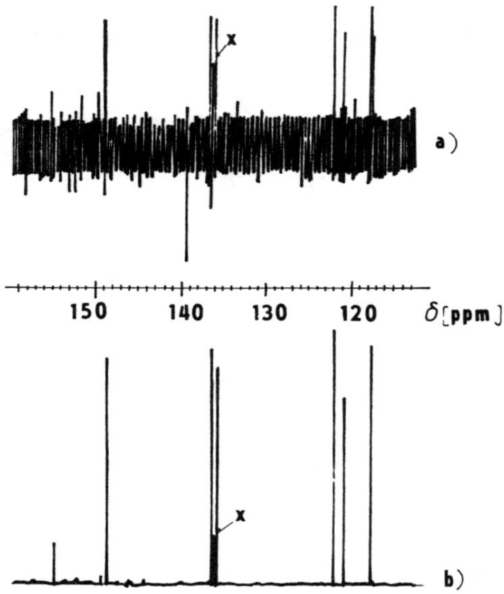

Figure 7.11 The MEM method during analysis of FT-NMR spectra: the measured spectrum 2-vinylpyridine in $CDCl_3$ **(a)** and the spectrum after application of the MEM method **(b)** — x labels the instrumentation artifact caused by imperfect signal transmission. (From DeMarco, A. and Wüthrich, K., *J. Magn. Reson.*, 24, 201, 1976. With permission.)

The delay of induction — FID, measured in the NMR experiment, can be decomposed to the form

$$M(t) = F(t) + u(t) \tag{7.75}$$

where $F(t)$ is the theoretical FID connected with the spectrum in the frequency region $X(\nu)$ by the relationship

$$F(t) = \int_{-\infty}^{+\infty} X(\nu)\exp(-i\,2\pi\nu t)\,d\nu \tag{7.76}$$

and $u(t)$ represents the noise influence characteristic of the given spectrometer. $X(\nu)$ can be decomposed to the convolution product (see Appendix D) of δ functions with the peak-shape function:

$$X(\nu) = \sum_{\upsilon=1}^{K}\int_{-\infty}^{+\infty} I_\upsilon(\nu-\nu')\,\delta(\nu'-\nu_\upsilon)\,d\nu' \tag{7.77}$$

where $I_\upsilon(\nu)$ is the function of the shape of the resonance curve, which can be approximated by the Lorentz function (Chapter 2.). Substituting Equation

7.77 into Equation 7.76 and using the convolution theorem (see Appendix D), the following results:

$$F(t) = \sum_{v=1}^{K} \left[\int_{-\infty}^{+\infty} I_v(\nu)\ \exp(-i\,2\pi\nu t)\,d\nu \right.$$

$$\left. \cdot \int_{-\infty}^{+\infty} \delta(\nu - \nu_v)\ \exp(-i\,2\pi\nu t)\,d\nu \right]$$

$$= \sum_{v=1}^{K} C_v(t) \int_{-\infty}^{+\infty} \delta(\nu - \nu_v)\ \exp(-i\,2\pi\nu t)\,d\nu \qquad \textbf{(7.78)}$$

where

$$C_v(t) = \int_{-\infty}^{+\infty} I_v(\nu)\ \exp(-i\,2\pi\nu t)\,d\nu \qquad \textbf{(7.79)}$$

is the correlation function for the resonance frequency ν.[25,26]

If $I_v(\nu)$ is the Lorentz function (the general assumption in NMR spectroscopy), $C_v(t)$ has an exponential character and Equation 7.78 is reduced to the form

$$F(t) = \sum_{v=1}^{K} \exp(-k_v t) \int_{-\infty}^{+\infty} \delta(\nu - \nu_v)\ \exp(-i\,2\pi\nu t)\,d\nu \qquad \textbf{(7.80)}$$

where k_v is the rate constant of the induction delay in the vth resonance.

If all spectral bands have the same halfwidths, i.e., $k_v = k$ for $v = 1, 2, \ldots K$, Equation 7.79 will change to the form

$$F(t) = \exp(-kt) \int_{-\infty}^{+\infty} \sum_{v=1}^{K} \delta(\nu - \nu_v)\ \exp(-i\,2\pi\nu t)\,d\nu$$

$$= \exp(-kt) \int_{-\infty}^{+\infty} N(\nu)\ \exp(-i\,2\pi\nu t)\,d\nu \qquad \textbf{(7.81)}$$

where the quantity $N(\nu)$ is known as the total spectral density:

$$N(\nu) = \sum_{v=1}^{K} \delta(\nu - \nu_v) \qquad \textbf{(7.82)}$$

If values k_v are different for individual bands, it is possible to use the value k corresponding to the narrowest band of a spectrum.[25] Similarly, Equation 7.77 and Equation 7.82 can be modified in the following way:

$$N(\nu) = \sum_{v=1}^{K} \int_{-\infty}^{+\infty} L_v(\nu - \nu')\, \delta(\nu' - \nu_v)\, d\nu' \tag{7.83}$$

Resonance frequencies will have effective halfwidths in a spectrum after deconvolution:

$$\frac{k'_v}{\pi} = \frac{(k_v - k)}{\pi} \tag{7.84}$$

corresponding to the Lorentz band L_v (ν), while k'_v is the rate constant of induction delay after deconvolution.

In general, the deconvolution can be performed using Equations 7.78 and 7.79. The individual bands must not have the shape of the Lorentz functions, and this is not likely because, in many NMR experiments applying various digital filters, different complicated forms of spectral bands appear.

Analysis of NMR spectra by the MESFD method is illustrated in Figure 7.12.

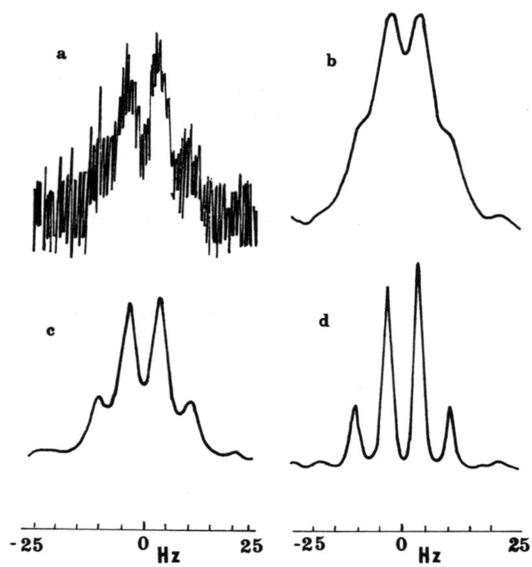

Figure 7.12 The MESFD method: the measured spectrum of proton resonance of the multiplet C-H of the amino acid alanine in D_2O **(a)**, the spectrum after broadening the curves by application of the Fourier transformation with broad Lorentz curves **(b)**, the spectrum after applying the MESFD method with Lorentz curves (Equation 7.83) **(c)**, and the spectrum after applying the MESFD method with general type of bands (Equations 7.78 and 7.79). (From Ni, F. and Scheraga, H. A., *J. Raman Spectrosc.*, 16, 337, 1985. With permission.)

REFERENCES

1. Farrar, T. C. and Becker, E. D., *Pulse and Fourier Transform NMR,* Academic Press, New York, 1972.
2. Shaw, D., *Fourier Transform NMR Spectroscopy,* Elsevier, Amsterdam, 1976.
3. Marshall, A. G., *Fourier, Hadamard and Hilbert Transforms in Chemistry,* Plenum Press, New York, 1982.
4. Goljer, I. and Liptaj, T, *New Methods in FT NMR Spectroscopy of Liquids,* VEDA, Bratislava, 1986.
5. Ernst R. E., *Advances in Magnetic Resonance,* Vol. II, Waugh, J. S., Ed., Academic Press, New York, 1966.
6. Ferrige, A. G. and Lindon, J. C., *J. Magn. Reson.,* 31, 337, 1978.
7. De Marco, A. and Wüthrich, K., *J. Magn. Reson.,* 24, 201, 1976.
8. Brereton, R. G., Garson, M. J., and Staunton, J., *J. Magn. Reson.,* 43, 224, 1981.
9. Gueron, M., *J. Magn. Reson.,* 30, 515, 1978.
10. Clin, B., De Bony, J., Lalanne, P., Biais, J., and Lemanceau, B., *J. Magn. Reson.,* 33, 457, 1979.
11. Lohman, J. A. B., *J. Magn. Reson.,* 38, 163, 1980.
12. Delsuc, M. A., Ni, F., and Levy, G. C., *J. Magn. Reson.,* 73, 548, 1987.
13. Kay, S. M. and Marple, S. I., *Proc. IEEE,* 69, 1380, 1981.
14. Gull, S. F. and Daniell, G. J., *Nature (London),* 272, 686, 1978.
15. Sibisi, S., Skilling, J., Brereton, R. G., Laue, E. D., and Staunton, J., *Nature (London),* 311, 406, 1984.
16. Sibisi, S., *Nature (London),* 301, 134, 1983.
17. Gull, S. F. and Skilling, J., *Proc. IEE(F),* 131, 646, 1984.
18. Laue, E. D., Skilling, J., Staunton, J., Sibisi, S., and Brereton, R., *J. Magn. Reson.,* 62, 437, 1985.
19. Livesey, A. K. and Skilling, J., *Acta Crystallogr.,* A41, 113, 1985.
20. Gull, S. F. and Skilling, J., *Indirect Imaging,* Roberts, J. A., Ed., Cambridge University Press, London, 1984.
21. Shore, J. E. and Johnson, R. W., *IEEE Trans. Inform. Tech.,* IT-26, 26, 1980.
22. Shore, J. E. and Johnson, R. W., *IEEE Trans. Inform. Tech.,* IT-29, 942, 1983.
23. Jaynes, E. T., *IEEE Trans. Systems Sci. Cybernetics,* SSC-4, 227, 1968.
24. Skilling, J. and Bryan, R. K., *Mon. Nat. R. Astr. Soc.,* 211, 111, 1984.
25. Ni, F., Levy, G. C., and Scheraga, H. A., *J. Magn. Reson.,* 66, 385, 1986.
26. Ni, F. and Scheraga, H. A., *J. Raman Spectrosc.,* 16, 337, 1985.

Appendix A

Foundations of Matrix Calculations[1]

A matrix $\bar{\bar{A}}$ is a set of $m \times n$ objects A_{ij}, $i = 1 \to m$, $j = 1 \to n$, in a rectangular arrangement of m rows and n columns:

$$\bar{\bar{A}} = \begin{vmatrix} A_{11} & A_{12} & \cdots & A_{1n} \\ A_{21} & A_{22} & \cdots & A_{2n} \\ \vdots & \vdots & \vdots & \vdots \\ A_{m1} & A_{m2} & \cdots & A_{mn} \end{vmatrix} \quad \textbf{(A.1)}$$

Quantities A_{ij} are called *matrix* $\bar{\bar{A}}$ *elements.* Matrix elements can take various forms. They can be numbers, matrices, functions, operators, vectors and so on. The first index i is the order of a row and the second index j is the order of a column, in which an element A_{ij} is placed. If all matrix elements A_{ij} are numbers it is called a number matrix, either real or complex, depending on whether numbers are only from a set of real numbers or if at least one of the matrix elements A_{ij} is a complex number.

The matrix defined above is a *two-dimensional matrix.* Multi-dimensional matrices are also used, but it is not necessary to describe them here.

If $m \neq n$, the matrix is called *rectangular,* or a matrix of the type (m, n). If $m = n$, it is called a *square* matrix, of order n. A matrix of the type $(n, 1)$ is called a *column matrix,* or a *column vector* $\vec{A}$, and a matrix of the type $(1, n)$ is called a *row matrix,* or a *row vector* $\vec{A}^T$. A square matrix of order 1 is called a *point matrix.*

Elements A_{ii} of a square matrix $\bar{\bar{A}}$ are called the *principal elements.* They form the *principal diagonal.*

The sum of principal matrix elements is called a *matrix trace:*

$$Tr\left(\bar{\bar{A}}\right) = \sum_{i=1}^{n} A_{ii} \tag{A.2}$$

A square matrix is an *upper triangular* or *lower triangular matrix,* if all its elements are zero below or above a diagonal, respectively, and at least one of the principal elements has a nonzero value.

A matrix having at least one nonzero principal element and all nondiagonal elements zero is a *diagonal matrix.* A diagonal matrix with all principal elements equal to 1 is called a *unit matrix* $\bar{\bar{E}}$:

$$E_{ij} = \delta_{ij} \tag{A.3}$$

where δ_{ij} is the Kronecker delta function for which

$$\delta_{ij} = \begin{Bmatrix} 1 & \text{for} & i = j \\ 0 & \text{for} & i \neq j \end{Bmatrix} \tag{A.4}$$

is defined.

A *zero matrix* is a matrix $\bar{\bar{O}}$ with all elements equal to zero:

$$O_{ij} = 0 \quad \text{for all } i \text{ and } j \tag{A.5}$$

A matrix $\bar{\bar{A}}^*$ is a *complex conjugate matrix* of a matrix $\bar{\bar{A}}$ if for all values i, j

$$\left(\bar{\bar{A}}^*\right)_{ij} = \left(A_{ij}\right)^* \tag{A.6}$$

is valid.

A matrix $\bar{\bar{A}}^T$ of the type (n, m) is a *transpose matrix* of a matrix $\bar{\bar{A}}$ of the type (m, n) if for all its elements

$$\left(\bar{\bar{A}}^T\right)_{ij} = A_{ji} \tag{A.7}$$

A square matrix $\bar{\bar{A}}$ is a *symmetric matrix* if

$$\bar{\bar{A}}^T = \bar{\bar{A}} \tag{A.8}$$

A matrix $\bar{\bar{A}}^+$ is a *Hermitian transpose matrix* if

$$\left(\bar{\bar{A}}^+\right)_{ij} = \left(\bar{\bar{A}}^{*T}\right)_{ij} = A_{ji}^* \tag{A.9}$$

A square matrix $\bar{\bar{A}}$ is a *Hermitian matrix* if

$$\bar{\bar{A}} = \bar{\bar{A}}^{+} \tag{A.10}$$

For elements of a Hermitian matrix, the following relationship is known:

$$\left(\bar{\bar{A}}^{T}\right)_{ij} = \left(\bar{\bar{A}}^{*}\right)_{ji} \tag{A.11}$$

All principal elements of the Hermitian matrix are from the set of real numbers.

The number

$$\det\bar{\bar{A}} = \sum_{k_i} (-1)^{r} \; A_{1k_1}, A_{2k_2}, \ldots A_{nk_n} \tag{A.12}$$

is called the *determinant of the nth degree* of a square matrix $\bar{\bar{A}}$. The summation runs through all permutations $k_1, k_2, \ldots k_n$ of numbers $1, 2, \ldots n$, and r represents the number of inversions of $k_1, k_2 \ldots k_n$ permutations.

A determinant of the $(n - 1)$ degree formed from a matrix $\bar{\bar{A}}$ by eliminating the ith row and the jth column is a *minor* or a *subdeterminant* conjugated to a matrix $\bar{\bar{A}}$ belonging to the element A_{ij}. It is labeled M_{ij}.

The *algebraic supplement* of an element A_{ij} in a matrix $\bar{\bar{A}}$ determinant is a minor M_{ij} multiplied by a number $(-1)^{i+j}$. It is labeled D_{ij} and

$$D_{ij} = (-1)^{i+j} \; M_{ij} \tag{A.13}$$

is given for it.

A determinant does not change its value if

1. Its rows are replaced with columns and its columns with rows so that

$$\det\left(\bar{\bar{A}}\right) = \det\left(\bar{\bar{A}}^{T}\right) \tag{A.14}$$

2. Any linear combination made from the other rows (columns) is added to any of its rows (columns).

If all elements of a row (column) are multipled by any number, the value of the determinant is equal to the product of the original determinant and the given number. The determinant of a matrix which is itself the product of matrices is equal to the product of the determinants of the factor matrices.

$$\det\left(\bar{\bar{A}}\bar{\bar{B}}\right) = \det\left(\bar{\bar{A}}\right)\det\left(\bar{\bar{B}}\right) \tag{A.15}$$

A matrix $\bar{\bar{A}}$ of the type (m, n) has an order of h if at least one of its minors of the hth order is nonzero but all its minors of orders higher than h (if they exist) are zero.

If the order of a square matrix is equal to its degree ($h = n$), it is called a *regular (nonsingular) matrix.* If $h < n$, the matrix is *singular.*

For two real matrices $\bar{\bar{A}}$ and $\bar{\bar{B}}$ the following relationships are known:

$$\bar{\bar{A}} = \bar{\bar{B}} \quad \text{if} \quad A_{ij} = B_{ij} \quad \text{for all } i, j \tag{A.16}$$

$$\bar{\bar{A}} \geq \bar{\bar{B}} \quad \text{if} \quad A_{ij} \geq B_{ij} \quad \text{for all } i, j \tag{A.17}$$

$$\bar{\bar{A}} \leq \bar{\bar{B}} \quad \text{if} \quad A_{ij} \leq B_{ij} \quad \text{for all } i, j \tag{A.18}$$

$$\bar{\bar{A}} \neq \bar{\bar{B}} \quad \text{if} \quad A_{ij} \neq B_{ij} \quad \text{for at least one pair of } i, j \tag{A.19}$$

For complex matrices only the equality is defined.

The *matrix equality* has the following properties:

1. reflexivity $\quad \bar{\bar{A}} = \bar{\bar{A}} \quad$ **(A.20)**

2. symmetry $\quad \bar{\bar{A}} = \bar{\bar{B}} \Rightarrow \bar{\bar{B}} = \bar{\bar{A}} \quad$ **(A.21)**

3. transitivity $\quad \bar{\bar{A}} = \bar{\bar{B}} \text{ and } \bar{\bar{B}} = \bar{\bar{C}} \Rightarrow \bar{\bar{A}} = \bar{\bar{C}} \quad$ **(A.22)**

A matrix $\alpha\bar{\bar{A}}$ is called the *product of a matrix* $\bar{\bar{A}}$ *and a number* α, which is formed from a matrix $\bar{\bar{A}}$ by multiplying all its elements by a number α:

$$\left(\alpha\bar{\bar{A}}\right)_{ij} = \alpha A_{ij} \tag{A.23}$$

A *sum of two matrices* $\bar{\bar{A}}$ and $\bar{\bar{B}}$ of the same type (m, n) is a matrix $\bar{\bar{C}}$, for which

$$\bar{\bar{C}} = \bar{\bar{A}} + \bar{\bar{B}} \tag{A.24}$$

is valid, if

$$C_{ij} = A_{ij} + B_{ij} \qquad \text{for all } i, j \tag{A.25}$$

For a sum of matrices the following properties are given:

$$\bar{\bar{A}} + \left(\bar{\bar{B}} + \bar{\bar{C}}\right) = \left(\bar{\bar{A}} + \bar{\bar{B}}\right) + \bar{\bar{C}} \qquad \text{associativity} \tag{A.26}$$

$$\bar{\bar{A}} + \bar{\bar{B}} = \bar{\bar{B}} + \bar{\bar{A}} \qquad \text{commutativity} \tag{A.27}$$

$$\bar{\bar{A}} + \bar{\bar{O}} = \bar{\bar{A}} \qquad \text{where } \bar{\bar{O}} \text{ is the zero matrix} \tag{A.28}$$

$$\left.\begin{aligned} k_1\bar{\bar{A}} + k_2\bar{\bar{A}} &= \left(k_1 + k_2\right)\bar{\bar{A}} \\ k\bar{\bar{A}} + k\bar{\bar{B}} &= k\left(\bar{\bar{A}} + \bar{\bar{B}}\right) \end{aligned}\right\} \qquad \text{distributivity} \tag{A.29}$$

The *product of a matrix* $\bar{\bar{A}}$ of the type (m, n) with a *matrix* $\bar{\bar{B}}$ of the type (n, p) is a matrix $\bar{\bar{C}}$ of the type (m, p). Its elements are defined by the relationship

$$C_{ij} = \sum_{k=1}^{n} A_{ik} + B_{kj} \qquad \text{for all } i, j \tag{A.30}$$

For the multiplication of matrices the following properties are defined:

$$\left(\bar{\bar{A}}\bar{\bar{B}}\right)\bar{\bar{C}} = \bar{\bar{A}}\left(\bar{\bar{B}}\bar{\bar{C}}\right) \qquad \text{associativity} \tag{A.31}$$

$$\left.\begin{aligned} \left(\bar{\bar{A}} + \bar{\bar{B}}\right)\bar{\bar{C}} &= \bar{\bar{A}}\bar{\bar{C}} + \bar{\bar{B}}\bar{\bar{C}} \\ \bar{\bar{A}}\left(\bar{\bar{B}} + \bar{\bar{C}}\right) &= \bar{\bar{A}}\bar{\bar{B}} + \bar{\bar{A}}\bar{\bar{C}} \end{aligned}\right\} \qquad \text{distributivity} \tag{A.32}$$

Generally, a product of matrices is not commutative, i.e., $\bar{\bar{A}}\,\bar{\bar{B}}$ is not always equal to $\bar{\bar{B}}\,\bar{\bar{A}}$. (If a product of matrices $\bar{\bar{A}}\,\bar{\bar{B}} = \bar{\bar{B}}\,\bar{\bar{A}}$ holds, the matrices $\bar{\bar{A}}, \bar{\bar{B}}$ *commute*, i.e., they are interchangeable). A *difference* $\bar{\bar{A}}\,\bar{\bar{B}} - \bar{\bar{B}}\,\bar{\bar{A}}$ is called a *commutator* of matrices $\bar{\bar{A}}$ and $\bar{\bar{B}}$. If

$$\bar{\bar{A}} = \bar{\bar{A}}_1 \cdot \bar{\bar{A}}_2 \cdot \bar{\bar{A}}_3 \ldots \bar{\bar{A}}_k \tag{A.33}$$

then

$$\bar{\bar{A}}^T = \bar{\bar{A}}_k^T \ldots \bar{\bar{A}}_3^T \cdot \bar{\bar{A}}_2^T \cdot \bar{\bar{A}}_1^T \tag{A.34}$$

A product of k ($k > 0$ is an integer) identical square matrices $\bar{\bar{A}}$ is called the *kth power* of $\bar{\bar{A}}$ and

$$\bar{\bar{A}}^k = \bar{\bar{A}} \cdot \bar{\bar{A}} \cdot \bar{\bar{A}} \ldots \bar{\bar{A}} \qquad (k\text{-times}) \tag{A.35}$$

For $k = 0$, $\bar{\bar{A}}^0 = \bar{\bar{E}}$, where $\bar{\bar{E}}$ is a unit matrix.

An *inverse matrix,* the inverse of a square matrix $\bar{\bar{A}}$ of the nth degree, is itself a square matrix $\bar{\bar{A}}^{-1}$ of the nth degree, for which the following holds:

$$\bar{\bar{A}}\bar{\bar{A}}^{-1} = \bar{\bar{A}}^{-1}\bar{\bar{A}} = \bar{\bar{E}} \qquad \textbf{(A.36)}$$

where $\bar{\bar{E}}$ is a unit matrix.

For an inverse matrix

$$\bar{\bar{A}}^{-1} = \frac{1}{\det\bar{\bar{A}}}\begin{vmatrix} D_{11} & D_{21} & \cdots & D_{n1} \\ D_{12} & D_{22} & \cdots & D_{n2} \\ \vdots & \vdots & \vdots & \vdots \\ D_{1n} & D_{2n} & \cdots & D_{nn} \end{vmatrix} \qquad \textbf{(A.37)}$$

where D_{ij} are algebraic cofactors of elements A_{ij} in a matrix $\bar{\bar{A}}$ determinant.

A matrix

$$\text{adj}\bar{\bar{A}} = \begin{vmatrix} D_{11} & D_{21} & \cdots & D_{n1} \\ D_{12} & D_{22} & \cdots & D_{n2} \\ \vdots & \vdots & \vdots & \vdots \\ D_{1n} & D_{2n} & \cdots & D_{nn} \end{vmatrix} \qquad \textbf{(A.38)}$$

is called an *adjugated matrix* to a matrix $\bar{\bar{A}}$. It can be written

$$\bar{\bar{A}}^{-1} = \frac{1}{\det\bar{\bar{A}}}\, adj\bar{\bar{A}} \qquad \textbf{(A.39)}$$

An inverse matrix of a matrix $\bar{\bar{A}}$ exists only if $\bar{\bar{A}}$ is a regular matrix. A matrix $\bar{\bar{A}}^{-1}$ is a regular matrix and

$$\left(\bar{\bar{A}}^{-1}\right)^{-1} = \bar{\bar{A}} \qquad \textbf{(A.40)}$$

An inverse matrix of a diagonal matrix is a diagonal matrix whose principal elements are equal to the inverse values of the principal elements of the original matrix.

A matrix inverse of the product of a finite number of regular matrices is equal to the product of inverse matrices in opposite order. If

$$\bar{\bar{A}} = \bar{\bar{A}}_1\bar{\bar{A}}_2\ldots\bar{\bar{A}}_k \qquad \textbf{(A.41)}$$

then

$$\bar{\bar{A}}^{-1} = \bar{\bar{A}}_k^{-1}\ldots\bar{\bar{A}}_2^{-1}\bar{\bar{A}}_1^{-1} \qquad \textbf{(A.42)}$$

To divide a matrix $\bar{\bar{B}}$ (a dividend) by a matrix $\bar{\bar{A}}$ (a divisor) means to find a matrix $\bar{\bar{X}}$ or $\bar{\bar{Y}}$, such that

$$\bar{\bar{B}} = \bar{\bar{A}}\,\bar{\bar{X}} \tag{A.43}$$

or

$$\bar{\bar{B}} = \bar{\bar{Y}}\bar{\bar{A}} \tag{A.44}$$

respectively.

A matrix $\bar{\bar{X}}$ ($\bar{\bar{Y}}$) is the left (right) quotient of matrix $\bar{\bar{B}}$ with matrix $\bar{\bar{A}}$. For a matrix $\bar{\bar{X}}$ ($\bar{\bar{Y}}$) the following is true:

$$\bar{\bar{X}} = \bar{\bar{A}}^{-1}\bar{\bar{B}} \tag{A.45}$$

$$\bar{\bar{Y}} = \bar{\bar{B}}\bar{\bar{A}}^{-1} \tag{A.46}$$

A regular real matrix $\bar{\bar{A}}$ is an *orthonormal matrix*, if

$$\bar{\bar{A}}^T\bar{\bar{A}} = \bar{\bar{E}} \tag{A.47}$$

or

$$\bar{\bar{A}}^T = \bar{\bar{A}}^{-1} \tag{A.48}$$

A regular complex matrix $\bar{\bar{A}}$ is a *unitary matrix* if

$$\left(\bar{\bar{A}}^T\right)^*\bar{\bar{A}} = \left(\bar{\bar{A}}^+\right)\bar{\bar{A}} = \bar{\bar{E}} \tag{A.49}$$

$$\bar{\bar{A}}^+ = \bar{\bar{A}}^{-1} \tag{A.50}$$

respectively.

Square matrices $\bar{\bar{A}}$ and $\bar{\bar{B}}$ of nth degree, with real or complex elements, are called similar matrices, if a regular matrix $\bar{\bar{P}}$ of the nth degree exists, so that

$$\bar{\bar{B}} = \bar{\bar{P}}^{-1}\bar{\bar{A}}\bar{\bar{P}} \tag{A.51}$$

The matrix

$$\bar{\bar{A}} - \lambda\bar{\bar{E}} \tag{A.52}$$

is called a *characteristic matrix* for the square matrix $\overline{\overline{A}}$ where λ is a real or complex number. A *characteristic polynomial* of a matrix $\overline{\overline{A}}$ is a determinant of a characteristic matrix $\overline{\overline{A}} - \lambda\overline{\overline{E}}$:

$$\det\left(\overline{\overline{A}} - \lambda\overline{\overline{E}}\right) = \det\begin{vmatrix} A_{11}-\lambda & A_{12} & \cdots & A_{1n} \\ A_{21} & A_{22}-\lambda & \cdots & A_{2n} \\ \vdots & \vdots & \vdots & \vdots \\ A_{n1} & A_{n2} & \cdots & A_{nn}-\lambda \end{vmatrix}$$

$$= (-1)^n\left(\lambda^n - p_1\lambda^{n-1} - p_2\lambda^{n-2} - p_{n-1}\lambda - p_n\right) = 0 \qquad \textbf{(A.53)}$$

Roots λ_i of this equation ($i = 1, n$) are called *eigenvalues* of a matrix $\overline{\overline{A}}$. An *eigenvector* $\vec{Y}_i$ corresponding to the eigenvalue λ_i is a nontrivial solution of the system

$$\left(\overline{\overline{A}} - \lambda_i\overline{\overline{E}}\right)\vec{Y}_i = 0 \qquad \textbf{(A.54)}$$

For evaluation of all n eigenvectors, it is necessary to solve systems of n linear equations (Equation A.52).

The number of linearly independent eigenvectors corresponding to the given eigenvalue is equal to its multiplicity. If all eigenvalues are different, only one eigenvector belongs to each one. Eigenvectors corresponding to different eigenvalues are orthogonal:

$$\vec{Y}_i^T\vec{Y}_j = 0 \qquad \text{for } i \neq j \qquad \textbf{(A.55)}$$

Characteristic values of a triangular matrix are directly equal to its diagonal elements. If a characteristic polynomial of a matrix $\overline{\overline{A}}$ of the nth degree has n single roots $\lambda_1, \lambda_2, \ldots \lambda_n$, then the system of elementary divisors of a characteristic matrix $\overline{\overline{A}} - \lambda\overline{\overline{E}}$ is $(\lambda - \alpha_1)(\lambda - \alpha_2) \ldots (\lambda - \alpha_n)$.

Matrices $\overline{\overline{A}}$ and $\overline{\overline{B}}$ of the same degree are similar only if their characteristic matrices $\overline{\overline{A}} - \lambda\overline{\overline{E}}$ and $\overline{\overline{B}} - \lambda\overline{\overline{E}}$ are equivalent, i.e., if they have the same elementary divisors. Therefore, similar matrices have the same characteristic polynomial and the same characteristic values.

Vectors $\vec{b}_i$, $i = 1, n$, are called *orthonormal,* if for all i and j

$$\vec{b}_i^T\vec{b}_j = \delta_{ij} \qquad \textbf{(A.56)}$$

A set of n orthonormal vectors $\vec{b}_i$ forms an *n-dimensional space.* Each vector $\vec{d}_k$ in such n-dimensional space can be expressed as a linear combination of those orthonormal vectors. Orthonormal vectors can be obtained by the *Schmidt orthogonalization procedure.*

Let us have any vectors $\vec{c}_i$, $i = 1, n$. Then a set of orthonormal vectors $\vec{b}_i$ will be obtained by the following method:

$$\alpha_1\vec{b}_1 = \vec{c}_1 \qquad \textbf{(A.57)}$$

where, for the normalization constant α_i

$$|\alpha_1|^2 = \frac{1}{\left(\vec{c}_1^{\,T}\vec{c}_1\right)} \tag{A.58}$$

$$\alpha_2 \vec{b}_2 = \vec{c}_2 - \vec{b}_1\left(\vec{b}_1^{\,T}\vec{c}_2\right) \tag{A.59}$$

is defined. Obviously, the vector $\vec{b}_2$ defined by this procedure is orthogonal to the vector $\vec{b}_1$. The normalization constant α_2 can be determined from the normalization condition:

$$|\alpha_2|^2 = \frac{1}{\vec{c}_2^{\,T}\vec{c}_2 + \left(\vec{b}_1^{\,T}\vec{c}_2\right)^2} \tag{A.60}$$

For any vector $\vec{b}_k$

$$\alpha_k \vec{b}_k = \vec{c}_k - \sum_{i=1}^{k-1} \vec{b}_i\left(\vec{b}_i^{\,T}\vec{c}_k\right) \tag{A.61}$$

is given.

Vectors $\vec{d}_i$ and $\vec{d}_j$ are called conjugate vectors to a positive definite matrix $\bar{\bar{A}}$ if

$$\vec{d}_i^{\,T}\bar{\bar{A}}\vec{d}_j = 0 \qquad \text{for } i \neq j \tag{A.62}$$

If $\bar{\bar{A}}$ is a unit matrix, the conjugate vectors will be orthogonal, so that the conjugation can be understood as a generalization of orthogonality. A set of n conjugate vectors forms an n-dimensional space, and any vector of this space can be expressed as a linear combination of conjugate vectors.

Conjugate vectors can be obtained using a procedure similar to the Schmidt orthogonalization method. Let us have any vector $\vec{d}_1$. Then a vector

$$\vec{d}_2 = \bar{\bar{A}}^T\vec{d}_1 - \frac{\vec{d}_1\bar{\bar{A}}^T\bar{\bar{A}}\vec{d}_1}{\vec{d}_1\bar{\bar{A}}\vec{d}_1}\,\vec{d}_1 \tag{A.63}$$

will be a conjugate vector to the vector $\vec{d}_1$. This procedure can be repeated using the same method, so that the vector $\vec{d}_3$ will be conjugated to vector $\vec{d}_1$, $\vec{d}_2$, etc., until the determination all of conjugate vectors $\vec{d}_i$ $i = 1, n$ is finished.

Appendix B

Systems of Linear Equations[2-3]

A system

$$\begin{aligned} A_{11}X_1 + A_{12}X_2 + \dots + A_{1n}X_n &= Y_1 \\ A_{21}X_1 + A_{22}X_2 + \dots + A_{2n}X_n &= Y_2 \\ \vdots \qquad\qquad \vdots \qquad\qquad &\ \ \vdots \\ A_{m1}X_1 + A_{m2}X_2 + \dots + A_{mn}X_n &= Y_m \end{aligned} \tag{B.1}$$

is called *a system of m linear equations in n unknown variables* and can be transformed into matrix form:

$$\bar{\bar{A}}\vec{X} = \vec{Y} \tag{B.2}$$

where a matrix $\bar{\bar{A}}$ of the type (m, n) is called a *system matrix* and $\vec{X}$ and $\vec{Y}$ are column vectors of the type $(n, 1)$. An extended matrix is a matrix where

$$\bar{\bar{B}} = \begin{vmatrix} A_{11} & A_{12} & \dots & A_{1n} & Y_1 \\ A_{21} & A_{22} & \dots & A_{2n} & Y_2 \\ \vdots & \vdots & \vdots & \vdots & \vdots \\ A_{m1} & A_{m2} & \dots & A_{mn} & Y_m \end{vmatrix} \tag{B.3}$$

A system is solvable only if the rank of a system matrix is equal to the rank of an extended system matrix.

A system of equations (Equation B.1) is called a *homogeneous system,* if a vector $\vec{Y}$ is a null vector. A system of n homogeneous equations in n unknowns

$$\begin{aligned} A_{11} X_1 + A_{12} X_2 + \ldots + A_{1n} X_n &= 0 \\ A_{21} X_1 + A_{22} X_2 + \ldots + A_{2n} X_n &= 0 \\ \vdots \qquad\qquad \vdots \qquad\qquad &\;\; \vdots \\ A_{m1} X_1 + A_{m2} X_2 + \ldots + A_{mn} X_n &= 0 \end{aligned} \tag{B.4}$$

always has a null (trivial) solution

$$\vec{O} = (0, 0, 0, \ldots 0) \tag{B.5}$$

If a system (Equation B.4) has a solution $\vec{\xi} = (\xi_1, \xi_2, \ldots \xi_n)$, it also has a solution $\alpha\vec{\xi} = (\alpha\xi_1, \alpha\xi_2, \ldots \alpha\xi_n)$, where α is any scalar number.

If vectors $\vec{\xi}^{(1)} = (\xi_1^{(1)}, \xi_2^{(1)}, \ldots \xi_n^{(1)})$, $\vec{\xi}^{(2)} = (\xi_1^{(2)}, \xi_2^{(2)}, \ldots \xi_n^{(2)})$, $\vec{\xi}^{(k)} = (\xi_1^{(k)}, \xi_2^{(k)}, \ldots \xi_n^{(k)})$ are solutions of a system (Equation B.4), then any linear combination

$$\vec{\xi} = \sum \alpha_1 \vec{\xi}^{(1)} \tag{B.6}$$

is also its solution.

If a system matrix of homogeneous linear equations (Equation B.4) has rank h, then this system has $n - h$ linear independent solutions and each succesive solution is the linear combination of such $n - h$ solutions.

Let a matrix of a system (Equation B.1) have a rank h and a solution $\vec{\eta} = (\eta_1, \eta_2, \ldots \eta_n)$. Let (Equation B.4) have $n - h$ independent solutions $\vec{\xi}^{(1)}$, $\vec{\xi}^{(2)}, \ldots \vec{\xi}^{(n-h)}$. Then all the solutions to Equation B.1 will be yielded when to any solution of Equation B.4

$$\sum_{l=1}^{n-h} \alpha_1 \vec{\xi}^{(1)} \tag{B.7}$$

the solution of the nonhomogeneous system (Equation B.1) is added, i.e., each solution of Equation B.4 can be written in the following form:

$$\sum_{l=1}^{n-h} \alpha_1 \vec{\xi}^{(1)} + \vec{\eta} \tag{B.8}$$

where α_1 are any numbers.

METHODS FOR SOLUTION OF SYSTEMS OF LINEAR EQUATIONS

Methods for solving a system of linear equations can be split into two groups:

1. *Exact methods* giving a finite algorithm for a solution (e.g., the Gauss method, the method of principal elements, the method of quadratic roots, etc.)

2. *Iterative methods,* which can obtain a solution of a predefined accuracy using convergent infinite processes (e.g., the simple iterative method, the Gauss-Seidel method, etc.)

Usually solutions are rounded off for the first type of method, so these methods are in fact only approximate. An estimation of roundoff error is rather difficult.

The effectiveness of the iterative methods strongly depends on the convenience of the choice of initial approximation and on the convergence speed.

SOLUTION OF LINEAR EQUATION SYSTEMS WITH AN INVERSE MATRIX

A system of n linear equations and of n variables

$$\begin{aligned} A_{11} X_1 + A_{12} X_2 + \ldots + A_{1n} X_n &= Y_1 \\ A_{21} X_1 + A_{22} X_2 + \ldots + A_{2n} X_n &= Y_2 \\ \vdots \qquad\qquad \vdots \qquad &\quad \vdots \\ A_{n1} X_1 + A_{n2} X_2 + \ldots + A_{nn} X_n &= Y_n \end{aligned} \tag{B.9}$$

can be written in a matrix form:

$$\overline{\overline{A}}\vec{X} = \vec{Y} \tag{B.10}$$

where $\overline{\overline{A}}$ is a system matrix, and $\vec{X}$ and $\vec{Y}$ are column vectors of unknown X_1, respective of terms of a right side Y_i. This system has solution only if

$$\det \overline{\overline{A}} \neq 0 \tag{B.11}$$

(*Cramer's rule*). Then an inverse matrix $\overline{\overline{A}}^{-1}$ to a matrix $\overline{\overline{A}}$ exists, and a solution of Equation B.10 will be

$$\vec{X} = \overline{\overline{A}}^{-1}\vec{Y} \tag{B.12}$$

i.e., an inverse matrix to a matrix $\overline{\overline{A}}$ must be found and the right side vector $\vec{Y}$ must be multiplied by it from the left side. This method is not effective when a matrix $\overline{\overline{A}}$ is close to a singular matrix. In that case, it is necessary to use special numerical methods of inversion.

The Gauss Method

The *Gauss method,* also called *the elimination method,* is the most widespread method for solving a system of linear equations. It is based on successive elimination of variables. However, during transformation of the equations, considerable numerical errors can accumulate. Therefore, only certain modifications of this method are used today.

The Gauss method searches for the pivot element and selects a coefficient A_{ij} of a matrix $\bar{\bar{A}}$ with a maximal absolute value (a principal element). Let it be an element A_{pq}. A row p including a principal element is called a principal row. Then values

$$m_i = \frac{A_{iq}}{A_{pq}} \quad \textbf{(B.13)}$$

are calculated for all $i \neq p$. An extended matrix of a system (Equation B.9) is transformed by the following procedure: a principal row multiplied by the m_i value is subtracted from each ith row ($i \neq j$). By eliminating a principal row, a new extended matrix of the system in which the qth column has zero values is yielded. This procedure is repeated until the matrix obtained consists of only two rows. This matrix is called the principal matrix.

Then all the principal rows formed after simple transformations are connected with the triangular matrix equivalent to the original matrix. Values of variables X_i are obtained by solving this system with the triangular matrix.

The Compact Gauss Method

This method represents a coefficient matrix $\bar{\bar{A}}$ in the form of a product of two triangular matrices $\bar{\bar{B}}$ and $\bar{\bar{C}}$

$$\bar{\bar{A}} = \bar{\bar{B}}\bar{\bar{C}} \quad \textbf{(B.14)}$$

as one of them (e.g., matrix $\bar{\bar{C}}$) would have values of 1 in the diagonal, and the original system of linear equations is equivalent to the equation system with the matrix $\bar{\bar{C}}$.

Let us label an extended matrix of an original system as $\bar{\bar{Z}}$, a matrix of a transformed system as $\bar{\bar{U}}$ (with a coefficient matrix $\bar{\bar{C}}$), and an extended system matrix with a matrix $\bar{\bar{B}}$ as $\bar{\bar{V}}$. Then

$$V_{ij}^{(2j-1)} = Z_{ij}^{(0)} + \sum_{k=1}^{j-1} V_{ik}^{(2k-1)} U_{kj}^{(2k)} \quad \textbf{(B.15)}$$

for $i = j, j + 1, \ldots n$, and

$$U_{jm}^{(2j)} = \frac{Z_{jm} - \sum_{k=1}^{j-1} V_{jk}^{(2k-1)} U_{km}^{(2k)}}{V_{jj}^{(2j-1)}} \quad \textbf{(B.16)}$$

for $m = j + 1, j + 2, \ldots n + 1$.

These equations are used alternately. An upper index labels an iterative cycle.

Values of unknown $X_1, X_2,..X_n$ are obtained by solving the equations:

$$X_i + \sum_{k=i+1}^{n} U_{ik}^{(2j)} X_k = U_{i\,n+1}^{(2i)} \tag{B.17}$$

The Quadratic Root Method

This method is used for systems of linear equations with a symmetric matrix $\bar{\bar{A}}$. In this case, a matrix $\bar{\bar{A}}$ can be divided into a product of transposed triangular matrices:

$$\bar{\bar{A}} = \bar{\bar{S}}^T \bar{\bar{S}} \tag{B.18}$$

where

$$\bar{\bar{S}} = \begin{vmatrix} S_{11} & S_{12} & \dots & S_{1n} \\ 0 & S_{22} & \dots & S_{2n} \\ \vdots & \vdots & \vdots & \vdots \\ 0 & 0 & \dots & S_{nn} \end{vmatrix} \tag{B.19}$$

and matrix elements S_{ij} are

$$S_{11} = \left(A_{11}\right)^{\frac{1}{2}} \tag{B.20}$$

$$S_{1j} = \frac{A_{1j}}{S_{11}} \tag{B.21}$$

$$S_{ii} = \left(A_{ii} - \sum_{m=1}^{n-1} S_{mi}^2\right)^{\frac{1}{2}} \quad \text{for } i > 1 \tag{B.22}$$

$$S_{ij} = \frac{A_{ij} - \sum_{m=1}^{i-1} S_{mi} S_{mj}}{S_{ii}} \quad \text{for } j > i \tag{B.23}$$

$$S_{ij} = 0 \quad \text{for } i > j \tag{B.24}$$

After finding a matrix $\bar{\bar{S}}$, the following matrix equation is solved

$$\bar{\bar{S}}^T \vec{Z} = \vec{Y} \tag{B.25}$$

and then unknowns X_i are obtained by solving

$$\bar{\bar{S}}\vec{X} = \vec{Z} \tag{B.26}$$

The solution of these equations is simple because the matrix $\bar{\bar{S}}$ is a triangular matrix.

The Orthogonalization Method

This method changes an original system of linear equations (Equation B.10) to a system of orthogonal equations, using the sequence of transformations:

$$\bar{\bar{A}}^{(s)}\vec{X} = \vec{Y}^{(s)} \tag{B.27}$$

In this new system, rows of a coefficient matrix $\bar{\bar{A}}^{(s)}$ are mutually orthogonal

$$\sum_{l=1}^{n} A_{il}^{(s)} A_{jl}^{(s)} = 0 \qquad \text{for } i \neq j \tag{B.28}$$

The transformation is performed by the following equations:

$$\lambda_{i\,m-1} = \frac{\sum_{k=1}^{n} A_{ik}^{(m-1)} A_{m-1\,k}^{(m-1)}}{\left[\sum_{k=1}^{n} A_{m-1\,k}^{(m-1)}\right]^2} \tag{B.29}$$

$$A_{ij}^{m} = A_{ij}^{(m-1)} - \lambda_{i\,m-1} A_{m-1\,j}^{(m-1)} \tag{B.30}$$

$$Y_i^{(m)} = Y_i^{(m-1)} - \lambda_{i\,m-1} Y_{m-1}^{(m-1)} \tag{B.31}$$

for $m = 2, 3, \ldots n$; $i = m, m+1, m+2, \ldots n$; and $j = 1, 2, \ldots n$. Then values of unknowns X_i are determined by

$$X_i = \sum_{k=1}^{n} \left\{ \frac{A_{kk}^{(n)} Y_k^{(n)}}{\sum_{s=1}^{n} \left[A_{ks}^{(n)}\right]^2} \right\} \tag{B.32}$$

for $i = 1, 2, \ldots n$.

The Simple Iteration Method

A system of linear equations (Equation B.9) can be transformed to

$$\begin{aligned} X_1 &= C_{11} X_1 + C_{12} X_2 + \dots + C_{1n} X_n + f_1 \\ X_2 &= C_{21} X_1 + C_{22} X_2 + \dots + C_{2n} X_n + f_2 \\ &\vdots \\ X_n &= C_{n1} X_1 + C_{n2} X_2 + \dots + C_{nn} X_n + f_n \end{aligned} \tag{B.33}$$

or into a matrix form:

$$\vec{X} = \bar{\bar{C}} \vec{X} + \vec{f} \tag{B.34}$$

If any vector $\vec{X}^{(0)}$ is used as an initial approximation, by replacing $\vec{X}^{(0)}$ in Equation B.34 a new vector

$$\vec{X}^{(1)} = \bar{\bar{C}} \vec{X}^{(0)} + \vec{f} \tag{B.35}$$

can be obtained. The above procedure is repeated until the desired difference between the vectors $|\vec{X}^{(m)} - \vec{X}^{(m-1)}|$ is obtained.

Matrix Inversion Methods

The Addition Method. Based on an original matrix $\bar{\bar{A}}$ and a unit matrix $\bar{\bar{E}}$, the method of mathematical induction of a matrix sequence $\bar{\bar{A}}^{(0)}$, $\bar{\bar{B}}^{(1)}$, $\bar{\bar{A}}^{(1)}$, $\bar{\bar{B}}^{(2)}$, $\bar{\bar{A}}^{(2)}$,... $\bar{\bar{B}}^{(n)}$, $\bar{\bar{A}}^{(n)}$ is formed and

$$\bar{\bar{A}}^{(0)} = \bar{\bar{A}} - \bar{\bar{I}} \tag{B.36}$$

$$B_{ij}^{(m)} = \begin{Bmatrix} A_{ij}^{m-1} & \text{for } i \neq j \\ \delta_{ij} & \text{for } i = j \end{Bmatrix} \tag{B.37}$$

$$A_{ij}^{(m)} = B_{ij}^{(m)} - \left[B_{im}^{(m)} A_{mj}^{(m-1)} \right] \cdot \left[1 + A_{mm}^{(m-1)} \right]^{-1} \tag{B.38}$$

for $i, j, m = 1, 2, \ldots n$. The inverse matrix $\bar{\bar{A}}^{-1}$ is defined by

$$\bar{\bar{A}}^{-1} = \bar{\bar{A}}^{(m)} \tag{B.39}$$

The Submatrices Method. This method inverts matrices of high order using an inversion of the lower order matrices.

Let us split a matrix $\bar{\bar{A}}$ of the type (n, n) into four submatrices:

$$\bar{\bar{A}}(n,n) = \begin{vmatrix} \bar{\bar{\alpha}}_{11}(r,r) & \bar{\bar{\alpha}}_{12}(r,s) \\ \bar{\bar{\alpha}}_{21}(s,r) & \bar{\bar{\alpha}}_{22}(s,s) \end{vmatrix} \quad \textbf{(B.40)}$$

where the types of individual matrices are in parentheses and $r + s = n$. Then for a matrix $\bar{\bar{A}}^{-1}$

$$\bar{\bar{A}}^{-1} = \begin{vmatrix} \bar{\bar{\beta}}_{11}(r,r) & \bar{\bar{\beta}}_{12}(r,s) \\ \bar{\bar{\beta}}_{21}(s,r) & \bar{\bar{\beta}}_{22}(s,s) \end{vmatrix} \quad \textbf{(B.41)}$$

can be written. The following equations are defined for the determination of submatrices $\bar{\bar{\beta}}_{ij}$:

$$\bar{\bar{\beta}}_{11} = \bar{\bar{\alpha}}_{11}^{-1} + \bar{\bar{X}}\,\bar{\bar{Z}}^{-1}\bar{\bar{Y}} \quad \textbf{(B.42)}$$

$$\bar{\bar{\beta}}_{12} = -\bar{\bar{X}}\,\bar{\bar{Z}}^{-1} \quad \textbf{(B43)}$$

$$\bar{\bar{\beta}}_{21} = -\bar{\bar{Z}}^{-1}\bar{\bar{Y}} \quad \textbf{(B.44)}$$

$$\bar{\bar{\beta}}_{22} = \bar{\bar{Z}}^{-1} \quad \textbf{(B.45)}$$

where

$$\bar{\bar{X}} = \bar{\bar{\alpha}}_{11}^{-1}\,\bar{\bar{\alpha}}_{12} \quad \textbf{(B.46)}$$

$$\bar{\bar{Y}} = \bar{\bar{\alpha}}_{21}\,\bar{\bar{\alpha}}_{11}^{-1} \quad \textbf{(B.47)}$$

$$\bar{\bar{Z}} = \bar{\bar{\alpha}}_{22} - \bar{\bar{\alpha}}_{21}\,\bar{\bar{X}} \quad \textbf{(B.48)}$$

This method is used in combination with other methods of submatrix inversion.

Ill-Conditioned Systems

Let us assume that a matrix $\bar{\bar{A}}$ in Equation (B.10) has some large elements. Let

$$\left(\bar{\bar{A}}^{-1}\right)_{ji} = \frac{D_{ij}}{\det \bar{\bar{A}}} \quad \textbf{(B.49)}$$

be one of them. The corresponding algebraic complement D_{ij} is greater than $\det\bar{\bar{A}}$. A small change in an element A_{ij} causes a relatively large error in an element $(\bar{\bar{A}}^{-1})_{ki}$, so a great relative error can appear at the calculation of a vector $\vec{X}$.

Let us suppose that the system of linear equations (Equation B.10) was solved by some of the above methods. Let us label an obtained solution $\vec{X}^{(1)}$ and introduce a residual vector

$$\vec{R}^{(1)} = \vec{Y} - \bar{\bar{A}}\,\vec{X}^{(1)} \qquad \textbf{(B.50)}$$

In absolute value the greater component of a vector $\vec{R}^{(1)}$ lies within an interval $[10^{(-k_1)}, 10^{(-k_1+1)}]$, where $k_1 \geq 0$. Absolute values of vector $\vec{X}$ components are assumed to be greater than absolute values of a vector $\vec{R}$ (this is always true except in some specially defined cases). Then let us solve (by any method) a system of linear equations:

$$\bar{\bar{A}}\,\vec{W} = 10^{(k_1)}\,\vec{R}^{(1)} \qquad \textbf{(B.51)}$$

and label the result $\vec{W}^{(1)}$. If

$$\vec{X}^{(2)} = \vec{X}^{(1)} + 10^{(-k_1)}\,\vec{W}^{(1)} \qquad \textbf{(B.52)}$$

is defined, a vector $\vec{X}^{(2)}$ will be improved against $\bar{\bar{X}}^{(1)}$. This procedure can continue so that for the $(i + 1)$th iteration, generally

$$\vec{X}^{(i+1)} = \vec{X}^{(i)} + 10^{(-k_i)}\,\vec{W}^{(i)} \qquad \textbf{(B.53)}$$

will hold. This iterative method is repeated until elements of a vector $\vec{X}$, yielded from succesive iterative steps, do not differ by a value higher than a chosen arbitrary value.

Appendix C

Regression Analysis

Many problems of molecular spectroscopy lead to a formulation in terms of a regression task. Usually the formulation of a certain functional relationship describing the physical nature of a studied system is a result of theoretical analysis. This functional relationship, called the *regression function,* represents a general functional characterization of the dependence between a dependent measured variable y (absorbance, transmittance, etc.) and an independent variable x (wavelength, wavenumber, incidental angle of radiation, etc.):

$$y = f(\vec{a}, x) \tag{C.1}$$

where f is the regression function consisting of a vector of unknown parameters $\vec{a}$

$$\vec{a} = \begin{pmatrix} a_1 \\ a_2 \\ \vdots \\ a_n \end{pmatrix} \tag{C.2}$$

where n is the number of unknown parameters. Regression analysis,[4-9] based on known (measured) pairs of values y_i, x_i, $i = 1, 2 \ldots N$, where N is the number of measurements, determines values and statistical characteristics of the best estimation of a vector $\vec{a}$. The most common method for solving for a regression model is the *least squares method,*[3] which obtains an estimate of a vector $\vec{a}$ by minimizing the objective function of the form

$$F(\vec{a}) = \sum_{i=1}^{N} W_i\left(y_i - f(\vec{a}, x_i)\right)^2 \tag{C.3}$$

where *weighting coefficients* W_i are proportional to a reciprocal value of a dispersion $s^2(y_i)$ of the value y_i, and fulfill a normalizing condition:

$$\sum_{i=1}^{N} W_i = N \tag{C.4}$$

It is assumed that values of an independent variable x_i are known precisely. In the case of a dependent variable y_i measurable with the constant dispersion, i.e., $s^2(y_i) = s^2(y)$ = constant unit weights W_i are used ($W_i = 1$). For the condition of a local extreme of a function $F(\vec{a})$, the following equation system results for an estimation of a vector $\vec{a}$:

$$\frac{\partial F(\vec{a})}{\partial a_k} = -2\sum_{i=1}^{N}\left[y_i - f(\vec{a}, x_i)\right]\frac{\partial f(\vec{a}, x_i)}{\partial a_k} = 0 \qquad \textbf{(C.5)}$$

for $k = 1, 2, \ldots n$, n is a number of unknown parameters.

If Equation C.5 is linear, the method is named *linear regression analysis;* in the opposite case the method is called *nonlinear regression analysis.*

Besides the least squares method, more seldom used regression procedures, the *maximum likelihood method* and the *method of bayesian estimation,* for example, are also known. These methods are described briefly at the end of this chapter

LINEAR REGRESSION ANALYSIS

In the case of linear regression analysis[4,7] a system of equations (Equation C.5) is linear, and for its solution some of the standard methods described in Appendix B can be used.

A general linear regression function can be written in the form

$$f(\vec{a}, x) = \sum_{k=1}^{n} a_k\, g_k(x) \qquad \textbf{(C.6)}$$

where $g_k(x)$ is any known function of x. Then Equation C.5 will be changed to the form

$$\sum_{k=1}^{n} a_k \sum_{i=1}^{N} g_k(x_i) g_l(x_i) = \sum_{i=1}^{N} y_i\, g_l(x_i) \qquad \textbf{(C.7)}$$

for $l = 1, 2, \ldots n$, or in the matrix form

$$\bar{\bar{G}}\, \vec{a} = \vec{b} \qquad \textbf{(C.8)}$$

where

$$G_{kl} = \sum_{i=1}^{N} g_k(x_i) g_l(x_i) \qquad \textbf{(C.9)}$$

$$b_l = \sum_{i=1}^{N} y_i\, g_l(x_i) \qquad \textbf{(C.10)}$$

for $k, l = 1, 2, \ldots n$.

The most frequent case of linear regression analysis is the *polynomial regression:*

$$f(\vec{a}, x) = \sum_{k=1}^{n} a_k \, x^{k-1} \tag{C.11}$$

In this case, for functions $g_k(x)$

$$g_k(x) = x^{k-1} \tag{C.12}$$

is defined. Then a matrix $\overline{\overline{G}}$ and a vector $\vec{b}$ have a simple form:

$$G_{kl} = \sum_{i=1}^{N} x_i^{k+1-2} \tag{C.13}$$

$$b_l = \sum_{i=1}^{N} y_i \, x_i^{1-1} \tag{C.14}$$

A specific characteristic of polynomial regression is usually the bad conditionality of the matrix $\overline{\overline{G}}$ that occurs when using polynomials of higher degrees. In this case, special algorithms for solving a system of linear equations (Equation C.7) (see Appendix B) must be used.

NONLINEAR REGRESSION ANALYSIS

The solution of a system of nonlinear equations (Equation C.5), which leads to nonlinear regression analysis, is generally involved, and the solution need not be unique. Therefore, many special algorithms for calculating regression parameters a_k by minimizing the objective function (Equation C.3) were developed. This function can be developed in the neighborhood of a point $\vec{a}'$ in the Taylor sequence:

$$F(\vec{a}) = F(\vec{a}') + \vec{g}^T (\vec{a} - \vec{a}') + \frac{1}{2}(\vec{a} - \vec{a}')^T \, \overline{\overline{G}}(\vec{a} - \vec{a}') + \ldots \tag{C.15}$$

where

$$\vec{g} = \begin{pmatrix} g_1 \\ g_2 \\ \vdots \\ g_n \end{pmatrix} \tag{C.16}$$

is the *vector of first derivatives* of a function $F(\vec{a})$ in points $\vec{a}'$:

$$g_j = \left[\frac{\partial F(\vec{a})}{\partial a_j} \right]_{\vec{a} = \vec{a}'} \tag{C.17}$$

and $\bar{\bar{G}}$ is a *matrix of second derivatives* of a function $F(\vec{a})$ in points $\vec{a}'$ (*the Hessian matrix*):

$$G_{jk} = \left[\frac{\partial^2 F(\vec{a})}{\partial a_j \, \partial a_k} \right]_{\vec{a} = \vec{a}'} \quad \textbf{(C.18)}$$

The first term of the series (Equation C.15) is constant. The second term is proportional to a gradient $\vec{g}$. A gradient $\vec{g}$ characterizes the direction in which the function (Equation C.3) grows speedily. The quadratic term describes parabolic properties, and it is the lowest member of this sequence. Opposite to $\vec{g}$, it can be supposed that $\bar{\bar{G}}$ is approximately constant, for a sufficiently narrow segment.

A large number of optimization algorithms have been developed. It is necessary to note that none of these methods leads uniquely to the global minimum if a function $F(\vec{a})$ has a sequence of local minima. At the end of this chapter some algorithms that do determine the global minimum of a function $F(\vec{a})$ are introduced.

Stepwise Methods of Minimization

The Grid Analysis Method. This method is based on the determination of values of a function $F(\vec{a})$ for a grid of uniformly distributed vector $\vec{a}$ components, and on the determination of its minimum. Localizing the minimum of a function of one variable to approximately 99% accuracy requires 10^2 evaluations of function $F(\vec{a})$. For n variables this number of required evaluations is on the order of 10^{2n}. Therefore, this method can practically be used only when the function $F(\vec{a})$ maximally involves two parameters requiring optimization. In the case of multivariate functions, the Monte Carlo method, calculating only a small random sample from the total grid points is used instead.

The Single Parameter Step. The method seeks the minimum with respect to one optimized parameter. This procedure is then repeated for all variables. After optimizing single parameters, this cycle is repeated until self-consistency is obtained. Because of its slow convergence, this method is not practical for multidimensional optimizations.

Rosenbrock's Algorithm. Rosenbrock's algorithm[10] consists of one-dimensional optimizations. After optimizing all parameters, a new set of orthogonal axes derived from the line connecting the start point and end point of a cycle is defined in parameter space. This vector involves the direction of all previous improvements and is considered a convenient direction for future improvement. This method is relatively stable, and it is applicable for narrow valleys of a function $F(\vec{a})$. However, its efficiency decreases as the number of optimized parameters increases.

The Simplex Method. The simplex method[11,12] is the most successful stepwise method. A simplex is an n-dimensional polyhedron specified by its $n + 1$ peaks P_i, $i = 1, 2, \ldots n + 1$. Figure C.1 shows the situation for the two-dimensional example. At the first step a functional value of $n + 1$ randomly selected points is evaluated. Let P_H and P_L be the points of the simplex in which the values of the function $F(\vec{a})$ have the highest and lowest value. For the center of mass of all points except P_H the following is valid:

$$\overline{P} = \frac{1}{n}\left(\sum_{i=1}^{n+1} P_i - P_H\right) \tag{C.19}$$

A new simplex is formed from the original simplex by replacing P_H with a point of a lower functional value $F(P_n)$ in the next step. The position of this point P^* is stated by reflecting P_H around P:

$$P^* = \overline{P} + \left(\overline{P} - P_H\right) \tag{C.20}$$

If $F(P^*) < F(P_L)$, a new point is given by the relationship

$$P^{**} = \overline{P} + 2\left(\overline{P} - P_H\right) \tag{C.21}$$

If $F(P^*) > F(P_L)$, a new point is determined by

$$P^{**} = \overline{P} - \frac{1}{2}\left(\overline{P} - P_H\right) \tag{C.22}$$

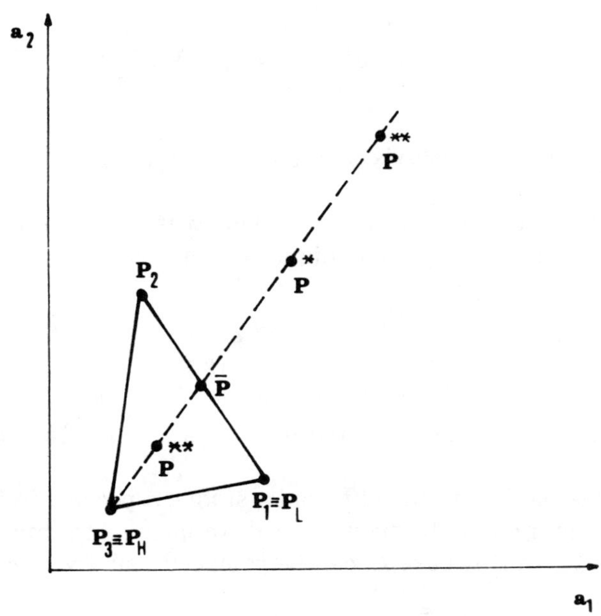

Figure C.1 The simplex method for the two-parametric example.

Then the better point of the points P^*, P^{**} replaces P_H. Unless neither point is better than P_H a new simplex is formed around P_L with dimensions reduced by a factor of 1/2.

Many modifications of this method using differently defined contractions or expansion factors are known. A quadratic interpolation along points P_H, P^*, and P^{**}, searching the minimum in this area, is also used.

A simplex algorithm is rather insensitive to local minima or a fine structure. The convenient convergence criterion is based on the difference $F(P_H) - F(P_L)$.

Gradient Methods

The Steepest Descent Method. If values of the first derivatives of a minimized function are known, it is advantageous to minimize in the direction of a negative gradient vector, i.e., in the direction of the steepest descent of a functional value. This method, first used by Cauchy more than 100 years ago, is the basis of the steepest descent method.

The method consists of a series of one-dimensional minimizations in the direction of a local gradient. Because a gradient direction is subsequently changed, this method requires many iterative steps.

The Newton Method. A general quadratic function specified by its functional values of first and second derivatives of the given point can be minimized in one step:

$$\vec{a}_{\min} = \vec{x}_0 - \bar{\bar{G}}^{-1}\vec{g}\Big|_{\vec{a}=\vec{a}_0} = \vec{x}_0 - \bar{\bar{V}}\vec{g}\Big|_{\vec{a}=\vec{a}_0} \tag{C.23}$$

where $\bar{\bar{V}}$ (the covariance matrix) is the inverse of the matrix of the second derivatives $\bar{\bar{G}}$.

Generally, for a nonquadratic function, when $\bar{\bar{G}}$ is the unconstant matrix, the following iterative technique can be used:

$$\vec{a}_{m+1} = \vec{a}_m - \lambda_m \vec{G}^{-1}\vec{g}\Big|_{\vec{a}=\vec{a}_m} \tag{C.24}$$

where λ_m is the step size. This method is called the Newton-Raphson method.[5,6,8,9]

Usually, the method is not sufficiently stable, especially in the area far away from the minimum. The divergence also appears if the matrix $\bar{\bar{G}}$ is not positive definite. In this case $\bar{\bar{G}}$ can be replaced using various methods:

1. By a unit matrix $\bar{\bar{G}} = \bar{\bar{I}}$, when a Newton step is the steepest descent step

2. By considering only matrix $\bar{\bar{G}}$ diagonal elements, if all of the elements are positive
3. By a matrix $\bar{\bar{G}} \pm \xi \bar{\bar{I}}$, where a number ξ is higher than the highest negative matrix $\bar{\bar{G}}$ rank.

The Method of Conjugate Directions. The Fletcher-Reeves theorem[13] states that a linear minimization sequence in each of the n-conjugate directions leads to minimization of a general quadratic function of n-variables:

$$F(\vec{a}) = F(\vec{0}) + \vec{g}^T \Big|_{\vec{a}=\vec{0}} \vec{a} + \frac{1}{2}\vec{a}^T \bar{\bar{G}} \Big|_{\vec{a}=\vec{0}} \vec{a} \tag{C.25}$$

The quadratic function (Equation C.25) and n-vectors $\vec{d}_i$ are conjugate (Appendix A) with respect to the Hessian matrix $\bar{\bar{G}}$:

$$\vec{d}_i^T \bar{\bar{G}} \vec{d}_j = 0 \qquad \text{for } i \neq j \tag{C.26}$$

Then the vectors $\vec{a}$ and $\vec{g}$ can be expressed as linear combinations:

$$\vec{a} = \sum_{i=1}^{n} \alpha_i \vec{d}_i \tag{C.27}$$

$$\vec{g} = \sum_{i=1}^{n} \beta_i \vec{d}_i \tag{C.28}$$

so Equation C.25 becomes

$$F(\vec{a}) = F(\vec{0}) + \left(\sum_i \beta_i \vec{d}_i^T\right)\left(\sum_j \alpha_j \vec{d}_j\right) + \frac{1}{2}\left(\sum_i \alpha_i \vec{d}_i^T\right)\bar{\bar{G}}\left(\sum_j \alpha_j \vec{d}_j\right) \tag{C.29}$$

After the modification, Equation C.29 can be expressed as

$$\begin{aligned} F(\vec{a}) &= F(\vec{0}) + \sum_i \sum_j \beta_i \vec{d}_i^T \vec{d}_j \alpha_j + \frac{1}{2}\sum_j \alpha_j^2 \vec{d}_j^T \bar{\bar{G}} \vec{d}_j \\ &= F(\vec{0}) + \sum_j \left(b_j \alpha_j + b_j' \alpha_j^2\right) \end{aligned} \tag{C.30}$$

where

$$b_j = \sum_j \beta_i \vec{d}_i^T \vec{d}_j \tag{C.31}$$

$$b'_j = \vec{d}_i^T \bar{\bar{G}} \vec{d}_j \tag{C.32}$$

are constants. Thus, the general quadratic equation (Equation C.25) is separated by this procedure into a set of independent one-parameter functions. Therefore, minimizations with respect to α_i will be independent.

The method of conjugate directions for a general nonquadratic function is stable, because it performs a series of linear optimizations in an independent direction and it guarantees convergence in a finite number of steps. A disadvantage is that n minimizations are required in a direction $\vec{d}_1$, and only one in a direction $\vec{d}_n$ for each iteration. This asymmetry is avoided in the version of Powell.[14]

The Method of Conjugate Gradients. Let us suppose that a function and its gradients are determined for two points $\vec{a}_0$ and $\vec{a}_1$. Then, let us define the differences:

$$\Delta\vec{a} = \vec{a}_1 - \vec{a}_0 \tag{C.33}$$

$$\Delta\vec{g} = \vec{g}_1 - \vec{g}_0 \tag{C.34}$$

For the quadratic function $F(\vec{a})$ with the Hessian matrix $\bar{\bar{G}}$, the following is defined:

$$\Delta\vec{g} = \bar{\bar{G}}\,\Delta\vec{a} \tag{C.35}$$

Then, any vector $\vec{d}_1$ orthogonal to $\Delta\vec{g}$ will be conjugated to a vector $\Delta\vec{a}$, because

$$\vec{d}_1^T\,\Delta\vec{g} = \vec{d}_1^T\,\bar{\bar{G}}\,\Delta\vec{a} = 0 \tag{C.36}$$

Equation C.35 shows possibilities for obtaining conjugate directions without knowing the Hessian matrix.

In the conjugate gradient method, one-dimensional minimizations are performed in conjugate directions, with each direction used only once per iteration. A first direction:

$$\vec{d}_0 = -\vec{g}_0 \tag{C.37}$$

is a vector of steepest descent in a point $\vec{a}_0$. Let the minimum in this direction be $\vec{a}$ with gradient $\vec{g}_1$. Then the next direction $\vec{d}_1$ is

$$\vec{d}_1 = -\vec{g}_1 + b\,\vec{d}_0 \tag{C.38}$$

while

$$\left(-\vec{g}_1^T + b\,\vec{d}_0^T\right)\bar{\bar{G}}\,\vec{d}_0 = \left(-\vec{g}_1^T - b\,\vec{g}_0^T\right)\left(\vec{g}_1 - \vec{g}_0\right) = 0 \tag{C.39}$$

$$\vec{g}_1^T\,\vec{g}_0 = 0 \tag{C.40}$$

By rearranging these equations for the constant b

$$b = \frac{\vec{g}_1^T \vec{g}_1}{\vec{g}_0^T \vec{g}_0} \tag{C.41}$$

is obtained. By placing into Equation C.38 for a direction $\vec{d}$, the following expression results:

$$\vec{d}_1 = -\vec{g}_1 + \frac{\vec{g}_1^T \vec{g}_1}{\vec{g}_0^T \vec{g}_0} \vec{d}_0 \tag{C.42}$$

Variable Metric Methods

The variable metric methods are based on the assumption that for a real nonlinear function $F(\vec{a})$, terms higher than the quadratic are small but not negligibly so, and the Hessian matrix changes only slowly.

In this case the matrix $\bar{\bar{G}}$ need not be calculated in each iterative cycle, but it is efficient to correct it using information from the previous iteration. These corrections are calculated using various approaches and procedures.

The method consists of the following steps:

1. The original point $\vec{a}_0$, its gradient $\vec{g}_0$ and an approximation to the $\bar{\bar{G}}^{-1}$ matrix, (as a rule, a unit matrix is used, $\bar{\bar{V}}_0$) are given.
2. The step

$$\vec{a}_1 = \vec{a}_0 - \bar{\bar{V}}_0 \vec{g}_0 \tag{C.43}$$

will be the minimum if $F(\vec{a})$ is the quadratic function and $\bar{\bar{V}}_0$ is the exact covariance matrix (in the general case, $F(\vec{a})$ is not the quadratic function). After performing the step (Equation C.43)

$$F(\vec{a}_1) = F\left(\vec{a} - \alpha \bar{\bar{V}}_0 \vec{g}_0\right) \tag{C.44}$$

is calculated.

3. The matrix $\bar{\bar{V}}_0$ is corrected using the following formula:

$$\bar{\bar{V}}_1 = \vec{V}_0 + f\left(\bar{\bar{V}}_0, \vec{a}_0, \vec{a}_1, \vec{g}_0, \vec{g}_1\right) \tag{C.45}$$

Subsequently, $\bar{\bar{V}}_0$, $\bar{\bar{g}}_0$, and $\bar{\bar{a}}_0$ are replaced by $\bar{\bar{V}}_1$, $\bar{\bar{g}}_1$, and $\bar{\bar{a}}_1$ and the second step is repeated until the convergence criteria are satisfied. The various methods differ only in the choice of function shape.

The Davidon Second Order Formula. Davidon[15] and Fletcher and Powell[16] proposed the following formula for a matrix $\bar{\bar{V}}_1$:

$$\bar{\bar{V}}_1 = \bar{\bar{V}}_0 + \frac{\vec{\delta}\,\vec{\delta}^T}{\vec{\delta}^T\,\vec{\gamma}} - \frac{\bar{\bar{V}}_0\,\vec{\gamma}\,\vec{\gamma}^T\,\bar{\bar{V}}_0}{\vec{\gamma}^T\,\bar{\bar{V}}_0\,\vec{\gamma}} \tag{C.46}$$

where

$$\vec{\delta} = \vec{a}_1 - \vec{a}_0 \tag{C.47}$$

$$\vec{\gamma} = \vec{g}_1 - \vec{g}_0 \tag{C.48}$$

The fundamental requirement of Equation C.45 is to satisfy the following equality:

$$\bar{\bar{V}}_1\,\vec{\gamma} = \vec{\delta} \tag{C.49}$$

For the quadratic function:

$$\vec{g}_{m+1} - \vec{g}_m = \bar{\bar{G}}\left(\vec{a}_{m+1} - \vec{a}_m\right) \tag{C.50}$$

is defined, and it can be shown that Equation C.45 is convenient for this requirement.

The First Order Formula. Several authors independently introduced a first-order formula to determine a matrix $\bar{\bar{V}}$:[6,8]

$$\vec{V}_1 = \bar{\bar{V}}_0 + \frac{\left(\vec{\delta} - \bar{\bar{V}}_0\,\vec{\gamma}\right)\left(\vec{\delta} - \bar{\bar{V}}_0\,\vec{\gamma}\right)^T}{\vec{\gamma}^T\left(\vec{\delta} - \bar{\bar{V}}_0\,\vec{\gamma}\right)} \tag{C.51}$$

It can be shown that for the matrix $\bar{\bar{V}}_1$ defined above, not only

$$\bar{\bar{V}}_1\,\vec{\gamma} = \vec{\delta} \tag{C.52}$$

but also

$$\bar{\bar{V}}_1\,\Delta\vec{\gamma} = \Delta\vec{\delta} \tag{C.53}$$

is valid, where $\Delta\delta$ is the step vector of previous iterations and $\Delta\gamma$ is the vector of gradient changes.[17]

Fletcher's Unified Approach. Fletcher[18] developed an approach that unifies the above first- and second-order equations.

The fundamental idea of this approach is based on the requirement that a new formula satisfies the basic relationship

$$\bar{\bar{V}}_1 \vec{\gamma} = \vec{\delta} \tag{C.54}$$

or

$$\vec{\gamma} = \bar{\bar{V}}_1^{-1} \vec{\delta} \tag{C.55}$$

The second order formula gives

$$\bar{\bar{V}}_1^{-1} = \left(\bar{\bar{E}} - \frac{\vec{\gamma}\, \vec{\delta}^T}{\vec{\delta}^T \gamma} \right) \bar{\bar{V}}_0^{-1} \left(\bar{\bar{E}} - \frac{\vec{\delta}\, \vec{\gamma}^T}{\vec{\delta}^T \vec{\gamma}} \right) + \frac{\gamma\, \vec{\gamma}^T}{\vec{\delta}^T \vec{\gamma}} \tag{C.56}$$

The defined relationship can be considered a mapping of $\vec{\delta} \to \vec{\gamma}$ since $\vec{\gamma} = \bar{\bar{V}}_1^{-1} \vec{\delta}$. The mapping of $\vec{\gamma} \to \vec{\delta}$, which produces a new formula where $\bar{\bar{V}}_1 \vec{\gamma} = \vec{\delta}$, can be obtained by replacing $\vec{\delta}$ with $\vec{\gamma}$. The new dual expression can then be written in the form

$$\bar{\bar{V}}_1 = \left(\bar{\bar{E}} - \frac{\vec{\delta}\, \vec{\gamma}^T}{\vec{\delta}^T \vec{\gamma}} \right) \bar{\bar{V}}_0 \left(\bar{\bar{E}} - \frac{\vec{\gamma}\, \vec{\delta}^T}{\vec{\delta}^T \vec{\gamma}} \right) + \frac{\vec{\delta}\, \vec{\delta}^T}{\vec{\delta}^T \vec{\gamma}} \tag{C.57}$$

Let us consider a class of equations including both the second order and dual expressions (Equations C.55 and C.56) as the special case.

Let us introduce the notation

$$\bar{\bar{V}}_1 = \bar{\bar{T}}\left(\bar{\bar{V}}_0 \right) \quad \text{for the second order formula}$$

$$\bar{\bar{V}}_1 = \bar{\bar{D}}\left(\bar{\bar{V}}_0 \right) \quad \text{for the dual formula} \tag{C.58}$$

and consider the general class of expressions:

$$\bar{\bar{V}}_\Phi = (1 - \Phi)\, \bar{\bar{T}} + \Phi\, \bar{\bar{D}} \tag{C.59}$$

where Φ is a parameter that determines the exact formula. Then for the first order formula:

$$\Phi(\text{1st order}) = \frac{\vec{\delta}^T \gamma}{\left(\vec{\delta}^T \vec{\gamma} - \vec{\gamma}^T \bar{\bar{V}}_0 \vec{\gamma} \right)} \tag{C.60}$$

is defined.

In Equation C.58 it is necessary to ensure a monotonic convergence $\bar{\bar{V}}$ for a quadratic function. Any formula for $\bar{\bar{V}}_{\Phi}$ with Φ value in the interval (0, 1) possesses this monotonic convergence property. This requirement need not always be satisfied for the first order formula. Therefore, Fletcher[18] proposed the following algorithm for determining the parameter Φ: if Φ(first order) < 0, then $\Phi = 0$, and the usual second order formula is obtained. If Φ(first order) > 1, then $\Phi = 1$ corresponding to the dual formula. If Φ(first order) is inside the interval (0,1), Φ is equal to its real value.

SPECIAL APPROACHES

All the methods introduced up to now have rather general applicability, although they assume approximately quadratic behavior in the vicinity of the minimum. However, special minimization techniques used for special physical situations are known as well.

Chi-Square Minimization

Minimization of the weighted sum of squares of deviations between measured and calculated values (Equation C.3) is often used, and it can be written in the form

$$F(\vec{a}) = \sum_{k=1}^{N} f_k^2(\vec{a}) = \sum_{k=1}^{N} \left[\frac{y_k^{\text{exp}} - y_k^{\text{mod}}(\vec{a})}{\sigma_k} \right]^2 \tag{C.61}$$

where $\vec{y}_k^{\text{exp}}$ and $\vec{\sigma}_k$ are measured values and their mean square deviations $\vec{y}_k^{\text{mod}}(\vec{x})$ are the values of y calculated from the regression relationship (Equation C.1):

$$y_k^{\text{mod}}(\vec{a}) = f(\vec{a}, x_k) \tag{C.62}$$

where $\vec{a} = (a_1, a_2, \ldots a_n)^T$ is the parameter vector.

The minimization function $F(\vec{a})$ yields the best estimate of the n-parameter of the vector $\vec{a}$ based on N measurements of the value Y with random errors σ.

Let us consider the second derivatives of a function $F(\vec{a})$:

$$G_{ij} = \frac{\partial^2 F}{\partial a_i \, \partial a_j} = \sum_{k=1}^{N} 2 \frac{\partial f_k}{\partial a_i} \frac{\partial f_k}{\partial a_j} + \sum_{k=1}^{N} 2 f_k \frac{\partial^2 f_k}{\partial a_i \, \partial a_j} \tag{C.63}$$

Often the second term of Equation C.63 is small compared to the first term, and is negligible. This is called linearization of y_k^{mod} as a function of $F(\vec{a})$. In the case of the linear least squares method, the second term of Equation C.63 is exactly zero, so $F(\vec{a})$ is a quadratic function of unknown parameters;

the whole minimization problem is reduced to the matrix inversion of the function $F(\vec{a})$ second derivatives, i.e., only one Newton step is made.

Generally, in the case of the nonlinear least squares method, linearization consists of the following approximation:

$$\frac{\partial^2 F}{\partial a_i\, \delta a_j} \approx \sum_{k=1}^{N} 2\frac{\partial f_k}{\partial a_i}\frac{\partial f_k}{\partial a_j} \quad \textbf{(C.64)}$$

This approximation has an advantage in easy calculation and in the fact that the approximated matrix of the second derivatives is always positive definite. The Newton minimization method is often used effectively in such approximations.

Maximum Likelihood Maximization

The increasing importance in experimental data fitting has given rise to the maximum likelihood method, which is an alternative to the least squares method. In this case maximization of the following function is performed:

$$\begin{aligned} F(\vec{a}) &= -\sum_{k=1}^{N} \ln f_k(\vec{a}) \\ &= -\sum_{k=1}^{N} \ln\left[y_k^{\text{exp}} - y_k^{\text{mod}}(\vec{a})\right] \end{aligned} \quad \textbf{(C.65)}$$

For the second derivatives of the function $F(\vec{a})$ the following can be found:

$$\frac{\partial^2 F}{\partial a_i\, \partial a_j} = \sum_{k=1}^{N} \frac{1}{f_k^2}\frac{\partial f_k}{\partial a_i}\frac{\partial f_k}{\partial a_j} - \sum_{k=1}^{N} \frac{1}{f_k^2}\frac{\partial f_k}{\partial x_i}\frac{\partial f_k}{\partial x_j} \quad \textbf{(C.66)}$$

Similarly, this relationship can be linearized by approximation:

$$\frac{\partial^2 F}{\partial a_i\, \delta a_j} \approx \sum_{k=1}^{N} \frac{1}{f_k^2}\frac{\partial f_k}{\partial a_i}\frac{\partial f_k}{\partial a_j} \quad \textbf{(C.67)}$$

The advantages of the given approximation are analogous to those of χ^2 minimization.

The Separable Calculations Method

It often happens that computation of the values of the function value $F(\vec{a})$ can be arranged so that a large part of the calculation depend only on a few parameters a_i. These parts will remain unchanged if the corresponding parameters have not changed since the previous evaluation. A special case is when the calculation can be separated into n parts, each depending only on one parameter. In this case, the method of subsequent variances of individual parameters can be used.

LOCAL AND GLOBAL MINIMA

Whenever the function has more then one local minimum, a new problem arises. First of all, it is necessary to specify which of the four following possibilities of minimization is required:

1. Finding any local minimum
2. Specifying the global minimum
3. Specifying a certain minimum (the physical solution), which must not be the global minimum
4. Finding all local minima

The first possibility is quite rare, but it is possible to realize it using any minimization procedure. The second is much more common. The methods ensuring this solution will be discussed below. The third possibility arises if approximate values of some parameters are known and what is sought is the situation where only one solution is not outside the "right" valley on an optimized hyperspace. The usual technique requires fixing the approximately known parameters first. After minimizing according to the other parameters, the complete minimization is started.

The fourth possibility of determining all the local minima is the most difficult. Although some of the methods described below can be helpful, up to now no exhaustive method for solving this problem is known. The only possibility (numerically very complex) is applying many starting parameter estimations equally distributed on an n-dimensional grid.

The Gelfand Algorithm

Gelfand[19] proposed the ad hoc stepwise method, which naturally provides the increases and decreases of a functional value in any step, while it has a general tendency to decrease a functional value.

From a starting point $\vec{a}_0$, a local minimization begins (e.g., the gradient method) until the function's differences between the following steps become sufficiently small (the point $\vec{x}_0$ in Figure C.2). Then we go back to the starting point, and by a "long" random step, we will take the point $\vec{a}_1$ in which another local minimization is performed until the point $\vec{x}_1$ is reached. Then a "precipitous" step is taken along a line of $\vec{x}_0 - \vec{x}_1$ up to the point $\vec{a}_2$. From this point, another local minimization is performed up to the point

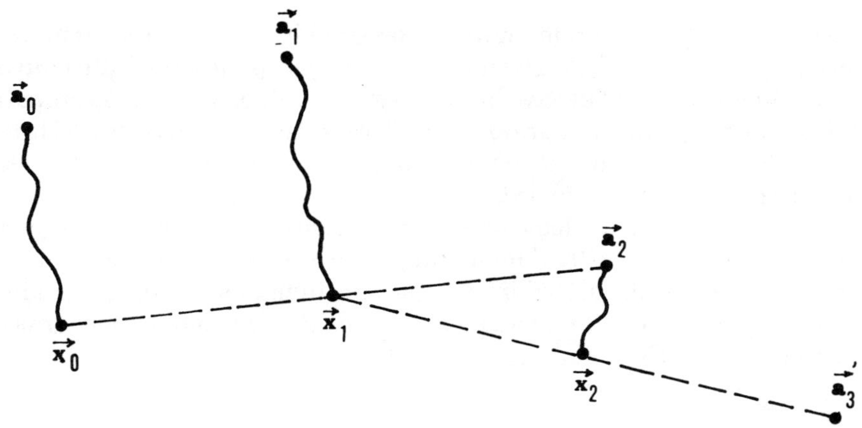

Figure C.2 The Gelfand algorithm for searching for a global minimum of a function $F(\vec{a})$.

$\vec{x}_2$ and another "precipitous" step from $\vec{x}_2$ to $\vec{a}_3$. The whole cycle continues in this way.

The "precipitous" steps are important to the determination, if the solution "rolls" over small ridges or skirts a high mountain. But no precise method for determining the necessary step length exists, and therefore it must be chosen experimentally.

The Goldstein–Price Method

Goldstein and Price[20] proposed the simple but elegant method that determines other local minima after one local minimum is found. It is based on the properties of the Taylor series of the minimized function. Let us assume that the function can be represented as the Taylor series around a local minimum $\vec{a}_1$ (in which first derivatives have null values):

$$F(\vec{a}) = F(\vec{a}_1) + \frac{1}{2}(\vec{a} - \vec{a}_1)^T \,\bar{\bar{G}}(\vec{a} - \vec{a}_1) + \text{h.t.} \qquad \textbf{(C.68)}$$

The higher terms (h.t.) involving third and higher derivatives are important now, because they specify other local minima. Now, let us perform the function transformation so that only the higher terms remain. Such a transformed function F_1 has the following form:

$$F_1(\vec{a}_1, \vec{a}) = -\frac{2\left[\vec{F}(a) - \vec{F}(a_1)\right]}{(\vec{a} - \vec{a}_1)^T \,\bar{\bar{G}}(\vec{a} - \vec{a}_1)} \qquad \textbf{(C.69)}$$

By means of this transformation, the minimum in a point $\vec{a}_1$ is removed and the way is cleared to search for other local minima generated by the higher terms in the function series near the point $\vec{a}_1$. Then the method continues by determining the local minimum of a function F_1. Because the quadratic form $(\vec{a} - \vec{a}_1)^T \,\overline{\overline{G}}\,(\vec{a} - \vec{a}_1)$ is always positive (for the positive definite matrix $\overline{\overline{G}}$), the function F_1 will always have a negative value, unless the estimate of the position of point $\vec{a}_1$ improves. In such cases the method is started from this improved point, and the original function F_1 is minimized locally until a new local minimum is yielded.

If the minimal value determined for the function F_1 is positive, it will also correspond to a new local minimum, but it will not be an improvement over $\vec{a}_1$. In this case the procedure can continue from this new point, forming a new function F_2 from the function F_1 (just as the function F_1 was formed from the function F).

STATISTICAL INFERENCE AND THE REGRESSION METHOD

The efficiency of a regression function, as well as the precision of regression parameter values, can be characterized by statistical inference. The differences between experimental data on one hand and values obtained by the regression function on the other are analyzed in this treatment.

The fundamental characteristic describing the precision of fitting experimental data by the regression function is the *standard deviation of an approximation:*

$$s_{\mathrm{apr}} = \left(\frac{F_{\min}}{N - n} \right)^{\frac{1}{2}} \tag{C.70}$$

where $F_{\min}$ is the minimal value of the sum of squared deviations (Equation C.3), N is the number of experimental points, and n is the number of optimized parameters a_i of the regression function. The value s_{apr}^2 characterizes the dispersion of experimental values around the regression function and is called the *residual dispersion.*

The *correlation* between the experimental values $y_i^{\exp}$ and the values y_i^{mod} obtained from the regression function (Equation C.61) is characterized by the *coefficient of correlation:*

$$\rho = \frac{\sum_{i=1}^{N} \left(y_i^{\exp} - \overline{y}^{\exp} \right) \left(y_i^{\mathrm{mod}} - \overline{y}^{\mathrm{mod}} \right)}{\left[\sum_{i=1}^{N} \left(y_i^{\exp} - \overline{y}^{\exp} \right)^2 \sum_{i=1}^{N} \left(y_i^{\mathrm{mod}} - \overline{y}^{\mathrm{mod}} \right)^2 \right]^{\frac{1}{2}}} \tag{C.71}$$

where the mean values are given by

$$\overline{y}^{\alpha} = \frac{\sum_{i=1}^{N} y_i^{\alpha}}{N} \qquad \alpha = \exp, \mathrm{mod} \tag{C.72}$$

Another variable characterizing the precision of fitting experimental data by the regression function is the *R-factor* (used especially for X-ray structural analysis):

$$R = \left[\frac{\sum_{i=1}^{N} \left(y_i^{\exp} - y_i^{\mathrm{mod}} \right)^2}{\sum_{i=1}^{N} \left(y_i^{\exp} \right)^2} \right]^{\frac{1}{2}} \tag{C.73}$$

The convenience of the regression equation can be tested by *Fisher's F-test:*

$$F = \frac{s_y^2}{s_{apr}^2} \tag{C.74}$$

where for the sample dispersion of experimental data the following expression is defined:

$$s_y^2 = \frac{\sum_{i=1}^{N} \left(y_i^{\exp} - \overline{y}^{\exp} \right)^2}{N - 1} \tag{C.75}$$

Such a defined criterion shows how many times lower the dispersion around the regression equation is compared to the dispersion around the mean value; the convenience of the regression equation is greater the higher F is than the critical value of Fisher's F-statistic $F_p\,(\nu_1, \nu_2)$ for the chosen significance level p and the number of degrees of freedom $\nu_1 = N - 1$ and $\nu_2 = N - n$.

The result of minimizing the object function (Equation C.3) is the best estimation of the parameters a_i. The standard deviations of the individual parameters can be estimated using the relationship

$$s(a_i) = \left[2\left(\overline{\overline{G}}^{-1} \right)_{ii} \right]^{\frac{1}{2}} s_{\mathrm{apr}} \tag{C.76}$$

where $(\overline{\overline{G}}^{-1})_{ii}$ is the diagonal element of the inverse matrix to the second derivative matrix of an object function for parameters at the point of its minimal value:

$$G_{ij} = \frac{\partial^2 F_{\min}\left(\vec{a} \right)}{\partial a_i \, \partial a_j} \tag{C.77}$$

The statistical significance of individual parameters of a regression model can be tested by the *Student t-test:*

$$t(a_i) = \frac{|a_i|}{s(a_i)} \tag{C.78}$$

Parameters with a value $t(a_i)$, which is lower than the critical value of the Student distribution with $N - n$ degrees of freedom on the chosen significance level, are considered to be statistically nonsignificant. The values of such parameters from the given experimental data set cannot in principle be determined. If the presence of a parameter in a regression model is required from physical theory, the experimental data set must be adequately extended to permit its estimation. In the opposite case (empirical regression models) it is necessary to keep the given range of the experimental data set, to simplify the regression function by eliminating the corresponding parameter, and improve it by another method.

The presence of linear bonds between parameter pairs is indicated by high values of *coefficients of correlation:*

$$\rho\left(a_i, a_j\right) = \frac{\left(\bar{\bar{G}}^{-1}\right)_{ij}}{\left[\left(\bar{\bar{G}}^{-1}\right)_{ii}\left(\bar{\bar{G}}^{-1}\right)_{jj}\right]^{\frac{1}{2}}} \tag{C.79}$$

Individual values of linearly correlated (or collinear) parameters may not be determined uniquely using regression analysis.

If any analytical function $g(\vec{a}, x)$ is formed by means of parameter vector $\vec{a}$, then its dispersion can be estimated by the relationship:

$$s^2\left[g(\vec{a}, x)\right] = \sum_{i=1}^{N} c_i^2 s^2\left(a_i\right) + \sum\sum_{i \neq j} c_i c_j \operatorname{cov}\left(a_i, a_j\right) \tag{C.80}$$

where the covariance between parameters is

$$\begin{aligned} \operatorname{cov}\left(a_i, a_j\right) &= 2\left(\bar{\bar{G}}^{-1}\right)_{ij} s_{\text{apr}}^2 \\ &= \rho\left(a_i, a_j\right) s\left(a_i\right) s\left(a_j\right) \end{aligned} \tag{C.81}$$

and coefficients c_i are defined by the following equation:

$$c_i = \frac{\partial g(\vec{a}, x)}{\partial a_i} \tag{C.82}$$

Appendix D

The Fourier Transformation

The *Fourier transformation* (References 21–24) $F(s)$ of a function $f(x)$ is defined by the following equation:

$$\mathcal{F}\{f(x)\} = F(s) = \int_{-\infty}^{+\infty} f(x)\, e^{-i2\pi xs}\, dx \qquad \textbf{(D.1)}$$

If the Fourier transformation is used for the function $F(s)$, the following is obtained:

$$\mathcal{F}\{F(s)\} = f(x) = \int_{-\infty}^{+\infty} F(s)\, e^{i2\pi xs}\, ds \qquad \textbf{(D.2)}$$

Substituting Equation D.1 into the expression in Equation D.2, the *Fourier integral theorem* results:

$$f(x) = \int_{-\infty}^{+\infty} \left[\int_{-\infty}^{+\infty} f(x)\, e^{-i2\pi xs}\, dx \right] e^{i2\pi xs}\, ds \qquad \textbf{(D.3)}$$

If a function $f(x)$ is discontinuous, the left side of Equation D.3 must be replaced by the relationship $1/2[f(x) + f(-x)]$, i.e., by the mean value of the limits from the right side and left side at the point of discontinuity.

The Fourier transformation of function $f(x)$ exists if:

1. An integral $\int_{-\infty}^{+\infty} f(x)dx$ exists
2. If all discontinuities of a function $f(x)$ are finite
3. If a function $f(x)$ has only a finite number of extremes

Any function $f(x)$ can be split into an odd and even part:

$$f(x) = E(x) + O(x) \qquad \textbf{(D.4)}$$

where

$$E(x) = \frac{1}{2}\left[f(x) + f(-x)\right] \tag{D.5}$$

is the even part and

$$O(x) = \frac{1}{2}\left[f(x) - f(-x)\right] \tag{D.6}$$

is the odd part of a function $f(x)$. Then the Fourier transformation $f(x)$ is reduced to the following form:

$$\mathcal{F}\{f(x)\} = 2\int_{-\infty}^{+\infty} E(x)\cos(2\pi xs)\,dx - 2i\int_{-\infty}^{+\infty} O(x)\sin(2\pi xs)\,dx \tag{D.7}$$

The following function

$$h(x) = f(x) * g(x) = \int_{-\infty}^{+\infty} f(u)\,g(x-u)\,du \tag{D.8}$$

is called the *convolution* of two functions $f(x)$ and $g(x)$, where * means the *convolution product.*

For the convolution product the following properties are known:

1. $f(x) * g(x) = g(x) * f(x)$ (commutativity)
2. $f(x) * [g(x) * p(x)] = [f(x) * g(x)] * p(x)$ (associativity)
3. $f(x) * [g(x) + p(x)] = f(x) * g(x) + f(x) * g(x)$ (distributivity)

The *self-convolution of a function* $f(x)$ is defined as a convolution of a function by itself:

$$f^*(x) * f(x) = \int_{-\infty}^{+\infty} f^*(x)\,f(x-u)\,du \tag{D.9}$$

The function

$$\gamma(x) = \frac{\int_{-\infty}^{+\infty} f^*(u)\,f(x+u)\,du}{\int_{-\infty}^{+\infty} f(u)\,f^*(u)\,du} \tag{D.10}$$

is called the *auto-correlation function* $\gamma(x)$ of a function $f(x)$.

A *cross correlation* $f(x)$ of two functions $g(x)$ and $h(x)$ is defined as follows:

$$f(x) = g(x) \otimes h(x) = \int_{-\infty}^{+\infty} g^*(u - x)\, h(u)\, du \tag{D.11}$$

The commutative law is not valid for cross correlation

$$g(x) \otimes h(x) \neq h(x) \otimes g(x) \tag{D.12}$$

but if $f(x) = g(x) \otimes h(x)$ then $h(x) \otimes g(x) = f(-x)$.

A function

$$\operatorname{sinc} x = \frac{\sin \pi x}{\pi x} \tag{D.13}$$

has the following properties

$$\operatorname{sinc} 0 = 1 \tag{D14}$$

$$\operatorname{sinc} n = 0 \qquad (\text{for a nonzero integer number n})$$

$$\int^{+\infty} \operatorname{sinc} x\, dx = 1 \tag{D.15}$$

Function $Si(x)$ is defined by the following relationship:

$$\int_{-\infty}^{+\infty} \operatorname{sinc} x\, dx = 1$$

$$Si(x) = \int_0^x \frac{\sin t}{t}\, dt \tag{D.16}$$

Therefore,

$$\int_{-\infty}^{+\infty} \operatorname{sinc} x\, dx = \frac{Si(\pi x)}{\pi} \tag{D.17}$$

and

$$\operatorname{sinc} x = \frac{d}{dx}\left(\frac{Si(\pi x)}{\pi}\right) \tag{D.18}$$

are valid.

The *Dirac delta function* is defined by the relationship

$$\delta(x) = \frac{1}{2\pi} \int_{-\infty}^{+\infty} e^{ikx}\, dk \tag{D.19}$$

and has the following properties:

$$\int_{-\infty}^{+\infty} \delta(x)\,dx = 1 \tag{D.20}$$

$$\int_{-\infty}^{+\infty} \delta(x)\,f(x)\,dx = f(0) \tag{D.21}$$

$$\int_{-\infty}^{+\infty} \delta(x-a)\,f(x)\,dx = f(a) \tag{D.22}$$

$$\int_{-\infty}^{+\infty} \delta(x)\,f(x-a)\,dx = f(-a) \tag{D.23}$$

$$\delta(ax) = \frac{1}{|a|}\delta(x) \tag{D.24}$$

$$\delta(x) = \delta(-x) \tag{D.25}$$

$$\int_{-\infty}^{+\infty} \delta(x')\,f(x-x')\,dx' = \int_{-\infty}^{+\infty} \delta(x-x')\,f(x')\,dx' = f(x) \tag{D.26}$$

and

$$\delta(x) * f(x) = f(x) * \delta(x) = f(x) \tag{D.27}$$

If $f(x)$ has a discontinuity at a point x, then

$$\delta(x) * f(x) = \frac{1}{2}\left[f(x) + f(-x)\right] \tag{D.28}$$

is defined.

For the derivatives of the function $\delta(x)$ the following properties are known:

$$\int_{-\infty}^{+\infty} \delta'(x)\,dx = 0 \tag{D.29}$$

$$\int_{-\infty}^{+\infty} x\,\delta'(x)\,dx = -1 \tag{D.30}$$

$$\int_{-\infty}^{+\infty} |\delta'(x)|\,dx = \infty \tag{D.31}$$

$$\delta'(x) * f(x) = \int_{-\infty}^{+\infty} \delta'(x' - a)\, f(x')\, dx' = f'(x) \tag{D.32}$$

$$\delta'(-x) = -\delta'(x) \tag{D.33}$$

$$\int_{-\infty}^{+\infty} \delta''(x)\, dx = 0 \tag{D.34}$$

$$\delta''(x) * f(x) = f''(x) \tag{D.35}$$

$$\delta^{(n)}(x) = (-1)^{(n)}\, n!\, x^{-n}\, \delta(x) \tag{D.36}$$

$$\delta^{(n)}(x) * f(x) = f^{(n)}(x) \tag{D.37}$$

$$\int_{-\infty}^{+\infty} \delta^{(n)}(x)\, f(x)\, dx = (-1)^{(n)}\, f^{(n)}(0) \tag{D.38}$$

The *similarity theorem* states that if $F(s)$ is the Fourier transformation of a function $f(x)$, then $|a|^{-1}F(s/a)$ is the Fourier transformation of a function $f(ax)$:

$$\begin{aligned}\int_{-\infty}^{+\infty} f(ax)\, e^{-i2\pi xs}\, dx &= \frac{1}{|a|}\int_{-\infty}^{+\infty} f(ax)\, e^{-i2\pi(ax)\left(\frac{s}{a}\right)}\, d(ax)\\ &= \frac{1}{|a|}\, F\left(\frac{s}{a}\right)\end{aligned} \tag{D.39}$$

The *summation theorem* states that if $f(x)$ and $g(x)$ have the Fourier transformation $F(s)$ and $G(s)$, then a function $f(x) + g(x)$ has the Fourier transformation $F(s) + G(s)$:

$$\begin{aligned}\int_{-\infty}^{+\infty} \left[f(x) + g(x)\right] e^{-i2\pi xs}\, dx &= \int_{-\infty}^{+\infty} f(x)\, e^{-i2\pi xs}\, dx + \int_{-\infty}^{+\infty} g(x)\, e^{-i2\pi xs}\, dx\\ &= F(s) + G(s)\end{aligned} \tag{D.40}$$

The *shift theorem* states that if a function $f(x)$ has the Fourier projection $F(s)$, then a function $f(x - a)$ has the Fourier projection $e^{-i2\pi as}F(s)$:

$$\begin{aligned}\int_{-\infty}^{+\infty} f(x - a)\, e^{-i2\pi xs}\, dx &= \int_{-\infty}^{+\infty} f(x - a)\, e^{-i[2\pi(x-a)s]} \cdot e^{-i2\pi as}\, d(x - a)\\ &= e^{-i2\pi as}\, F(s)\end{aligned} \tag{D.41}$$

The *modulation theorem* states that if a function $f(x)$ has the Fourier projection $F(s)$, then a function $f(x)\cos(\omega x)$ has the Fourier projection $1/2[F(s - \omega/2\pi) + F(s + \omega/2\pi)$:

$$\int_{-\infty}^{+\infty} f(x)\cos(\omega x)\, e^{-i2\pi xs}\, dx$$

$$= \frac{1}{2}\left[\int_{-\infty}^{+\infty} f(x)\, e^{-i\omega x} \cdot e^{-i2\pi xs}\, dx + \int_{-\infty}^{+\infty} f(x)\, e^{-i\omega x}\, e^{-i2\pi xs}\, dx\right]$$

$$= \frac{1}{2}\left[\int_{-\infty}^{+\infty} f(x)\, e^{-i2\pi x(\frac{s-\omega}{2\pi})}\, dx + \int_{-\infty}^{+\infty} f(x)\, e^{-i2\pi x(\frac{s+\omega}{2\pi})}\, dx\right]$$

$$= \frac{1}{2}\left[F\left(\frac{s-\omega}{2\pi}\right) + F\left(\frac{s+\omega}{2\pi}\right)\right] \quad \textbf{(D.42)}$$

The *convolution theorem* states that if functions $f(x)$ and $g(x)$ have the Fourier projections $F(s)$ and $G(s)$, then the convolution of $f(x) * g(x)$ has the Fourier projections $F(s)\,G(s)$:

$$f(x) * g(x) = \int_{-\infty}^{+\infty}\left[\int_{-\infty}^{+\infty} f(x')\, g(x - x')\, dx'\right] \cdot e^{-i2\pi xs}\, dx$$

$$= \int_{-\infty}^{+\infty} f(x')\left[\int_{-\infty}^{+\infty} g(x - x')\, e^{-i2\pi xs} \cdot dx\right] dx'$$

$$= \int_{-\infty}^{+\infty} f(x')\, G(s)\, e^{-i2\pi x's}\, dx'$$

$$= G(s)\int_{-\infty}^{+\infty} f(x')\, e^{-i2\pi x's}\, dx' = F(s)\, G(s) \quad \textbf{(D.43)}$$

Hence

$$\mathcal{F}\{f * g\} = \mathcal{F}\{f\}\,\mathcal{F}\{g\} \quad \textbf{(D.44)}$$

$$\mathcal{F}\{fg\} = \mathcal{F}\{f\} * \mathcal{F}\{g\} \quad \textbf{(D.45)}$$

are valid.

Rayleigh's theorem states that the integral of the absolute square value of the function $f(x)$ is equal to the integral form of the absolute square value of its Fourier transformation:

$$\int_{-\infty}^{+\infty} |f(x)|^2 \, dx = \int_{-\infty}^{+\infty} |F(s)|^2 \, ds \qquad \textbf{(D.46)}$$

The *power theorem* states

$$\int_{-\infty}^{+\infty} f(x)\, g^*(x)\, dx = \int_{-\infty}^{+\infty} F(s)\, G^*(s)\, ds \qquad \textbf{(D.47)}$$

The *self-correlation theorem* states that if a function $f(x)$ has the Fourier projection $F(s)$, then its self-correlation function has the Fourier projection $|F^2(s)|$:

$$\int_{-\infty}^{+\infty} |F(s)|^2 \, e^{i2\pi xs} \, ds = \int_{-\infty}^{+\infty} F(s)\, F^*(s)\, e^{i2\pi xs} \, ds$$

$$= f(x) * f^*(-x) = \int_{-\infty}^{+\infty} f(u)\, f^*(u - x)\, du$$

$$= \int_{-\infty}^{+\infty} f^*(u)\, f(u + x)\, du \qquad \textbf{(D.48)}$$

The *derivative theorem* states that if a function $f(x)$ has the Fourier projection $F(s)$, then $f'(x)$ has the Fourier projection $i2\pi sF(s)$:

$$\int_{-\infty}^{+\infty} f'(x)\, e^{-i2\pi xs} \, dx = \int_{-\infty}^{+\infty} \lim \frac{f(x + \Delta x) - f(x)}{\Delta x} e^{-i\pi xs} \, dx$$

$$= \lim \int_{-\infty}^{+\infty} \frac{f(x + \Delta x)}{\Delta x} e^{-i2\pi xs} \, dx$$

$$- \lim \int_{-\infty}^{+\infty} \frac{f(x)}{\Delta x} e^{-i2\pi xs} \, dx$$

$$= \lim \frac{e^{-i2\pi \Delta xs} F(s) - F(s)}{\Delta x} = i2\pi sF(s) \qquad \textbf{(D.49)}$$

The derivative of a convolution product of two functions is equal to the convolution product of one function with the derivative of the second function:

$$[f(x) * g(x)]' = f'(x) * g(x) = f(x) * g'(x) \qquad \textbf{(D.50)}$$

The function $f(x)$ integral from $-\infty$ to $+\infty$ is equal to its Fourier projection at the origin:

$$\int_{-\infty}^{+\infty} f(x)\,dx = \left.\int_{-\infty}^{+\infty} f(x)\,e^{-i2\pi xs}\,dx\right|_{s=0}$$

$$= \{f(x)\}_{s=0} = F(0) \qquad \textbf{(D.51)}$$

An integral

$$M_f^{(n)} = \int_{-\infty}^{+\infty} x^n\,f(x)\,dx = (-i2\pi)^{-n}\int_{-\infty}^{+\infty}(-i2\pi x)^n\,f(x)\,dx$$

$$= \frac{F^{(n)}(0)}{(-i2\pi x)^n} \qquad \textbf{(D.52)}$$

is called the n-th moment of function $f(x)$.

The second moment of a function $f(x)$ is also called the *moment of inertia,* and for it

$$M_f^{(2)} = \int_{-\infty}^{+\infty} x^2\,f(x)\,dx = -\frac{F''(0)}{(4\pi)^2} \qquad \textbf{(D.53)}$$

holds.

The variance σ^2 of a function $f(x)$ is called a *mean-square value:*

$$\sigma^2 = <(x - <x>)^2> = \frac{\int_{-\infty}^{+\infty}(x - <x>)^2\,f(x)\,dx}{\int_{-\infty}^{+\infty} f(x)\,dx}$$

$$= -\frac{F''(0)}{4\pi^2\,F(0)} + \frac{[F'(0)]^2}{4\pi^2[F(0)]^2} \qquad \textbf{(D.54)}$$

where the brackets < > mean the mean value.

The variance of a convolution product $f(x) * g(x)$ is equal to the sum of the variances of $f(x)$ and $g(x)$:

$$\sigma_{f*g}^2 = \sigma_f^2 + \sigma_g^2 \qquad \textbf{(D.55)}$$

The value

$$\frac{\int_{-\infty}^{+\infty} f(x)\,dx}{f(0)} \qquad \textbf{(D.56)}$$

is called the equivalent width of any function $f(x)$. If $f(0) = 0$ is the numerator of Equation D.56, the function $f(x)$ has no equivalent width for the purposes of spectroscopy.

References to Appendices

1. Bodewig, E., *Matrix Calculus,* Interscience Publishers Inc., New York, 1959.
2. Hildebrand, F. B., *Introduction to Numerical Analysis,* McGraw-Hill, New York, 1956.
3. Ralston, A., *A First Course in Numerical Analysis,* Blaisdell Publishing Co., New York, 1958.
4. Lawson, C. L. and Hanson, R. J., *Solving Least Squares Problems,* Prentice-Hall, Inc., Englewood Cliffs, New Jersey, 1974.
5. James, F., *Function Minimization,* Proc. 1972 CERN Computing and Data Processing School, Pertisau, 1972.
6. Kowalik, J. and Osborne M. R., *Methods for Unconstrained Optimization Problems,* Elsevier, New York, 1968.
7. Bunday, B. D., *Basic Optimization Methods,* Edward Arnold, London, 1984.
8. Dixon, L. C. W., *Non-Linear Optimization,* English University Press, London, 1972.
9. Box, M. J., Davies, D., and Swann, W. H., *Non-Linear Optimization Techniques,* Oliver and Boyd, Edinburgh, 1969.
10. Rosenbrock, H. H., *Comput. J.,* 3, 175, 1960.
11. Nelder, J. A. and Mead, R., *Comput. J.,* 7, 308, 1965.
12. Yarbro, L. A. and Deming, S. N., *Anal. Chim. Acta,* 73, 391, 1974.
13. Fletcher, R. and Reeves, C. M., *Comput. J.,* 7, 149, 1964.
14. Powell, M. J. D., *Comput. J.,* 6, 163, 1963.
15. Davidon, W. C., *Comput. J.,* 10, 406, 1968.
16. Fletcher, R. and Powell, M. J. D., *Comput. J.,* 6, 163, 1960.
17. Powell, M. J. D., *Rank One Methods for Unconstrained Optimization, Appearing in Integer and Non-LinearProgramming,* Adabic, J., Ed., North-Holland Publishing Co., Amsterdam, 1970.
18. Fletcher, R., *Comput. J.,* 13, 317, 1970.
19. Gelfand, I. M. and Tsetlin, M. L., *Soviet Phys.,* 6, 192, 1961.
20. Goldstein, A. A. and Price, J. F., *Math. Comput.,* 25, 569, 1971.
21. Papoulis, A., *The Fourier Integral and its Applications,* McGraw-Hill, New York, 1962.

22. Marshall, M. G., Ed., *Fourier, Hadamard, and Hilbert Transforms in Chemistry,* Plenum Press, New York, 1982.

23. Gaskill, J. P., *Linear Systems, Fourier Transforms and Optics,* John Wiley & Sons, New York, 1978.

24. Bloomfield, P., *Fourier Analysis of Time Series,* John Wiley & Sons, New York, 1976.

Index

D

E

F

G

H

I

J

K

L

O

P

Q

R

S

T